Nature and Experience in the Culture of Delusion

Also by David W. Kidner

NATURE AND PSYCHE: RADICAL ENVIRONMENTALISM AND THE POLITICS OF SUBJECTIVITY

Nature and Experience in the Culture of Delusion

How Industrial Society Lost Touch with Reality

David W. Kidner
Nottingham Trent University, UK

First published 2012 by
PALGRAVE MACMILLAN

Palgrave Macmillan in the UK is an imprint of Macmillan Publishers Limited, registered in England, company number 785998, of Houndmills, Basingstoke, Hampshire RG21 6XS.

Palgrave Macmillan in the US is a division of St Martin's Press LLC, 175 Fifth Avenue, New York, NY 10010.

Palgrave Macmillan is the global academic imprint of the above companies and has companies and representatives throughout the world.

Palgrave® and Macmillan® are registered trademarks in the United States, the United Kingdom, Europe and other countries.

ISBN 978–0–230–30848–0

This book is printed on paper suitable for recycling and made from fully managed and sustained forest sources. Logging, pulping and manufacturing processes are expected to conform to the environmental regulations of the country of origin.

A catalogue record for this book is available from the British Library.

A catalog record for this book is available from the Library of Congress.

10 9 8 7 6 5 4 3 2 1
21 20 19 18 17 16 15 14 13 12

Printed and bound in Great Britain by
CPI Antony Rowe, Chippenham and Eastbourne

Contents

Preface

Douglas Adams' novel *A Hitch-Hiker's Guide to the Galaxy* begins with the stealthy approach of huge inter-stellar Vogon spaceships, unnoticed by the world's combined military defences – a source of obliteration that emerges, as we say, 'out of the blue'. This seems an apt parable for current threats to human being, which also emerge from within systems we have not evolved to recognise. Unlike the Vogon spaceships, however, the threats that face us are not simply external, since they arise out of the same systems that we ourselves are intimately and often unknowingly entangled with. For similar reasons, and despite their passion and often heroism, environmentalists have generally been no more successful than Arthur Dent in stemming industrialism's accelerating destruction and commodification of the natural world. This is often taken to indicate the failure of environmental activism; but it is more realistically understood as a measure of the power of the forces ranged against the natural order and our difficulties in identifying these forces.

What has befallen the world since the advent of industrialism is a catastrophe. Despite the torrent of denials, evasions, excuses, compartmentalisations, talks about 'changing cultures', 'contested natures', and 'discursive mediation', I know of no other intellectually honest reading of the history of the colonisation and destruction of bioregional ecosystems, cultures, and languages. When academic writing ignores, conceals, intellectualises, justifies, or celebrates this history, it simply reveals its infection by the forces of colonisation.

This book is the story of a conquest – one that is changing the lives of every creature on earth, while saving its most drastic consequences for future generations. Despite its momentous character, the process of conquest is barely reported; indeed it has become a taken-for-granted aspect of living, corroding the health of cultures and individuals from within as it simultaneously restructures both the natural world and our experience of that world to fit the emerging industrial order. Indeed, it is highly significant that 'reality' has become a dirty word in many quarters today, especially but not exclusively within academia. In a reaction to scientism that is as simplistic as the understanding it tries to replace, some writers tell us that 'reality' is simply a product of culture and language. If this is the case, then the decline of the natural world becomes merely another cultural 'myth', with no more significance or emotional import than tales of dragons or fairies. For example, one academic writer tells us that she is "interested in" the "tale" of disenchantment that "defines a particular range of ethical problems calling for worry and redress".[1] Another asks: "Why not accept what may have been wrongly regarded as abnormal and terror-filled – the Enlightenment nightmares of machinic dehumanisation – as what is actually the normal, ordinary, or everyday?"[2]

The 'real world' consists not only of sensorily apparent physical properties but also of emergent ecological and subjective properties, particularly the tendency to self-organise, which is the basis of the whole of life. This is a world that contains structure, and also tendencies towards structure – a process dimension. It is itself a troubling sign that this book needs to begin by affirming that humans, like other animals, have evolved within a natural world, and that our physical and psychological characteristics reflect our evolutionary dependence on a world that is both symbolic and material. Social reductionism is as prevalent as biological reductionism; and what James Hillman has referred to as social scientists' "paralysing ... obsession with language and communication"[3] is as much a symptom of the deeper problem this book addresses – namely, the dissociation between physical being and symbolic form – as others' more scientistic preoccupations. Both these aspects of being are essential parts of the real world; and the dead end of idealist influence over many disciplines has to be challenged in all its forms. As Roy Bhaskar and others have shown very clearly, this has nothing to do with 'naïve realism'.

It is the glory and the bane of human being that our cognitive powers allow us to envision situations and creations that do not at the moment exist. This capability can be the means of creativity and invention; or it can be the road to delusion and insanity. My focus here concerns the process whereby whole civilisations – as I argue – can lose touch with reality and become effectively insane. When an individual loses touch with reality, we call it 'psychosis'. When a civilisation loses touch with reality, we call it 'normality'.

Academic etiquette dictates that I should not merely review the evidence and offer an analysis of it but also complete the book by offering solutions to the problems I have outlined, thereby bringing a comfortable sense of closure to my argument. This is the academic equivalent of the Hollywood 'happy ending'. Most of the 'solutions' currently on offer, however, will have the effect of enabling industrialism, rather than the natural order, to survive a little longer. Not only is the story I have to tell unfinished but the ending is one that is largely beyond the power of human consciousness to foretell. While solutions have become more urgently required over the past half-century, our ability to provide them has faded as industrialism has tightened its grip. My aim is more modest than making predictions or offering solutions, although it is a necessary precursor to both: to provide what is so far largely – and in certain disciplines completely – absent: namely a reasonably realistic assessment of our situation. Solutions, if they exist, will emerge out of a more complete awareness of our situation rather than from a continuing effort by consciousness to retain its delusory grip on our fate.

One of the most prominent of our problematic assumptions is that knowledge is best arranged into more or less separate disciplines, ensuring that a systemic understanding of industrial society as a whole – always an ambitious aim – has remained elusive. Most science – even recent science – has understandably recoiled from the systemic complexity of industrial society, preferring to focus on more detailed and fragmentary types of knowledge. There are exceptions: the work of Robert Ulanowicz and Stuart Kauffman in the natural sciences; the emergence of biosemiotics, catalysed by Gregory Bateson's pioneering writings, and greatly extended by Jesper Hoffmeyer and John Deely; the emergence of complexity theory and its application to ecosystems. Nevertheless, mainstream science, while brilliantly effective in specific domains, has failed to throw much light on the social, economic, and technological systems within which we live – and especially on their integration to generate the phenomenon I refer to as 'industrialism'. The consequence is that while we know a great deal about specific processes, we have almost no conception about where all this is leading or about the larger frameworks that provide the backdrop to our daily lives. Industrialism therefore involves a multiplicity of connections that are normally overlooked because of the splintering of knowledge into disciplines and specialities. Much the same could be said of the system it opposes and consumes – the natural order. There are peculiar difficulties, then, in describing these systems or the conflict between them in the linear stream of language that normally constitutes a book. The approach I have taken here is therefore more akin to a painting (or an ecosystem), each fragment or section contributing a piece of the whole, and relating to many other fragments.

Another tenet of academic etiquette – one that I also ignore – demands that I define my terms clearly before beginning my argument. This particularly applies to the term 'industrialism', as some of my critics have noted. But clear definition assumes that the important players in the drama have already been identified, and their shape and constitution agreed upon; so like chess pieces, it is merely their interaction that determines outcomes. Systemic interactions, however, are based not only on such 'bottom up' determinations but also on *redefinition* and *transformation* of component entities according to their interactions. Furthermore, 'clear definition' assumes that entities and ideas can be precisely translated into language. None of this is the case in our present situation: our character and boundaries, which were debatable even prior to the industrial era, are being transformed as we are uprooted from our natural context and transplanted into a quite different order, and our thought and language are themselves extensively colonised by industrialism. What is needed is that we sensitively and reflexively *feel* our way towards identifying the shape and dynamics of the industrial system, using a somewhat broader range of faculties than is generally considered proper.

Despite these misgivings, it is useful to have an approximate starting point for understanding; and in this provisional spirit I will characterise industrialism as that colonising system which is re-ordering the world and its constituents to fit the emerging imperatives towards commoditisation and capital accumulation, consuming ecological relationship as it does so. But this is already inadequate – not least because it implies that industrialism, like the Vogon spaceships, is *external* to human subjectivity; whereas it might be better regarded as an outgrowth of the historically emerging confluence of technological vision and economic dominance, exploiting vulnerabilities that open up only after a certain complexity of symbolic organisation has been reached.

One commentator on an early draft of this book suggested that I should use the term 'capitalism' to refer to this colonising system. While I would agree that capitalism is at the heart of the industrial system, the term carries baggage that I prefer to do without. For example, capitalism implies private ownership of the means of production; but as Vaclav Havel pointed out, the socialist republics of the Cold War era shared many essential characteristics with the capitalist world. Industrialism is broader and more pervasive than even Marx's pioneering insights recognised, and can be visualised as a sort of 'gravity well' into which everything tends to roll, resulting in a general distortion of the landscape of value, meaning, and relation.

I would like to thank my colleagues Matt Connell, Neil Turnbull, and Nigel Edley for covering some of my teaching while I worked on this book; and to Eugene Hargrove for permission to include brief extracts from my various papers in *Environmental Ethics*. Thanks also to Andy Fisher, Joel Kovel, Jack Manno, Hal Mansfield, Ugur Parlar, Eileen Patzig, and Gill Wyatt for their diverse criticisms and suggestions over the years; and particularly to C, for her less easily definable but nevertheless essential contribution.

1
Symbolism Breaks Free

External storage and the human brain

How is it that a tall, large-brained type of early human, living on the African plains many millennia ago, began to make tools and to use symbols in ways that would eventually revolutionise life on the planet? And how did these early developments set in motion trains of events that today seem to be hurtling towards an endgame over which we seem to have little control? In order to understand the initial stages of this process, we need to glance backwards, with Merlin Donald, into our evolutionary history:

> The last stage in the encephalisation of hominids came with the arrival of homo sapiens, about 200,000 to 100,000 years ago. The final increase in brain size entailed a further increment of about 20% in overall brain volume. At the same time, there was a continuous acceleration of the rate of cultural change. ... However, around 100,000 years ago ... toolmaking became gradually more refined, until it was revolutionised in the Mesolithic and Neolithic cultures. The correlation with brain size breaks down during this period; cultural change, once it began to accelerate, proceeded without any further change in brain size or, as far as can be determined, brain structure. ... Ritual, art, myth and social organisations developed and flourished in rapid succession. A new cognitive factor had obviously been introduced into the equation. The human capacity for continuous innovation and cultural change became our most prominent characteristic.[1]

The crucial question, as Donald suggests, is this: if brain size and organisation were not changing during the explosion of culture, then what was? Donald offers us a crucial insight when he suggests that with the advent of printing and, later, computers, there emerged an "external symbolic system" (ESS). Within this system, memories, or 'exograms', are "virtually unlimited in both format and capacity. ... systems of exogram storage are infinitely expandable, lending themselves to virtually any system of access, cross-indexing, cataloguing and organisation."[2] Narratives today do not need to be constantly retold in order to survive: they can be stored in books, DVDs, and hard drives, thereby reducing the demands on the brain, which learns a different function: how to *access* information that is stored *externally*. Knowledge can survive indefinitely; and this is the basis of the 'ratchet effect' – Michael Tomasello's term for our capacity to preserve, gain access to, and then modify earlier techniques or scientific models.[3] Uniquely among animals, we can store our inventions and techniques so that other humans can experiment with them and, sometimes, improve them. Rather than having to re-invent a technique over and over again with each new generation, reliable means of recording and disseminating discoveries allow us to stand on the shoulders of Newton, Galileo, and Darwin. The downside of this, of course, is that building on earlier discoveries embeds certain initially insignificant decisions into the foundations of technological society; and these may have implications that only much later come back to haunt us.

To return to our main theme, once a certain level of technological capability has been reached, the groundwork has been laid for the emergence of complex forms of symbolic organisation that, while they were originally rooted in the human brain, need not be confined to it. As neural complexity increased, the relation between consciousness and the external environment became more tenuous, and humans began to be able to imagine things and situations that were not, or not yet, physically real. The evolution of human neural organisation beyond a certain level of complexity was therefore a 'gateway event' that made possible entirely new technological systems, opening up possibilities hitherto unknown in nature. "To an extraordinary degree", suggests Donald, "this makes the human mind externally programmable", pointing to "the scripting of individual awareness by culture to enhance our understanding of the universe".[4]

These new technological systems extended human power enormously; but with increasing technological sophistication, together with the harnessing of technology to economic purposes, the question arises as to whether the power is actually human at all, or whether we are becoming increasingly peripheral adjuncts to an autonomous industrial system that is neither human nor natural. Our anthropocentric and individualist conceits that technology is purely an expansion of *human* capacities may be blinding us to the gradual shift in the locus of power; and as Edwin Hutchins remarks, the way we segregate what is 'inside' the mind from what is 'outside' "creates the impression that individual minds operate in isolation and encourages us to mistake the properties of complex sociocultural systems for the properties of individual minds."[5] In much psychology, for example, all mind-like properties are condensed into the individual, as in Tooby and Cosmides' claim that "what mostly remains, once you have removed from the human world everything internal to individuals, is the air between them."[6]

In the case of industrial society, consciousness has been drastically reshaped to fit an environment that is very different to that in which our species evolved, so that as Donald puts it in a later book, "[m]odern conscious experience has drifted very far from the natural streams of events that we imagine to be the normal sources of experience. We now have entire industries that deliberately design and manufacture experience."[7] The implication is that experience is nowadays dominated by ideological systems that owe little or nothing to organic life; and as Donald remarks, the "same shared technology can also be used to deceive, confuse, regiment, and even prevent thought".[8] In particular, consciousness may lose its grounding in our own embodiment, existing only within abstract technological systems; and Katherine Hayles has suggested that given "market forces already at work, it is virtually … certain that we will increasingly live, work, and play in environments that construct us as embodied virtualities." Hayles' response to this is to accept the situation as a *fait accompli*, putting an 'interpretive spin on it' and seeing information and physical being as "complementary rather than antagonistic".[9] But once we accept that reality is bearable only through such 'interpretive spins', then Pandora's Box has been opened, and the replacement of nature by industrialism is a foregone conclusion.

The question, then, is who or what is doing the programming? Of course, there are a large number of ideological systems around the world, each of which programmes its new recruits in its own specific way. Consider Richard Dawkins' example:

> A beautiful child close to me, six and the apple of her father's eye, believes that Thomas the Tank Engine really exists. She believes in Father Christmas, and when she grows up her ambition is to be a tooth fairy. She and her school-friends believe the solemn word of

respected adults that tooth fairies and Father Christmas really exist. This little girl is of an age to believe whatever you tell her. If you tell her about witches changing princes into frogs she will believe you. If you tell her that bad children roast forever in hell she will have nightmares. I have just discovered that without her father's consent this sweet, trusting, gullible six-year-old is being sent, for weekly instruction, to a Roman Catholic nun. What chance has she?[10]

Madness, in earlier ages and in non-industrial cultures, has been thought of in terms of possession by alien entities; and the notion that the mind is 'programmable' sits easily with such views. Furthermore, social reinforcement will cause beliefs to have a 'centripetal' tendency, to adapt Robert Ulanowicz's term: in other words, they will be dragged towards consensual views in a manner analogous to the way interstellar dust coalesces to form stars; and because this is a *social* process rather than one that necessarily depends on interaction with the physical world, it is easy to understand the emergence of consensual delusions that are incompatible with the outside world. The remarkable thing about this is not that we are 'programmable', but the extent to which we are unaware of this situation, comfortably deluded into beliefs about our (individual or collective) 'control' over our situation. A thin veneer of conscious individual agency covers up the systemic workings of industrial societies, concealing the real loci of power.

Emergent empathy and the external world

According to Thomas Wynn,[11] around 300,000 years ago hominids attained the capacity to think in a 'decentered' way. Wynn comes to this conclusion by examining the stone axe-heads and scrapers produced by the early inhabitants of the Olduvai Gorge in what is now the Serengeti Plain in Tanzania, concluding that the greater sophistication of later stone tools arose from an 'operational' ability to conceptualise the tool as it would be after several stages of shaping and trimming. Here, Wynn draws on Piaget's notion of 'decentration', which indicates the ability to transcend the 'egocentricity' that relates everything to the perspective of the individual concerned. Decentred thought enables us not only to view things from the standpoint of other individuals but also to reach beyond a human perspective altogether in order to access *systems* of thought such as science. Thus subjectivity breaks out of its initial confines in the human mind, encompassing what is beyond our own boundaries.

But is the notion that subjectivity can extend into the world simply a metaphor, or is there a deeper truth here? From a perspective that conceives of thought as necessarily a property of the individual brain, the idea of subjectivity as extended beyond the brain is merely a poetic indulgence. According to the widely taken-for-granted separation of the 'individual' from the 'environment', we neurologically model the external situation, and so become cognitively capable of visualising another's viewpoint; and so the evolution of the large hominid brain accounts for our sophisticated symbolic capacities. But the difficulty with this supposedly close relation between brain size and symbolic capacities is that the changes in brain size and in particular, the growth of the frontal lobes to approximately their contemporary size resulted in few technological innovations for at least 50,000 years. This suggests that while brain enlargement may have been *necessary* for technological development, it was not *sufficient*; and for Gregory Feist[12] this indicates that while the *capacity* of the brain may have been relatively stagnant during this period of technological latency, changes were occurring in neural connectivity and organisation. This, of course, is an untestable hypothesis, since, as Feist notes, there is no fossil record of such changes; and it is one that conveniently maintains the Cartesian assumption that the brain is an island of intelligence within an ocean of mere matter.

There is, however, another – and I think more plausible – possibility: namely, that the changes necessary for the appearance of technology involved not so much the brain in isolation, but rather the ways the brain interacted with the world outside – through the development of symbolic frameworks, most notably language. Of course, it is widely recognised that the evolution of thought involved a dialectical process involving artifacts and ways of thinking; but I am suggesting something more than this. As brain organisation was influenced by cultural and linguistic forms, and later by the materialised consequences of these forms such as tools and other artifacts, a dynamic may have developed that in effect extended subjectivity into the outside world, so that the relation between thought and the artifactual realm formed a widening sphere that possessed systemic qualities. Thus a tool does not *only* enable us to enact our material intentions; it also begins to shift the locus of intelligence beyond the brain and, increasingly, beyond human subjective awareness – and the more complex the tool the more marked will this effect be. Because of this developing dialectic between intelligence and the external world, technology cannot be seen as developing directly and simply 'out of the brain', but rather as involving a compounding process which reached further and further out into the world, reshaping it in increasingly radical ways. The time necessary for the feedback loops involved to take effect is somewhat analogous to the phenomenon of 'turbo lag' in turbocharged cars; and it may account for the time lapse between the appearance of hominid brains similar to our own and the symbolic and technological capacities which eventually appeared. Intelligence did not merely reshape the world; it also, increasingly, reincarnated itself *in* the world.

This symbolic 'evolution', then, is not *just* symbolic: it also has a crucial *material* dimension. As Terrence Deacon explains:

> Stone and symbolic tools, which were acquired with the aid of flexible ape-learning abilities, ultimately turned the tables on their users and forced them to adapt to a new niche opened by these technologies. Rather than being just useful tricks, these behavioural prostheses for obtaining food and organising social behaviours became indispensable elements in a new adaptive complex. The origins of 'humanness' can be defined as that point in our evolution where these tools became the principle sources of selection on our bodies and brains.[13]

There is a long tradition in archaeology which has overlooked these developments, because the relation between physical reality and conceptual structure has been viewed as operating in a single direction. That is, we diagnose conceptual functioning on the basis of artifacts, but ignore the influence of the artifacts on conceptual functioning; and more generally, we tend to be blind to their systemic integration. As Colin Renfrew notes, "without artifacts, material goods, many forms of thought simply could not have developed ... material culture is not only reflective of social relations and of cognitive categories; it is to a large extent constitutive of these also."[14] For example, the original cars were patterned after the horse-drawn coach, with the engine simply replacing the horse; and the upright stance and large, spindly wheels were only slowly superseded. In the century during which the modern car emerged, it is highly unlikely that any significant changes in brain structure have occurred; but while the human brain may have been *neurologically* capable of conceiving of a Porsche Carrera at the beginning of the twentieth century, this also required a lengthy process of conceptual refinement involving numerous intervening and sometimes physically realised forms. A certain form of intelligence, then, escaping from its original location within the brain, colonises the world, driving out the natural order as it does so.

Mind as an emergent property

An emergent property is one which results from the systemic interaction of the components of a system, and cannot be understood as simply due to the summative functioning of the parts of that system. For example, family therapists have long used concepts such as triangulation, enmeshment, and scapegoating that are not rooted in individual functioning; and a football team, a termite mound, an ecosystem, or industrial capitalism also possess properties that cannot be explained or understood simply by examining the parts of the system. Over time, emergent properties may lead to the crystallisation of more complex *physical* structures, so that the termite mound, for example, is the physical manifestation of a certain form of organisation. The evolution of life itself can be seen as dependent on emergent properties, as matter tends to organise itself into more complex, lower entropy forms. In a classic experiment, for example, Stanley Miller showed that a simple mixture of water, methane, ammonia, and hydrogen – a mixture designed to replicate the earth's primordial atmosphere – when subjected to electrical discharges intended to simulate lightning, generated complex hydrocarbons including amino acids.[15]

Although emergence is so common in artificial and natural systems that it is the rule rather than the exception, science has often struggled to recognise it. Indeed, science itself is traditionally built on the isolation of factors in controlled settings, thereby eliminating the emergent properties of complex interactions between many variables; but as scientists are now becoming more aware, in the real world this sort of complete functional isolation is rare. For example, for decades we assumed that the exhaust pipes and chimneys of the industrial world simply expelled gases such as carbon dioxide unproblematically into the atmosphere; but today we are uncomfortably aware that the resulting changes in atmospheric CO_2 concentration have systemic effects that are coming back to haunt us. Whether we like it or not, we are caught up in systems over which we have little control, and which may eventually act to eliminate the source of any imbalance: us, in other words.

It is sobering to consider the direction these changes are headed in. If – as Boyd and Richerson argue – complex brains are evolutionary responses to the 'deteriorated' and increasingly variable environmental conditions of the past several million years,[16] then humans' current effects on the climate, allied to the forthcoming exhaustion of fossil fuel reserves and other environmental challenges, will introduce a phase wherein two types of species will survive: firstly, the weedy, opportunistic 'r-selected' species which can quickly adapt to any available niche, and secondly, those of our descendants who are most techno-logically sophisticated and therefore capable of retreating from a hugely challenging environment into a manufactured 'world', allowing them, at least for a while, to cling to survival. The asymptote of this dialectic between environment and brain-power would seem to be pure cogni-tion existing within a humanly uninhabitable world – the endpoint of our symbolic 'retreat' from the natural order.

But I am getting ahead of myself; and my immediate concern is how subjectivity can be understood as emerging from the interaction of the brain and the rest of the world. Are we to assume – as most cognitivists do – that the brain is separate from the world, connected to it only selectively, and that it 'models' the 'environment', thereby allowing us to behave appropriately? In some respects, this assumption works well, but it is not a complete explanation, for the brain can also behave as part of a larger system which includes aspects of the environment and, sometimes, other people. This is the 'distributed cognition' hypothesis put forward by Edwin Hutchins.[17]

Abandoning the belief that subjectivity is localised *only* in the brain does not necessarily entail following Marvin Minsky's suggestion that we will soon be able to download the contents of consciousness onto a computer disc, leaving the body behind entirely[18] – an idea popularised in William Gibson's novel *Neuromancer*.[19] Such fantasies of disembodied subjectivity, whether they take technologically advanced forms such as this or more traditional, religious dreams of the soul ascending to Heaven, or both at once,[20] represent the persistent Western imperative to transcend our mate-riality and our ecological embeddedness – an imperative that is implicit in the drive towards the autonomy of symbolic structure. The anchoring of subjectivity in embodiment is precisely what makes us human, as Lakoff and Johnson have argued;[21] but what they do not add is that embodi-ment is itself embedded in the world. Subjectivity can reach out into the world through a decentred empathic and sensory awareness, so that while it is normally centred in the brain, it is not restricted to it. The notion of subjectively real entities that are not physically localised is not a new one: we do not ask where 'the family' is; and 'the family' can still exist even when its members are widely scattered. Nor do we try to localise the termite 'psyche' described by Eugene Marais[22] or seek out the 'invisible hand' of the market in the vaults of a bank. Likewise, the force of gravity acts at a distance, yet there is no connective medium between two plan-etary objects. But when it comes to subjectivity, it seems, we cling to the assumption that subjectivity is just an epiphenomenon of neural activity.

Wynn views the appearance of stone tools of increasing sophistication as evidence of evolving intelligence – which, of course, it is. But the appearance of tools also suggests that subjectivity was beginning to expand outwards from the brain. If a stick can be regarded as extending the awareness of a blind man beyond his bodily boundaries,[23] cannot a stone axe be similarly understood? And need it be simply manufactured items that function in this way? Could not watching a bird flying or a rock rolling down a scree slope or driving a car on a racetrack involve a similar extension of subjectivity? Rather than being a point-focus of awareness, can subjectivity be understood as somewhat flexibly emanating from the individual? If so, then it may not only be memory and calculating ability that migrate outwards into systems that extend beyond our physical boundaries. If, as Levi-Strauss suggested, 'animals are good to think with', then are cars, rivers, and forests also good to think with? Does subjectivity inhabit the environment, just as the environment assimilates and shapes subjectivity?

If so, then human nature is dependent not only on our embodiedness, but also on our 'enworldedness'. Many tribal peoples report a kind of empathy with other creatures that goes beyond imagination into something like telepathy; and supportive evidence for this can found not merely in the more fanciful and popular accounts, but also in reputable ethnographies.[24] One way of understanding this is to recognise that such peoples do not embrace the clear division between (intelligent, structuring) mind and (passive, structureless) world that we do, so that – in Tim Ingold's words – the world is "saturated with powers of agency and intentionality".[25] Subjectivity can be viewed as emergent from the entire system of mind-in-the-world rather than simply as an epiphenomenon of neural interaction.[26] Or – given that mind and world derive from the splitting of an original unity – perhaps there is something in Schelling's notion that, as Robert Richards translates it, "the ideas that constituted nature's creations were not captives of individual minds, but stood beyond self and nature, though were realised in both. Hence the solution to the puzzle of Goethe's epigram that 'an unknown, lawlike something in the object corresponds to an unknown, lawlike something in the subject.'"[27]

We might glimpse here an understanding of one of the more conceptually elusive aspects of human psychology – creativity. Since the pioneering research of Frank Barron in the 1960s,[28] few new ideas have been forthcoming, although there has been a plethora of mechanistic models which attempt to assimilate creativity to our cognitive preconceptions. True, the 'associative conception' of creativity recognises that creative thinking involves unusual connections; but since the majority of unusual connections are meaningless, the question remains: how do we recognise that tiny proportion of unusual connections which open our minds to new and insightful structure? And this is where the notion of metaphor comes in; for if, like Goethe, we see nature as poetry, and similarly guided by unspoken, analog rules and syntax, then a bone in one creature may be metaphorically related to 'the same' bone in another. As Bateson once remarked, a "metaphor is not just pretty poetry ... [it] is in fact the logic upon which the biological world has been built;"[29] and our ability to predict patterns across natural systems may be one manifestation of a creative metaphorical capacity which also exists in embodied form in the rest of the world. Certainly, it would be surprising if our mental capacities had not evolved to recognise naturally occurring forms and processes. As Karl Popper pointed out, talking about the 'nose' of a dog is anthropomorphising the animal; but this anthropomorphism is not just a flight of fancy, because the noses of dogs and humans are homologous organs, and their common ancestry means that they are related in functional, not just conceptual or linguistic, ways.[30] This is one of the reasons why it is inaccurate to say that mind *models* reality: mind and reality, like noses, are also complementary parts of a whole to which the reductionistic tendencies of science blind us. That is, if we begin from the presumption that mind is separate from the world, we constrict ourselves to a realm of concepts and language, and so inevitably find ourselves having to create models of the world; but if we understand mind as being the expression of a wider embodiment, then creativity and metaphor become essential ways of participating in the world. As Iain McGilchrist remarks, metaphor "is language's cure for the ills entailed on us by language."[31] In recognising subjectivity's rootedness in embodiment, then, we define ourselves as more than minds transported by dumb bodies; and as Richards puts it, when "Goethe, Humboldt, or Darwin rambled ... they discovered not only the divine beauty of their surroundings but their own emerging selves as well."[32] Although such experiences are mysterious to the type of thinking that resolutely maintains the boundary between the world and the mind, in them one can catch a glimpse of the creative person's sense that what 'they' create *comes from outside themselves.*

If this is the case, then the history of consciousness cannot be viewed simply as involving the development of better, more 'intelligent' ways of manipulating the world, nor even as an emergent property of the interaction of billions of neurons. Both these notions *begin* from the assumption that subjectivity is *separate* from the world; but perhaps, like the tree and the woodpecker's beak, there is a preexisting complementarity between subjectivity and world, shaping what Jacob von Uexkull called the 'Umwelt' – that is, the world-as-we are-aware-of-it, a sort of sub-set of physical reality. Put differently, subjectivity and the 'environ-ment' are not two separate systems which then jostle with each other to create a not-too-uncomfortable alliance; rather, our embodied and psychological being may be pre-fitted to interact with our surroundings. Our conceptualisations are sometimes directly anchored to the world and sometimes float free from it; and perhaps creativity occurs 'on the edge of autonomy',[33] as we play on the fringes of what is conceptually recognisable, dipping our toes into the immense unsuspected realm of nature that extends beyond these conceptual shores.

Unfortunately, the notion of the Umwelt has been used to argue that each species is, as it were, sealed within its own, self-constructed envi-ronment which is impenetrable by other species.[34] But this is to assume that the subjectivities of creatures have no common roots, either within their own physical being or that of the world they engage with – which is demonstrably incorrect on both counts. If I watch a blackbird looking for worms beneath the surface of my lawn, it is easy to empathise with her as she bends her head to one side to listen, then digs beneath the surface to extract a worm. Such empathy is not a fanciful anthropomor-phic construction on my part, but a totally unsurprising consequence of the commonalities between the world shared by the blackbird and myself, as well the similarities between our sensory systems. Jack Turner, discussing the apparently pointless habit of pelicans in soaring high above mountain ranges, argues that empathy goes well beyond scientific understanding:

> It is no more odd to say that pelicans love to soar and do so in *ecstasy* than it is to say what we so commonly say of human love and ecstasy: that our heart soars. … When I see white pelicans riding mountain thermals, I feel their exaltation, their love of open sky and big clouds. … I believe the reasons they are soaring over the Grand Teton are not so differ-ent from the reasons we climb mountains [and] sail gliders into great storms … Indeed, in love and ecstasy we are closest to the Other, for passion is at the root of all life and shared by all life.[35]

Information is not neutral

If the mind reaches out into the world, the other side of the coin is that characteristics of the world impregnate the mind; and in a natural context, this will create a benign dynamic that orients us as participants in the natural order. But if the contexts we inhabit are formed through a sort of metastatic evolution that diverges from the natural order, our cognitive capacities become the vehicle of our allegiance to a quite different 'world'. The ESS is unlikely to be simply a passive, neutral storehouse of information, waiting patiently to be accessed by active, intelligent, minds: it may well have emergent properties of which we are quite unaware. Most obviously, information is often selectively stored within various agencies, corporations, media, and government departments, each of which have their own organisational dynamics. What is recorded also reflects a number of decisions that are themselves unrecorded and implicit – concerning what is worth recording, the interests and priorities of researchers and reporters, and what is con-sistent with the ideological landscape. These prior decisions filter the potentially indefinitely large amount of information available, reducing it to a much smaller quantity of ideologically sieved material and promoting the internal consistency of the entire knowledge system. As well as including consciously and intentionally recorded informa-tion, therefore, this system also embodies a more unwitting type of record: that of the priorities, political influences, and taboos of the era, and therefore of the internal dynamics of the prevailing social structure. The ESS, then, can be understood not so much as neutral terrain that we can wander through at will, but rather as an array of well-established paths that have been established over centuries, some of which become main roads while others become overgrown and impassable. We can explore those paths that are currently maintained, but we have largely forgotten others, and we are not free to choose a completely different path. Changing the metaphor, the 'ratchet' does not simply prevent knowledge being lost; it also maintains a certain cultural momentum that operates in a specific *direction*.

This interpretation of external storage has its limitations, of course. Those with originality and courage can indeed wander away from the established paths of accumulated information, although it may take the insights of an Einstein or the moral commitment of a Gandhi to signifi-cantly change the direction of society. But such individuals are rare, and 'new' ideas are more often variations on old ones, confirming the appar-ently inevitable direction of our technological future. 'New' cars possess more speed, acceleration, fuel efficiency, and so on; but few people can envision a quite different transport system – or a context in which we have less need to travel. The result is that certain decisions made centuries ago now seem unchallengeable aspects of 'reality', especially if technologically materialised as 'part of the landscape'. Unfortunately for us, while the natural order can afford us a certain amount of leeway in that direction, humouring our inventiveness, our technologically created 'worlds' are not worlds at all, but clearings in the much larger forest of natural reality which we ultimately depend on. The general failure to recognise this reflects a potentially fatal lack of wisdom.

Given that the complexity and depth of this landscape often exceeds our cognitive ability to penetrate it, it inevitably embodies dynamics and directions that we are unaware of. We incorporate established attitudes, along with associated bundles of supporting information – mainly from a media which has its own unexamined ideological biases and commercial pressures. This is the sort of world most of us are born into; and given the plasticity of the infantile human brain, our cognitive structures quickly become consistent with it, building ideology into our cognitive processes. Brain development involves competition for neural resources;[36] and learning the skills necessary to access 'pre-digested' information from external sources is generally more economical than 'first principle' thinking, causing a capacity for critical, independent thinking to fade. In any case, given the sheer quantity of information we are assailed by in the industrial world, we become, in Kenneth Gergen's terms, 'saturated selves';[37] and under these conditions, the temptation to specialise narrowly and to retreat from the world's complexity into a socially reinforced, symbolic, 'parallel world' of cognitive categories may be irresistible. As Deacon remarks, "prefrontal overdevelopment has made us all savants of language and symbolic learning," so that we "tend to apply our one favoured cognitive style to everything."[38] Although the autistic character portrayed by Dustin Hoffman in 'Rain Man' – able to assess, with amazing accuracy, the number of matches spilled from a matchbox, but incapable of mastering everyday social routines – is an extreme example, we are nevertheless similarly distracted from crucial issues. For example, in a seminar in which we discussed the alleged shrinkage of political freedom in the UK, one of my students volunteered that he didn't have time to think about such things because he was too busy. "Busy with what?', I enquired. "Oh – things like what to have for dinner, whether I need to fill up with petrol", he answered. The loss of meaning in this intoxication with the minutiae of everyday life, and the loss of engagement with the fundamental issues that order our lives, is different only in degree from that experienced by the eighteenth century prodigy Jedediah Buxton who, when asked whether he had enjoyed a performance of Shakespeare's Richard III, could only reply that "there were 5202 steps during the dances and 12,445 words spoken by the actors" – numbers that turned out to be precisely accurate.[39]

This illustrates a general characteristic of industrial society: we tend to sacrifice the weighing and balancing of an integrated array of cognitive, sensory, and embodied capacities in favour of a few intensely developed symbolic abilities. For example, a pipeline engineer will be an expert in his or her particular speciality, but may be less familiar with aesthetics, environmental ethics, or the impact of transport on climate change. In effect, we export this type of awareness or decision to higher up the chain of command, or – more likely – to the implicit functioning of the system itself. To take a specific example, Brian Martin has studied the inquiry into a proposal to lay three oil pipelines through an area of metropolitan Sydney, showing how the inquiry systematically narrowed the frame of reference of the debate to local, economic, and technical practicalities. Conversely, the inquiry ignored broader and more fundamental issues such as whether growth in oil consumption was necessary or desirable, the character of community participation in the inquiry, and the adequacy of current distribution policies. Martin demonstrates that such refusals to consider the broader context of the proposal weighed against environmentalists' objections to the pipelines, since these objections were often persuasive only when located with a broader ecological and cultural context. Within a narrower perspective which assumed the need for an increasing volume of oil, the inevitability of the growth of air transport, and the primacy of economic justifications, then the project seemed perfectly justifiable.[40] Similarly, at the 2009 trial of the 29 climate change protesters who hijacked a train carrying coal to Drax power station, judge Justice Spencer banned any consideration of climate change.[41] This illustrates how the most important and fundamental issues are often sidelined during policymaking and legal proceedings, precedence being given instead to detailed points of law and policy that are consistent with an unchallenged social context. Put differently, government policy, the definition of 'acceptable' behaviour, and the supposedly independent legal system all begin from the same assumptions and embody the same compartmentalisations that affect other areas of society, and so in effect constitute complementary components of an integrated system that repels any challenges to its internal functioning. Any particular decision may appear perfectly rational when supported by a selective web of consistent beliefs and laws; and it is only when the entire web is examined in the broader context of the natural world that its unrealistic character and its divergence from natural processes become apparent.

Similarly, a leading textbook on road design encapsulates its ideological assumptions in the first two sentences:

Everybody travels, whether it be to work, play, shop, do business, or simply visit people. All foodstuffs and raw materials must be carried from their place of origin to that of their consumption or adaptation, and manufactured goods must be transported to the marketplace and the consumer.[42]

This is perfectly reasonable and unremarkable within a context that normalises the ideological baggage carried by such terms as 'raw materials', 'consumer', 'everybody', and 'marketplace'; but seen from a broader, ecological perspective which views 'business as usual' as problematic, it becomes alarming and questionable. To the extent that the ESS incorporates such ideological slants, it encourages an accelerating momentum in one particular direction – that of the expansion of industrial capitalism – regardless of the ultimate destructiveness of this direction.

By 'out-sourcing' information storage from our brains to the ESS, we are also out-sourcing the decision-making processes, ideological preferences, and much of our power to influence events. With the emergence of global capitalism, the ESS has become an ideological force that is all the more powerful because of its perceived universality. While the slogan 'there is no alternative to capitalism' sounds overtly political, the view that there is no alternative to the sort of information that is widely viewed as 'factual' is often tacitly accepted, and is therefore ultimately more pernicious. The finding that many of us are feeling impotent in the face of global political and economic developments, leading to unprecedented levels of apathy and alienation,[43] suggests that our intuition is ahead of rational consciousness in recognising that we have lost control of events as capitalism suffuses personality and social life. The metaphor of the 'social construction of the person' is ironically becoming more accurate as other sources of the self fade, although unfortunately constructionists take this metaphor as a generally applicable model rather than an index of a pathological process of recent origin.

The development of symbolic abilities

These current trends have their roots in evolutionary developments which reach back into prehistory. As Terrence Deacon argues, the growth of the prefrontal cortex of the human brain was not simply an addition to an already complete organ: rather, the emergence of language in particular and symbolic abilities in general represented a movement away from relatively direct representation of the world towards a system in which symbols primarily register their relation to other symbols within what Deacon refers to as "a closed logical system".[44] In William James' words, "the intellectual life of man consists almost wholly in his substitution of a conceptual order for the perceptual order in which his experience originally comes."[45] As this system grows, it develops its own character and momentum, drifting away from the natural world and eventually, as I will argue, becoming hostile to it. Mostly, we inhabit the border-zone between reality and fantasy: we read a novel or watch a film, becoming absorbed in it before returning to the physical realities of our daily existence. But over a longer timescale, those physical realities may themselves, to a greater or lesser extent, be manifestations of fantasy. The car and the freeway, and the field and the cow, have evolved out of originally natural contexts into the materialised fantasy that is modern life. Such materialised fantasies are more difficult to walk out of than a cinema; and experiencing them as temporary phenomena, emerging from a specific cultural background and soon to disappear along with their inhabitants, requires both a broad historical perspective and a firmly embodied sense of one's own identity as a particular sort of natural creature.

As Deacon argues, following Charles Sanders Pierce, the greater associational capacity of the cortically enlarged brain allows us not merely to represent aspects of the outside world 'iconically' (that is, based on resemblance between sign and object), or 'indexically' (through a causal linkage or a connection in time and space), but also 'symbolically'; that is, we can impose socially agreed patterns and relations on the world rather than being directly tied to the world through our sensory system. This is because symbolism allows us to abstract properties from the world as opposed to simply representing sensorily discrete entities as they present themselves to us. For example, we can pose the class of red things, or round things, or tools, or those that are made of copper, or those that need a 12-volt battery to operate, so that these qualities become basic components from which we can reconstruct the world in novel ways. Likewise, we can also mentally decompose an entity into its components: a sample of petroleum can be understood as made up of varying proportions of a range of hydrocarbons, which can themselves be conceptualised as composed of their constituents, carbon and hydrogen. Similarly – for better or worse – we can categorise people according to a range of characteristics: whether they are socialists, or evangelical Christians, or brown, or HIV positive, or female, and so on.

As the last three of these examples illustrate, not all such categorisations are arbitrary social judgements, and some are more directly related to physical and biological reality than others. However, it is clear that even if we restrict ourselves to those categorisations that are firmly rooted in physical reality, this is far from guaranteeing that the social systems derived from these classifications will be benign or ecologically sustainable. For example, categorising people according to whether they are male or female may lead to the sort of repressive social systems experienced in Britain in the nineteenth century – or perhaps even to the sort of society so depressingly well portrayed by Margaret Atwood in *The Handmaid's Tale*; and industrial society, although firmly based on penetrating chemical insights, is nevertheless in various ways unsustainable. Conversely, the belief among one Native American group that their religious practice enables the sun to rise each day, while inaccurate if interpreted literally, may nevertheless foster a sense of meaning and cultural strength. Social organisation is related to psychological or physical characteristics only in distant and indirect ways; and for a society to be sustainable it needs to be congruent with physical reality not only at the elemental level, but also at the higher-order organisational levels. While industrial society fulfils the first of these criteria through its mastery of chemical and physical processes, it is notably lacking so far as the second criterion is concerned. At the very least, sustainability would depend on recognising the extent of our ignorance in this latter area, together with an appropriate degree of humility.

A major root of our problems, therefore, lies in our tacit belief that specialised symbolic meanings are somehow more real than, and so can replace, overarching physical and ecological organisation. Our 'control and mastery' of nature is clear at elemental levels; but once we reach the levels of complexity involved in ecosystems, we are painfully inept and ignorant. In the past, because of the limited reach of knowledge, we necessarily lived within the constraints of the natural world; but as symbolic systems have increased in sophistication and power, we have gradually substituted artificial environments for our immersion in nature. In effect, we have prioritised one particular type of central nervous system function – the maintenance of autonomous symbolic systems – and relegated other, 'older' ones such as sensory representation to minor roles, rather than *integrating* new and old within one system. Since the structures mainly involved with the 'older' functions are those which developed during our ancestors' long evolution in intimate relationship with the natural world, it should not surprise us that our current maladjustment to the natural world is parallelled by our view of these older parts of the brain as 'primitive' and 'ancient' – the implication being that civilisation has moved beyond these quaint, simple, adaptations. Blinded by our technologically enabled self-sufficiency, we forget that the techno-logical realm flourishes only within a particular sort of biosphere which is vulnerable to the unintended consequences of that same technology.

Our ability mentally to deconstruct the world thus becomes a means towards its *physical* deconstruction. The first member of our species to 'see' a rock as a potential axe-head was the forerunner of later individuals for whom the transformation of the world has been a taken-for-granted basis of life. Rather than accepting the phenomenal reality of the world as largely given to us, we instead view it as made up of smaller, more 'real' elements and components – atoms, molecules, cells, and so on – constituting a cornucopia of 'raw materials' which await their transformation into useful commodities through industrial processes. Cognition generally operates in terms of these components, not with the entire phenomenal reality: we can count trees, but be rela-tively blind to the forest as a whole. We view the phenomenal reality of the world simply as the current, incidental arrangement of these components – not as a reality we inhabit and are defined by, but as a provisional terrain, an opportunity for us to use, transform, exploit. This cognitive attitude is enshrined in the philosophical dictum somewhat loosely referred to as the 'naturalistic fallacy', which claims that what *is* should not be taken as a guide to what *ought to be*. Applied to human institutions, this rightly suggests that the status quo is not inherently virtuous; but applied to the natural world, it dismisses the cumulative value of evolved life and reinforces the industrialist assumption that the world can be reshaped to fit industrialist needs.

These effects brings us face to face with the potentially disruptive impli-cations of symbolic power that has escaped from its earthly constraints. For example, as Deacon suggests, one way of understanding the discon-nection of words from the world is to view language as "an independent life form that colonises and parasitises human brains, using them to reproduce".[46] Thus humans become the unwitting agents of a system they originally used to survive. As Deacon conservatively puts it:

> By imagining language as a parasitic organism, we come to appreci-ate the potential for conflicting reproductive interests, where some language features might occur at the expense of the host's adaptations, and the possibility that many features have more to do with getting passed on from generation to generation than with conveying information.[47]

This insight is one that can be generalised to include other symbolic abilities; for language is not a self-sufficient entity which alone parasitises the brain, but rather one component of a well integrated whole that in its present form also involves economics, the legal system, much science and technology, the media, and education. In short, the entire infrastructure of global industrialism can be viewed as an integrated, rapidly evolving system that has grown out of a parasitic relationship with the human brain. While this relationship may have originally been a symbiotic one, it is today increasingly one-sided as the symbolic tail wags the dog of embodied human life; and in many respects, the health of the economic system seems to have human costs which are increasingly obvious.[48]

Even in the early days of language, the development of symbolic abilities may have been less oriented around meeting human needs than our contemporary anthropocentric hubris might assume. Deacon suggests that the prefrontal cortex evolved *in response to* symbolic demands, as opposed to the alternative scenario in which symbolic skills *followed* the evolution of the large hominid brain. This suggests that language in particular and symbolic skills in general were already operating systemically enough to exert evolutionary pressures of their own. Viewing language as a parasitic organism also opens up the possibility that, like any organic structure, symbolic capacities can *evolve*. And while both language and the human brain have evolved, language, according to Deacon, has done most of the evolving; and those features of language that are most easily learned are strongly selected for, resulting in the extraordinary ease with which children acquire their native tongue.

If the realm of symbols has developed to become a quasi-autonomous system that transcends the limits of the human brain, there is considerable potential for humans to be torn between symbolism and our more basic inherited propensities. From Freud onwards, theorists and psychotherapists have pointed to conflict between our embodied propensities and the cultural systems we inhabit as a cause of psychological maladjustment – a conflict captured most clearly in Carl Rogers' concept of 'incongruence'. Given that it is our sensory awareness that anchors us into the natural world, this incongruence implies a growing alienation between the symbolic function and the natural order, both within the person and more generally. A current irony is that the symbolic realm is still rooted in and therefore dependent on the human brain, itself part of the natural order; although with the increasing reach of technology into genetics and biology, this is a situation that is likely to change, with potentially enormous consequences. To paraphrase the title of

Bill Joy's much-read paper, 'the future doesn't need us',[49] although with historical hindsight it seems likely that the symbolic precursors of technological dominance may have grasped control over human destiny much earlier than is generally suspected.

How did technological symbolism achieve the 'escape velocity' that propelled us away from an ecologically defined role on the earth? As we were colonised by symbolic structures, we seem to have been driven by a desire to separate ourselves from our earthly origins, accentuating the distance between ourselves and other creatures. As Deacon puts it, by

> by failing to appreciate the constitutive role of lower forms of reference – iconic and indexical reference – this perspective kicks the ladder away after climbing up to the symbolic realm and then imagines that there never was a ladder. This leaves symbolic reference ungrounded and forces us to introduce additional top-down causal hypotheses, such as the existence of an ephemeral soul or the assumption that there can be forms of computation of mental language that are intrinsically meaningful.[50]

It is not obvious how this cognitively demanding emphasis on complex symbolic processing got started, given that a partial, fragmentary language would have been of little practical use in the early stages of its emergence. However, if one considers symbolic processing in general rather then simply in terms of language, it is not difficult to see how a distinctively human capacity for such processing would have had an immediate payoff. The ability to see a stone as a potential tool if carefully chipped to reveal the knife-like blade hidden within it has clear benefits for survival, as does the capacity to see a reed as an arrow-shaft. It would not be surprising if this ability to 'see' something as potentially something else developed in specific domains before becoming generalised to others; so already we can begin to see a building momentum to symbolic ability that would carry it over hurdles towards less immediately advantageous domains. We can also see how these initial stages of symbolic organisation would import certain implicit orientations to the world, the consequences of which would only much later become apparent. As Abraham Maslow's well-known aphorism has it, "if the only tool you have is a hammer, everything begins to look like a nail".

We should also recognise that initially unremarkable social decisions may lead to fateful unintended consequences. In the technological realm, for example, cars triumphed over railways despite the greater energy efficiency of the latter. Likewise, those West African peoples who developed open-field yam farming, thereby creating the large and potentially waterlogged clearings favoured by breeding malaria-carrying mosquitoes,[51] had little idea of the health implications of their actions; nor did the pioneering oilmen of British Petroleum, opening up the Saudi oilfields in the early twentieth century, foresee any climatic problems related to burning fossil fuels. *Homo sapiens* is largely blind to the long-term consequences of technological development.

Contemporary and early human brains

Our growing absorption into the symbolic order is consistent with findings that hominid brain size rapidly increased during the Middle Pleistocene epoch before *decreasing* by at least 100 mls in contemporary humans, accompanied by a corresponding decrease in overall body size.[52] While it is not difficult to understand why brain size increased at the same time as human communication and cognition grew in sophistication and complexity, how can we account for the *decrease* in brain size over the past 100,000 years, from a maximum of around 1460 grams to a present-day size of approximately 1300 grams?[53] This becomes doubly perplexing when we consider that far from stagnating, culture and technology have developed out of all recognition during this period.

It is only perplexing, however, if we assume that the sophistication of culture and technology directly reflects the sophistication of the brain. One of the most significant aspects of the evolution of culture, as we saw above, is that memory, according to Merlin Donald, became increasingly *external*. And as Deacon points out, as cultures grew in complexity, there was a parallel expansion of the need for enormously complex communicative 'circuitry' – "well beyond any reasonable hope of housing in one body,"[54] as Deacon suggestively puts it. This is consistent with the view that intelligence, as well as information storage, became more external. Even though he recognises that cultures resemble ecologies,[55] Donald explicitly rejects this hypothesis, arguing that "books and visual symbols are dead things which, left to themselves, sit on shelves. The ESS is not the field of action, the process of thought occurs in the individual biological mind."[56] Furthermore, "intelligence is isomorphic with our conscious experience of the world."[57]

But Donald's assertion seems increasingly questionable in the context of the contemporary world. Whereas the use of stone tools by the earliest anatomically modern humans demanded skilled hands-on knowledge of chipping techniques to produce a cutting edge, the tools we use in the contemporary world are often *designed* to require minimal individual skill or understanding of the manufacturing processes they entail. Using an MP3 player, sat nav, or computer, for example, is not cognitively demanding. The technology used to *create* these devices, of course, is highly sophisticated, involving a large number of designers, each of them specialists in their fields; and it will also draw on technological expertise accumulated over generations, stored in books and on computers, and physically realised in manufacturing processes that draw on the accumulated skills of generations of scientists. However, any individual need only possess a narrow range of specialist skills, allied with access to stored resources, greatly reducing the cognitive demands. Similarly, collecting our food from the supermarket requires no special knowledge of wild animals or the location of nut-bearing trees; and keeping warm depends on the flick of the central heating switch rather than the capacity to collect firewood and make a fire, often in difficult conditions. In short, while early modern humans' manufacture and use of tools and implements demanded considerable cognitive sophistication, there is now an enormous gulf between the skills and knowledge required to use a modern appliance and those needed to manufacture it.

Curiously, we tend to regard intelligence as valid only if it includes an *awareness* of the calculations involved, but in fact these are far from the most demanding calculations we make. Our bodies are enormously intelligent about how to fight off infection, heal wounds, how to retain our balance when we run, and so on; but because these are not conscious processes, the intelligence they imply tends to be disregarded. Likewise, plants are enormously intelligent in resisting diseases and assimilating the appropriate amounts of water and nutrients, and salmon manage to navigate back to the mouth of a river after roaming in the ocean – a feat that would be quite beyond us. If we relax the anthropocentric criterion that intelligence has to include an awareness of itself, then intelligence is all around us in the world. Furthermore, our tacit insistence that the prototype of all intelligent entities is the human brain conceals the intelligence of natural and industrial systems. Our *conscious* intelligence is actually rather limited and narrowly focused, and we pay a heavy price for insisting that nothing can be intelligent unless we can perceive it as such; for one of the consequences is that we remain blind to the gathering systemic power of industrialism.

The emphasis of 'Western' intelligence since the Enlightenment seems to have been away from an embodied, practical relation to the world and towards a more detached, manipulative style, almost as if intelligence today primarily involves the abstract manipulation of symbols and is only secondarily concerned with real-world consequences.

Although the intelligence of symbolic systems is permeating and transforming the world, our own mental capacities become more individually focused as we lose touch with natural realities; and personal awareness and understanding of what is outside ourselves is being replaced by the ability to purchase technological power. For example, Albert Borgmann tells us that in the eighteenth century, two methods were used to calculate longitude. One was the 'lunar distance method', which required an "intimacy with astronomy". The other was John Harrison's chronometer. As Borgmann comments: "Here is an example of how the progress of information technology yields information more spontaneously and easily while at the same time it disengages us from reality and diminishes our expertise".[58] Today, of course, we don't even need to be able to use a chronometer, now that GPS is available.

Contemporary navigational technologies, then, do not simply 'make it easier to get around'; they also undermine our sense of being located in a world that, as we significantly phrase it, *makes sense*. As Borgmann notes elsewhere, for the Blackfeet,

> mountain peaks and rivers were large-scale landmarks; buttes, trees, and cairns were more particular points of orientation. ... The word *orientation*, of course, derives from the rising sun (*sol oriens*) and reminds us of the role heavenly bodies once played in pointing the way. We need to understand, however, that sacred sites ... were not just precursors of navigational technology, of instruments. Rather the great monuments served to establish something like an ordered space. For the Blackfeet, the continental divide in what is now Glacier National Park was the backbone of the world, the ridge that gave reality its shape and stability. Chief Mountain was the lofty point where Blackfeet boys learned to comprehend the world and their place in it.[59]

Thus while navigational technologies have obvious advantages over more dated means of navigation, they also, more subtly, import an enormous loss of meaning, undermining what R. D. Laing referred to as our 'ontological security'.[60] More generally, as we increasingly relate to the world through screens and newsprint rather through embodied action, we lose our multiple anchorages in the world; and our situation becomes similar to that of the occupant of John Searles' 'Chinese Room',[61] in which an English-speaking person is locked in a room and receives a sequence of Chinese characters ('questions') to which he has to give an appropriate answer. As he speaks no Chinese, he has no idea what the 'questions' are, but he does have a compendium of the appropriate answers to any given 'question'. All he has to do, then, it to find an incoming Chinese character in his compendium, note the corresponding 'answer', which he then writes on a piece of paper and puts through the slot linking the room with the outside world. Searle's point is that the person can give the 'correct' answers even though he has no comprehension of either the 'question' or the 'answer'.

A notable feature of the 'Chinese Room' is that meaning resides elsewhere than in the individual's awareness. By analogy, many people in the industrialised world carry out tasks such as manipulating figures, getting from A to B, or responding to requests by phone, email, or letter, while having very little idea about the overall functioning of the organisations they are part of – and even less of the entire industrial system. The higher-level, more intrinsically meaningful decisions are made either by a tiny minority of powerful individuals, or else reflect the intrinsic functioning of the system. For example, many share dealers in major stock exchanges make their decisions according to pre-established, often computerised, algorithms, so that if share prices rise above or fall below particular values, this will automatically trigger buying or selling. By ceding decision-making power to software, we become operatives within systems we neither comprehend nor control. As Donald insightfully puts it, the "external memory field is really a sort of cultural Trojan Horse into the brain" which can "play our cognitive instrument, directing our minds to predetermined end states along a set course."[62] The ESS, as I suggested above, is much more than a mere storage facility; it is an active, expanding, and intelligent system with tendencies and objectives we are largely oblivious to.

In a sense, this externalisation of what we still fondly refer to as 'human' intelligence is a logical – indeed inevitable – step. As we noted above, greater intelligence, all other things being equal, requires greater brain size; and to maintain the same degree of functional integration, the number of connections between neurons must increase geometrically with the numbers of neurons. For this reason alone intelligence is distributed in highly complex systems; and in the case of industrialism, the individual human brain forms one tiny, individual locus of intelligence. With tongue only partly in cheek, Merlyn Donald suggests that "human evolution might be reconceived as the Great Hominid Escape from the Nervous System."[63] In other words, cognitive activity has broken out of the individual mind, if such ever existed, to colonise the cultural realm beyond, whereupon it forms a system whose emergent properties transcend those of the brains that were its starting point. In a sense, this takes us back to our cognitive beginnings, when we existed within a (natural) order whose overall functioning we understood hardly at all. There is, however, a crucial difference between our embeddedness in the natural and in the industrial systems: in the former case, the system is that which has allowed life to evolve over billions of years, whereas in the latter case, the system is one which has been immensely destructive of life.

While we generally assume that the type of intelligence demonstrated within the industrial system is simply an extension of human ability, we need to distinguish, as Edwin Hutchins reminds us, "between the cognitive properties of the sociocultural system and the cognitive properties of a person who is manipulating the elements of that system."[64] Increasingly, as IT systems in particular become more sophisticated, human intelligence becomes relatively trivial: just as the role of the fireman on an old-fashioned steam locomotive is simply to keep shovelling coal into the furnace, so the role of humans in IT systems is to input data to a computer. If Copernicus, Darwin, and Freud have each dethroned humans from our supposed centrality in the universe, the fourth – and probably final – dethronement is the still-nascent realisation that 'our' intelligence and technology are not ours at all, but the fruits of an external system which uses the human brain for purposes we cannot fathom. In Hutchins' terms, the intelligent systems we use are not so much extensions of human intelligence, but systems *"from which the human actor has been removed."*[65] Specifically,

> The computer was not made in the image of the person. *The computer was made in the image of the formal manipulations of abstract symbols. And the last 30 years of cognitive science can be seen as attempts to remake the person in the image of the computer.*[66]

Just as Descartes methodologically invalidated human sensory processes, so the computer model of human intelligence limits human powers to abstract manipulation. Although the physical world contains an abundance of symbols – think of pawprints, smoke, birds flying in a particular direction as an indicator of water or of the presence of a predator – today, we tend to assume that symbols are in the head.[67] Intelligence, Descartes declared, is about *thought*, not *sensation*; and because of this assumption, which is reinforced by computer models of the person-as-calculator, then our sensory apparatus becomes redundant:

> Remember that the symbols were outside, and the apparatus that fell off is exactly the apparatus that supported interaction with those symbols. When the symbols were put inside, there was no need

for eyes, ears, or hands. Those are for manipulating objects, and the symbols have ceased to be material and have become entirely abstract and ideational.[68]

In this situation – and despite the superficial individuality which serves to conceal our assimilation into industrialism as an ideological system – the crucial boundary is no longer that which separates the person from others or from the ideological system as a whole, but rather that which separates the coalescing *industrial* system, which includes industrialised personhood, from the *ecological* realities of the world. In its contemporary forms, the symbolic representation of embodied nature is an act of violence which transforms it into the 'raw materials' for industrial production; and as Deacon notes, nature gives us few reasons to recode it into symbol-learning problems. In "order for a set of objects to serve as symbol tokens, they ... need to be associated with one another in a pattern that maps to a closed logical system"[69] – a rare occurrence in nature; and this lack of relation between symbols and natural patterns "has cut language off from forces that shape biological evolution".[70] In summary, says Deacon, "symbolic reference itself is the only conceivable selection pressure for such an extensive and otherwise counterproductive"[71] learning emphasis; and recent hominid evolution seems to have been driven as much by the internal dynamics of symbolic systems as by our interaction with the physical world.

It is likely, as Deacon remarks, that Neanderthals "were fully modern and our mental equals. They had a brain size slightly above modern values".[72] What they lacked, however – and what separates their lives from our own – is access to the entire external system of industrialist symbolic order and its material realisation through technological power. References to 'the modern individual', by way of contrast to some presumed historical form of personhood, greatly overestimate the significance of individual characteristics and anthropocentrically ignore the properties of trans-human structures – ideological, political, and technological. Consistently with this view, Deacon argues that the extinction of the Neanderthals was not due to their being out-competed by *homo sapiens*, but rather their lack of immunity to viruses that the more mobile and inter-bred *homo sapiens* had developed resistance to. The irony may yet be that it is, speaking metaphorically, another 'virus' – that of industrialism – that does for *homo sapiens*. In Deacon's terms

though the invention of durable icons may not indicate any revolution in human biology, it was the beginning of a new phase of cultural evolution – one that is much more independent of human brains and speech, and one that has led to a modern runaway process which may very well prove to be unsustainable into the distant future. Whether or not it will be viewed in future hindsight as progress or just another short-term, irreversible, self-undermining trend in hominid evolution cannot yet be predicted. That we consider this self-undermining process as advancement, and refer to the stable, successful, and until recently, sustainable foraging adaptation of *homo erectus* as 'stagnation', may be a final irony to be played out by future evolution.[73]

If symbolic reference has developed its own evolutionary pressures, this is a situation within which subsequent human life must be contextualised. The large size of the human brain relative to body mass means that there are substantial brain areas that are not directly related to motor functioning and sensory decoding, and these are therefore free to develop more indirect and culturally variable forms of representation. Deacon argues that the immaturity of the human brain at birth, as compared to that of other mammals, allows infants quickly to learn the specific symbolic skills required by their particular culture. This specialisation, he suggests, is comparable to the remarkable compensatory skills developed by some individuals with congenital brain defects: we are in effect 'idiots savants' of language – although, of course, true to our anthropocentric hubris, we accentuate the 'savant' part and ignore the 'idiot'. This immaturity and ability to specialise, however, make us peculiarly dependent on the properties of our cultural environments. As Clifford Geertz famously remarked,

> As our central nervous system – and most particularly its crowning curse and glory, the neocortex – grew up in great part in interaction with culture, it is incapable of directing our behaviour or organising our experience without the guidance provided by systems of significant symbols ... To supply the additional information necessary to be able to act, we were forced, [during earlier eras], to rely more and more heavily on cultural sources – the accumulated fund of significant symbols. Such symbols are thus not mere expressions ... of our biological, and social, existence; they are prerequisites of it. Without men, no culture, certainly; but equally, and more significantly, without culture, no men.
>
> We are, in sum, incomplete or unfinished animals who complete or finish ourselves through culture.[74]

Despite the fig-leaf of our supposed individual autonomy, then, we are in fact dependent on the industrial system that currently dominates our cultural context, not only materially but also psychologically. Much of our capacity to 'make sense' of the world comes through the way we embed ourselves in the cultural and sub-cultural forms around us. This does not mean that we passively soak up and reproduce these forms: we may modify or even reject them, opening up alternative types of meaning and structure.

Deacon points out that this sort of behavioural flexibility in response to varying cultural contexts loosens "the linkage between structure and function".[75] In other words, behaviour is no longer solely determined by our physical characteristics, especially those of the brain; and in fact one of the main traits selected for will be behavioural *flexibility*. Although rapid cultural and technological change initially depended on the evolution of the enlarged cortex, this dependency was rapidly inverted as cultural changes generated their own selection pressures. Thus the human brain increasingly becomes one component within an emerging technological system; and the forces driving that system are less and less dependent on, and less and less responsive to, human needs or limitations. Whereas ecological structure is generally marked by a sort of dynamic balance between individual species and ecosystemic functioning, so that the structure of the whole system tends to create niches within which a diversity of species-specific behaviours can flourish, power within the industrial system seems to be much more clearly located within those higher-order organisational strata that set the agenda for the lower, more consciously accessible levels of functioning. This is not to say that we can always locate power within certain organisations; rather that possibilities of behaviour and self-definition are filtered through an implicit industrialist lens – a view that is consistent with the Foucauldian understanding of power as suffused throughout a system.

This neo-Cartesian split between those organisational levels that are controlling and those that are controlled is something of a recurring motif in industrial society, reappearing in the relations between (controlling) people and the machines we operate, between human and non-human creatures, in the denigration of nature to the position of 'raw materials', in the prioritisation of economic factors over all else, and most obviously in the hardware-software distinction. But these are only the most identifiable manifestations of a type of organisation that is itself developing well beyond the stage at which humans are the major players; and underlying all of them is the separation of symbolism and materiality that replaces the much more intertwined relation between these that exists in the wild world. As a particular form of symbolic organisation, having evolved in relative freedom from material constraints, returns to restructure the world, the more playful and diverse interplay between symbolism and materiality we find in the wild world will be replaced by a narrower, technologically consistent form of organisation.

Humans, then, long ago crossed the fateful threshold beyond which we can pose the possible as a transformation of the present. Not only can we see a stone as a knife: we see 'timber' converted to charcoal or houses, mountains containing iron ore converted into machinery, flowing water converted into electrical power. Once these transformations have occurred in the mind, the way is open for the development of a technology that will realise them in physical reality, in an accelerating dialectical process that is both material and symbolic. As Deacon puts it, "implicit in [the] stone tools and social-ecological adaptations [of our ancestors] were the seeds of future human characteristics. [All these changes] were caused by *ideas*."[76] Dreams of power and control, for the first time in evolutionary history, become realisable through the deconstruction and reconstruction of the world; and it is ironically these dreams that have led to our own disempowerment in the midst of a technologically defined world to which we have to adapt – although, ultimately, this adaptation is itself a sort of extinction. In this way, what began as a co-evolutionary change involving the human body and cultural practices achieves a momentum that sweeps us towards the cliff of our own subjugation and possibly extinction.

Cross-cultural considerations

Outside the industrial realm, symbolism has remained much more closely integrated with materiality, and non-industrial peoples typically regard the natural world as their primary source of knowledge. Just as we tend to use calculators, books, and computers as aids to thinking, so tribal peoples, as we noted above, find that animals are "good to think"' with – that is, they see knowledge as existing *in the world* rather than in some humanly constructed repository of information moored loosely alongside it. Among such peoples, then, the 'book of nature' is more than a distant metaphor, and the senses are essential pathways to knowledge rather than misleading distractions from it.

Tim Ingold has compared the type of learning that emerges from direct involvement with the world with that which takes place through books, computers, and formal teaching, later to be *applied to* the world. As an example of the former type of learning, Ingold describes how his father took him for walks in the countryside

> pointing out on the way all the plants and fungi ... that grew here and there. Sometimes he would get me to smell them, or to try out their distinctive tastes. ... If I would but notice the things to which he directed my attention, and recognise the sights, smells, and tastes that he wanted me to experience because they were so dear to him, then I would discover for myself much of what he already knew.[77]

What is happening here? The sort of learning Ingold describes can be understood as an introduction to a realm which otherwise might remain hidden, alien, to subjective awareness. As one becomes physically familiar with this realm, it becomes suffused by subjectivity, in rather the same way that subjectivity becomes shared between intimate friends. As Tebu, a Batek from the forests of Malaysia observed, "our souls live upon the trees. The forest is the veins and tendons of our lives."[78] In contrast, industrialist styles of learning involve not an intimate familiarity with a part of the world, but rather a sort of formal separation from it, involving such symbolic abstractions as energy flows, chemical reactions, or trophic levels. What we have here, then, are two rather different approaches, which in turn are associated with two distinct forms of knowledge, as a study by Robert Ryan illustrates.

Ryan studied the attitudes of various groups towards three natural areas in Michigan, finding that, on the one hand, those people who actively *used* the areas for recreational purposes developed a 'place-specific attachment', favouring minimal management and 'letting nature take its course'. "To many of them", Ryan states, "any tree or shrub, regardless of species, has a place in these natural areas and should be allowed to grow undisturbed".[79] In contrast, staff and volunteer restorationists favoured active intervention strategies involving cutting exotic plants and using herbicides, expressing very different attitudes:

> staff and volunteer restorationists expressed a more *conceptual* attachment; they were attached to a particular *type* of natural landscape such as a prairie rather than to a specific place. For these study participants, seeking another site would be an acceptable option if the one they were working on might change in a negative manner. This lack of place dependency suggests a substitutability of natural areas that is not shared by the other users.[80]

This suggests two very different epistemologies, one prioritising symbolic understandings, the other sensory, embodied relations. Drawing on the distinctions observed by Ryan, we can, rather provisionally and speculatively, elaborate these two epistemologies as follows:

Table 1.1 Polarised understandings of our place in the world

Conceptual epistemology	Embodied epistemology
Fundamental discontinuity between industrial humanity and the natural world.	Humanity is a constituent of the world.
Tendency towards actively changing the world.	Tendency towards living within the existing world.
Attachment to ideas, principles, abstractions.	Attachment to things-as-they-are.
Subjectivity as located within symbolic systems.	Subjectivity as reaching out into the world.
Symbolism as separate from the world.	Symbolism as part of the world.
Understanding.	Empathy and resonance.
Control and mastery.	Living within the existing world.
Only individual humans are intelligent.	World as intelligent.
Internal consistency of thought a priority.	The world takes priority.
Physical world as largely structureless.	Physical world as a source of order.

Others have suggested various similar schemes. Seymour Epstein, for example, has suggested that "people apprehend reality in two fundamentally different ways, one variously labelled intuitive, automatic, natural, nonverbal, narrative, and experiential, and the other analytic, deliberative, verbal, and rational." The first of these ways of knowing is "associated with feelings and experience", and the second with "intellect".[81] Epstein goes on to suggest that this scheme is compatible with those made by other researchers, including Bruner's distinction between propositional and narrative representations,[82] Rogers' discussion of the effects of 'incongruence' between the 'self concept' and 'experience',[83] and Tversky and Kahneman's proposal that there are two common forms of reasoning, which they term 'natural' and 'extensional'.[84] There is also a strong resonance here with the types of functioning that Iain McGilchrist has identified with the left and right cerebral hemispheres.[85]

The divergence I am proposing between forms of knowledge that owe their primary allegiance to an internally consistent symbolic realm, and those that are deeply embedded in the materiality of the world through sensory input and physical engagement, has been developing over many centuries; and consequently it has moved beyond a contrast between psychological styles towards becoming sedimented into the character of society as a whole. Keep and Mayhew, for example, exemplify the different social values attached to symbolic and sensory knowledge in the contrast they draw between the "abstract, theoretically based bodies of high-level knowledge of the type ... required by symbolic manipulators and analysts", and the 'tacit' knowledge required by those they refer to as "lower level" workers.[86] It is not surprising that in societies that are built on the dominance of symbolism over embodied structure, this dominance manifests itself not only in the relations between humanity and the non-human world, but also in the priority given to thought over feeling in the individual personality as well as in the social hierarchies that structure occupational status.

As an example of these two ways of knowing, take our experience of a beautiful landscape. We may well experience feeling of awe, wonder, humility, and so on, feeling that we are in the presence of something spiritual or divine. Now note what thought does to these experiences. We may 'know', for example, that a beautiful sunset is caused by dust particles in the air, somewhat diluting our sensory appreciation. Thus the two aspects of our experience are torn apart – with momentous implications for how we treat the world. While our embodied experience immerses us in the landscape in a felt way, opening our senses as ways of directly encountering it, thought diverts us into a world of symbolic relations, allowing us dispassionately to analyse and manipulate it.

As Ryan's study shows, these two modes of relation to landscape may conflict, both politically and intra-personally. Writ large, these diverging epistemologies are expressed in quite different ways of living. On the one hand, we can live in ways that accept and value naturally occurring diversity, accepting symbolic power as an articulation of naturally occurring relation; or we can isolate symbolism from the world, developing internally consistent symbolic models that extend our power *over* the world. Industrial humanity has clearly chosen the latter option – although individually, as Ryan's study shows, we may retain our sensory attachments as a sort of privatistic choice that is not reflected in economic and political realities. The industrialist symbolic project has permeated almost all aspects of our lives, often replacing direct experience of nature with digitally mediated experience, concealing natural order beneath administrative order, and muting our own embodied experience as natural creatures.

Through accidents of birth, location, and exposure to the wisdom of others, we can recapture our almost-lost experience of embodiment in a natural landscape, as Ingold's account makes clear. Ingold describes this approach as "an education of attention",[87] pointing out that "when the novice is brought into the presence of some component of the environment and called upon to attend to it in a certain way, his task ... is not to decode it. It is rather to discover for himself the meaning that lies within it".[88] Note: 'The meaning that lies within it', not the meaning that can be attached to it. Meaning in other words, exists *in the world*, ready to be discovered by the sensorily aware person. When we succeed in this quest, our behaviour does not need controlling by some separate symbolic guidance system, as Alan Drengson's account of Arne Naess's approach makes clear:

> To develop this sense of extended self is a natural process of maturing, and it does not destroy our ego but helps to moderate earlier tendencies to be self centred in the narrow sense. When we care for our place and others, we come to identify with their needs and well being, and we have a greatly enhanced and larger sense of community and interdependence. Our well being and that of our community are closely aligned. Thus, Naess says, we naturally and spontaneously care for our place and seek to protect it. For this we do not need a moral axiology, set of rules or enforcements held over us to force us to act.[89]

This attunement to the landscape is often inverted by cultural theorists who balk at the notion that knowledge may be found outside a community of words, books, and language. "Astonishingly", Ingold remarks, "meanings that people claim to discover *in* the landscape are attributed to the minds of people themselves and are said to be mapped *onto* the landscape",[90] thus protecting the fashionable Western orthodoxies of symbolic priority and cultural construction. As Vassos Argyrou points out, such theorisations serve to denigrate the thought of the indigenous informant and to elevate that of the anthropologist, who claims "to know the truths about native life ... to which the natives whose life it is are oblivious."[91]

Thus the inhabitants of the industrialised world have developed a different sort of knowledge – a knowledge that, while penetratingly accurate in a limited sense, gains power over nature by casting us adrift from it. Our conscious understanding of intelligence is of an *individual*, quintessentially *human* quality that we then apply *to* the world in order to manipulate it, rather than a *quality of the world* that we learn to participate in. As the French physiologist Claude Bernard put it, the scientist "sees only his idea and perceives only organisms concealing problems which he intends to solve."[92] While impressively perceptive in some respects, this form of intelligence has a crucial deficiency: it is blind to the emergent intelligence of the industrial system it contributes to, and even more ominously, it is blind to its own blindness.

Symbolism takes the reins

The loosening of the relation between brain structure and function allows the 'evolution' of industrialism to occur enormously faster than would be the case if every technological development was anchored to corresponding neurophysiological changes. Just as a computer can run many types of software from word processing programmes to games, so the brain, once it developed to a certain point, became capable of internalising a diversity of symbolic systems which can then 'evolve' while brain structure remains constant. Human symbolism, beyond a certain basic level of cerebral organisation, need not be an *extension* of evolution, but rather can lay the groundwork for evolution to be over-taken by faster and more powerful technological processes, including those which consciously and directly manipulate genetic codes. Thus change is occurring not so much through the natural mutation and selection of species, but through a somewhat analogous, but much more rapid, process in the realm of *symbolic* forms and their technological materialisations. While our participation in this process, through cognition, production, and consumption, brings us material living standards undreamt of by previous generations, it also introduces equally enormous vulnerabilities, since we often have little control over or conscious awareness of the direction of the system we contribute to.

But as I indicated above, there is no *necessary* divorce of symbolism from sensory awareness, or of knowledge from nature; so the splitting of the world that is the basis of industrialism does not become inevitable once symbolic capacity passes a certain point, as the viability of other cultures confirms. In fact, given the unsustainability of industrialism, this splitting is probably best regarded as a cultural dead-end with a limited life-span, not least because the deviation of symbolic structure from the embodied character of the world imports a blindness to those aspects of the world that remain unsymbolised. As Deacon puts it, we

> cannot help but see the world in symbolic categorical terms, dividing it up according to opposed features, and organising our lives according to themes and narrations ... we find pleasure in manipulating

the world so that it fits into a symbolic Procrustean bed, and when it does fit and seems to obey symbolic rules, we find the result comforting, even beautiful.[93]

When combined with our accumulated technological expertise, the next step seems predictable. Deacon's view is worth noting carefully:

> Though the evolution of brains has been about systems for modelling and predicting events in the world, the evolution of symbolic abilities has not just amplified this ability ... it has also introduced an insidiously inverted modelling tendency. The symbolic capacity seems to have brought with it a predisposition to project itself into what it models ... We are not just applying symbolic interpretations to human words and events; all the universe has become a symbol.[94]

Our symbolic capacity does not simply become disengaged from the world as we sense it. It also influences our sensory mechanisms and cognitive processes so that we see, hear, and think about the world in ways that are consistent with the evolving symbolic system, so that the symbolic tail begins to wag the sensory dog as we assimilate the world to fit our conceptual systems. This outward spread of conceptual structure continues like a spreading infestation as technology remakes the world to fit the industrialist symbolic system. For example, an estate agent may view landscapes in terms of their potential for 'development', just as an eighteenth century missionary might have viewed natives in terms of their potential for conversion. This assimilation is not a matter of cumulative inaccuracies in the way we conceptualise the world; rather, it is driven by the increasingly systematic behaviour of the emerging symbolic system as it responds to its own internal dynamics and seeks to assimilate everything to its conceptual schemes. While this is not an inevitable result of the growth of increasingly mediated and weakened responsiveness to sensory input, it is at least compatible with it. As Deacon puts it:

> We are not just a species that uses symbols. The symbolic universe has ensnared us in an inescapable web. Like a 'mind virus', the symbolic adaptation has infected us, and now by virtue of the irresistible urge it has instilled in us to turn everything we encounter ... into symbols, we have become the means by which it unceremoniously propagates itself throughout the world.[95]

This symbolic 'infection' is quickly followed by a *material* transformation, as our models *of* the world become models *for* its technological transformation, and any reference to an earlier, more natural world is dismissed as 'nostalgia' for an 'Edenic golden age' which allegedly 'never existed'.[96] While this suspicion of Edenic 'myths' is sometimes justifiable when it refers to the overblown romanticism of some popular nature writers, it is often employed in academic writing to imply that there never was and never will be a healthier nature, or people who lived more consistently with such a nature. The possibility of a nature that could be healthier and more diverse is unwelcome to those who identify with their roles in the industrial system, and so is energetically rejected.

What is the crucial step in this sequence that allowed the development of technological society? Was it the ability to think hypothetically, and so transcend the existing world? Was it the *ethical* decision to do so? Was it the capacity to selectively focus on certain properties, and to ignore others? Or was it the long-standing hostility of certain Middle Eastern nomads towards the wilderness, deeply sedimented within mythology and recorded within scripture,[97] that licensed the wholesale transformation of the world? While each of these may have played an essential part in the emergence of industrialism, it is in the systemic interaction of all these and more that the roots of a materialised delusional system arises – a topic I will explore in the following chapters.

2
The Natural and the Industrial

Torn between two worlds

Humans live on the cusp of a fundamental upheaval in the history of life on earth, as our symbolic capacity has brought with it the potential to transform, overwhelm, and perhaps destroy the evolutionary process out of which it originated. Yet despite the revolutionary character of this change, it is neither discontinuous with what has gone before nor easily perceived, largely because we do not observe it from some external point, but are ourselves swept up and defined within it. Having largely lost our reference points in the natural order, industrialism itself has become our source of reference, so that destruction becomes 'development', technocracy becomes 'culture', and both become normalised as 'just the way things are'. However, because symbolism grew out of embodiment and is still sometimes congruent with it, the relation between them is not a simple one of conflict and displacement, but rather of gradations of representation, distortion and subversion. Symbolism can enhance our relation to the natural order as well as undermine it; and this book should not be taken as a critique of symbolism per se, but only as pointing to the vulnerabilities that it introduces. Consequently, the realm of the 'human' includes all these sometimes consistent, sometimes opposing, tendencies, so that the current character of industrial humanity is one of tension and contradiction.

Some writers deal with – or rather fail to deal with – this complexity simply by expanding the category 'nature' to include *any* human activity. For example, John Gray suggests that "cities are no more artificial than the hives of bees. The internet is as natural as a spider's web".[1] Likewise, J. Baird Callicott argues that "if man is a natural, a wild, an evolving species not essentially different in this respect from all the others ... then the works of man, however precocious, are as natural as those of beavers, or termites, or any of the other species that dramatically modify their habitats." Despite acknowledging *homo sapiens'* unique cultural capacities, he considers that "the cultural works of man are evolutionary phenomena no less than are other massive structures created by living things like, say, coral reefs. They are, one and all, natural in that sense of the word".[2] Similarly, Steven Vogel suggests that "our artifacts are natural, every one of them", "whether what is being produced is an oak savanna or a tomato patch or a nuclear power plant or a baby."[3]

However, the emergence of technology cannot be viewed as an evolutionary development in the usual sense of the term: while it *depends on* a certain degree of evolved symbolic capacity, from that point onwards it is often disengaged from and destructive to the further evolution of organic life. As industrialised humans, we participate in both natural and industrial processes, making our situation thoroughly ambiguous; but it is becoming clear that given the extent and speed of industrialism's destruction of the natural order, it is – quite literally – a dead end. As Gregory Bateson remarked, *"the creature that wins against its environment destroys itself."*[4]

The conflation of culture and industrialism plays a thoroughly deceptive role in this black drama, disguising industrialism's widespread *destruction* of culture. Holmes Rolston, for example, while arguing for a culture that is "congenial to the values achieved over evolutionary natural history and ... ecological values" suggests in the same book that "[l]egitimate human demands for culture cannot be satisfied without the sacrifice of nature. That is the sad truth".[5] But while this may be a 'sad truth' for industrial 'culture' – if we can swallow such an oxymoron – it is certainly not the case for indigenous societies, as we will shortly see. Kate Soper, like Rolston, assumes the inevitability of this zero-sum game involving nature and culture: we are, she suggests, "both members, like other animals, of a natural species but at the same time quite unlike them in the urge we have to cultural transcendence". Ultimately, she views humans' 'cultural' character as taking priority over our evolved nature, asking whether we can "find more ecologically benign ways of *not* being natural but remaining assertively and distinctively human".[6] Other writers have gone even further in asserting the elimination of natural by symbolic structures, claiming that "nature is a construction of culture".[7] Here, the ontological priority of nature is reversed, bracketing off the entirety of evolutionary history.

The tensions within and between these various viewpoints suggest pervasive confusions about the character of culture and humanity. Are we natural beings, or has the explosion of symbolic capability radically redefined us? Within academia, these tensions are often sidestepped by the division of knowledge into the natural and the social sciences; but we live in *one* world, and it is clear that accepting such arbitrary fault lines in our understanding conceals some of the most crucial issues that face us today. In this chapter, I will explore these tensions, and suggest a way of theorising humanity's place in the world that offers a more coherent understanding.

How fragmented understandings abet industrialism

Anthropological definitions of culture have generally connoted an integrated system of beliefs and meanings which relate human life constructively to the natural world. In fact, culture is as vulnerable to industrialist colonisation as the natural order; and today the term is often used to refer to almost any recognisable lifestyle or form of social organisation, abandoning the integrative and evaluative connotations of earlier usage to produce a laundered meaning so vague that it can be fitted into industrialist accounts without protest. The traditional notion that culture operates in ways that are coherent and systemic, relating our symbolic functioning to our biological and ecological characteristics, giving meaning and purpose to people's lives, and locating the present within a temporal pattern which makes sense of the past and leads into a coherent future – all this is forgotten, mainly due to a fear of being seen to evaluate some societies or sections of society more positively than others.

While this is understandable given past prejudices, something very important has been lost: the understanding of culture's function, and consequently the valuing and safeguarding of cultural integration. For if the term is used simply to denote any loosely connected or geographically coincident group of peoples or interest groups, we lose the socially integrative meanings associated with traditional cultures. It is a short step, argues Russell Jacoby, from considering common sense as a cultural system[8] to contemplating "the 'culture' of drug addicts, soccer moms, or fans of Star Trek".[9] Likewise, the postmodern focus on the diversity of workers' 'life-worlds' has blinded us to the way that they are all structured through a common capitalist framework, leading to what Alf Hornborg sees as "a kind of moral-political paralysis."[10] Thus theorists' detailed focus and failure to engage with the 'big picture' condemns them to acting *within* the larger structures that they fail to address, a form of tacit collusion that leaves the most important questions of our era critically unscathed. Just as an elemental view of physical reality conceals the divergence between the natural and industrial systems, so reducing cultures to small 'chunks' deflects attention from orienting political contexts. Moments of joy or pathos, when taken in isolation, look much the same whether they are part of a political documentary or a 'feel-good' Hollywood film; and studying the 'cultures' of Liverpool football fans or Hackney drug dealers tells us nothing about the larger patterns and forces that orchestrate their lives.

Furthermore, the original meaning of 'culture' as 'rooted in the soil', thereby connecting us to nature,[11] has been replaced by its *opposite*, so that 'culture' – as we saw above – is now often *contrasted to* 'nature'. This is largely because viewing any human characteristics as partly 'natural' is associated in many social theorists' minds with overtones of eugenics and biological determinism, frightening them into taking refuge in an uncontroversial and value-free realm of individual choice and unrooted symbolism, and giving social theory its current ethereal ambience. But letting go of the necessary connections between nature and culture impoverishes both, as Gary Nabhan indicates:

> Camps abandoned, wells gone dry, tinajas drained or spoiled by livestock. Sheep populations corralled into smaller and smaller areas, where they are more vulnerable to birth defects rising out of shallow gene pools or to decimation by exotic diseases. And seasonally migratory bands of desert people being corralled as well, told to stay put on reservations or being enslaved to cotton farmers. What is being lost is more than a chunk of desert nature. More than a waning of native culture. What is being lost is a capacity for a long, deep relationship between wild animals and cultural traditions ... I am worried that as desert sheep slip out of sight, then out of mind, then out of dreams, a vacuum is created not only among desert people but among all people.[12]

Just as a culture separated from nature becomes vacuous and ungrounded, so nature unarticulated by culture is seen as alien, hostile, and meaningless. Today, the term 'nature' has lost its critical and inspirational leverage, becoming a synonym for 'the environment', or indicating the sum of 'natural resources' – an ecologically blank canvas awaiting humanity's imprint. What has been lost from the meanings both of 'nature' and 'culture' is the sense of *structure* that each derived partly from the other: we have abandoned the capacity to differentiate between a system of meaning and a mere assortment. This is entirely consistent with the requirements of industrialism, which *needs* 'human resources' rather than families and cultures, and 'raw materials' rather than functioning ecosystems. Current understandings of both 'nature' and 'culture' are reductive and therefore ideologically loaded, reflecting the colonisation of theory and discourse by industrialism, and bringing nature writing into line with the commercial realm. For example, the focus on biodiversity has diverted us from the necessary emphasis on *relationship* as fundamental to ecology. As Nabhan observes:

> the focus of efforts to conserve biodiversity has often been on species rather than on their interactions. In many cases, ignorance of biotic interactions has led to the decline of a particular plant or animal species that has lost its mutualists, even though it occurs within a formally protected area such as a national park or forest.[13]

In both the cultural and natural spheres, then, there is a persistent blindness to larger structures, so that the reductionist understandings of nature and culture fit comfortably with the reductionist needs of industry; and this blindness is related to industrial humans' greater trust in cognition, with its focus on specific concepts, definitions, and categories – rather than on embodied awarenesses unfettered by verbal meanings or categorical compartmentalisations. Although, as we will see below, these predominantly cognitive understandings often coincide with those of native peoples, a crucial difference is that the latter supplement detailed taxonomies with concepts representing *relationship*. As Shepard Kretch argues in his discussion of Native Americans' relationships to wildlife, for example, what was being preserved "was not a herd ... but one's economically vital, culturally defined ... and ritually expressed relationship with the buffalo".[14]

Tim Ingold usefully distinguishes between technological society, which "sets out to establish the epistemological conditions for society's control over nature by maximising the distance between them", and non-industrial societies which

> through their tools and techniques ... strive to *minimise* this distance, drawing nature into the nexus of social relations, or 'humanising' it. This 'drawing in' has as its object to establish the conditions not of control but of a kind of mutualism.[15]

Much of contemporary 'cultural' life is not so much cultural as *corrosive* to culture. Features such as consumerism, media that are enslaved by corporate interests, the commodification of spiritual belief, and the vacuous celebration of celebrity have all been associated with the fading of the cultural realm, along with meaning and well-being. As Philip Cushman puts it, commercial 'culture' "is a kind of mimicry of traditional culture for a society that has lost its own;"[16] and as Cushman and others have argued, the loss of culture as a system of social integration, resulting in an 'empty self' which lacks foundations, is *necessary* for the flourishing of a commercial realm which exploits insecurities, neediness, and material inequalities. The phrase 'industrial culture', then, is a mystifying oxymoron which tries to paper over industrialism's *destruction* of culture.

In indigenous societies, however, the relation between culture and nature is entirely different; and there are strong arguments for regarding the behaviour and traditions of many (but not all) *indigenous* peoples as contributing to ecosystemic health. In one study, for example, Gary Nabhan compared two wildlife sites – an O'Odham village and a US Park Service wildlife sanctuary, finding that since the human inhabitants had been excluded from the sanctuary, the wildlife had been disappearing. Visiting the two sites on consecutive days three times during one year, ornithologists identified more than 65 species at the village, and 32 at the wildlife sanctuary. As one O'odham farmer explained, "That's because those birds, they come where the people are. When people live and work in a place, and plant their seeds and water their trees, the birds go live with them. They like those places, there's plenty to eat and that's when we are friends to them."[17]

Not surprisingly, Nabhan suggests that "conserving wildlands with minimum human manipulation is at odds with indigenous beliefs";[18] and he comments that the "O'odham elders I know are still active participants in the desert without ever assuming that they are ultimately 'in control' of it".[19] Rather, the desert is recognised as having its own structure and being, into which human culture and life are integrated. It is "a place of songs ... the place where nightmares hide ... a place of power" as one O'odham poet puts it. Describing how they used to walk to the Sea of Cortez to collect salt, one O'odham elder explains, "the ocean birds would begin singing in their dreams. That's where our songs come from."[20] Nabhan explains: "This is where 'inner' and 'outer' become not a duality but a dynamic – like every breath we take. We are inspired by what surrounds us; we take it into our bodies, and we respond with expression. What we have inside us is, ultimately, always of the larger, wilder world".[21] O'odham culture, then, as for the majority of tribal cultures, is an *extension* of the natural order, not a replacement of it; and human action is therefore ecosystemic *in its very nature* – in contrast to the industrialised world, where despite our best conscious intentions, our actions are embedded within a system which is inherently *destructive* of ecosystems. Not only our cultural values, but also our economic and infrastructural contexts (think of the road transport system) make ecologically attuned living almost impossible. In looking for the roots of industrial humans' destruction of the natural order, we need to look not at the biological characteristics we share with other creatures, but rather at the character of the symbolic and material systems that have colonised us.

Consequently, to argue that wilderness affected by the human use of fire, cultivation, or hunting is for that reason no longer natural is to take an *implicitly* industrialist perspective in assuming an incompatibility between culture and nature. While the activities of humans in the industrialised world *are* likely to be damaging, this is much less likely to be the case among tribal groups. The dualistic opposition between so-called pristine and humanised landscapes may represent the effects of industrialism reasonably accurately; but we cannot generalise it to indigenous peoples who, in many cases, have been living in a sustainable and ecologically benign way for thousands of years, and among whom human welfare and ecosystemic health are profoundly related.[22] We would not argue that the presence of bears or beavers rendered an area 'no longer wilderness'; so why do so because humans live there in ecologically benign ways? The crucial distinction, therefore, is not between the *human* and the *nonhuman*, but between the *industrial* and the *nonindustrial*. The wild world *may* include humans, who can play as ecologically benign a role as any other creature, extending the natural order without trying to replace it. But equally, humans are uniquely vulnerable to being colonised by rampantly destructive ideological systems which alienate us from our own embodiment and resonances with wildness.

But our 'programming' by culture is not simply a passive process of inscribing cultural forms onto a receptive biological substrate; and most of us will be familiar with the deep feeling of rightness or wrongness that accompanies certain activities or beliefs. So, too, was Archie Belaney: born in the English south coast town of Hastings and later adopting the quasi-Native American name 'Grey Owl', Belaney clearly felt that the stifling middle-class urban lifestyle of his childhood experience was somehow inconsistent with some deeper current within himself, and that his adopted lifestyle in the Canadian wilderness, equally, was 'right'.[23] The ridicule which such attempts to live a natural lifestyle tend to encounter indicates the fear that motivates our attempted separations from nature – a fear that was also behind the witch-burnings of antiquity as well as the widespread animal killings which still occur today even in supposedly 'civilised' societies. But residual feelings of connection to nature may also be less rare than we might assume: Peter Kahn found that children from Texas, Alaska, Portugal, and the Brazilian Amazon share a strong affinity for the health of the natural world[24] – an affinity that is often discouraged in mainstream educational practice, which generally emphasises an anthropocentric focus on economically significant activities and 'objective' scientific understandings, as we will see in the next chapter.

Distinguishing the industrial and natural domains

The splitting apart of culture and nature, however, is only one of the ways that industrialism diverges from non-industrial ways of life.

In the natural world, the boundary regions between ecosystems ('ecotones') are sometimes particularly rich in diversity.[25] In contrast, in the industrialised world boundaries tend to reflect ecological *disjunctions* rather than loci of relation. The fence that separates my garden from a school playground, for example, does not serve to connect these two areas; nor does the plastic sack into which I put my unrecyclable rubbish integrate what I put in the sack with what is outside it. Likewise, the boundaries between fields clearly separate species of crop, as well as separating the crop from non-agricultural land, policed by an armory of pesticides and herbicides which destroy 'unwanted' species. In the industrial realm, boundaries invariably function to fragment rather than to integrate: the industrial world, at root, is created by reducing and simplifying the natural order according to a conceptual template so that it becomes a collection of distinct entities. As Jack Manno puts it, the commodification of nature "depends to a great extent on detaching things from their surroundings, ecological and cultural contexts".[26] What Manno refers to as 'commoditization' destroys relationship – "with each other, with our communities, with the earth", so that these relationships "grow ever simpler as the forces of commoditization operate."[27] As a result, the meanings associated with the landscape become shallower, more skeletal, more isolated from each other; and the industrialised world becomes a semantic desert containing only those meanings we impose on it, so that knowledge and landscape merge within a seamless industrialised whole. Compare, for example, the meaning of a field of sugar beet with that of an area of primal forest. The former has a single function, and is aesthetically unimpressive and invariant from place to place; while the latter is complex, elusive, multifaceted, variable across locations, and mostly beyond the power of cognition to order and classify.

Industrialism's destructiveness towards natural entities is not simply an unfortunate side-effect, but is inherent in its intrinsic method, which involves the reduction of natural entities – sometimes to their elements – so that they can be reconstituted as commodities. The planting of crops, for example, is necessarily preceded by the elimination of unwanted species; and the extraction of ore requires that it be separated from 'waste'. Something similar occurs in the social and psychological realms, where subjectivity is drained from shared social networks, becoming fragmented into individual desires and preferences, leaving arid areas of mechanistic bureaucracy and allowing the formation of a pool of decontextualised, biddable 'workers'. All these processes hasten the falling away from the industrial landscape of those natural, cultural, and aesthetic qualities that contribute towards human meaning and ecological health.

This simplified relationality is reflected in the relatively small number of roles played by industrial as opposed to natural entities. The relation between my vacuum cleaner and my living-room carpet, for example, is a straightforward one; and it would make no sense to talk of them as parts of an 'ecology' of relations. In contrast, while the silver birch in my garden provides nesting places for numerous birds and other creatures, this is not its sole purpose. It also provides leaves for my compost heap, shade during the summer, protection for the fish in my pond, food for caterpillars, a suitable environment for the snowdrops that appear each February, and so on. The vacuum cleaner has only one purpose. It does not provide a habitat for my toothbrush; nor does it prey on the cushions in my sitting room; and electrical cables do not parasitically entwine themselves around its plastic pipes. Apart from providing a home for a few thousand dust mites, it is ecologically inert. We might classify it as a member of the 'family' of vacuum cleaners, which itself is a member of the category 'housefold appliances', giving the impression that families of artifacts are of the same type as genera in the natural world. But this analogy is misleading, since my vacuum cleaner will not joyfully interact with the world around it, and it will beget no offspring as part of any long-term process of evolution. When it is not in use, it sits inert in my cupboard, serving no function; and when it has lived its useful life the most we can hope for is for a few components to be recycled. So while we classify natural entities and artifacts in similar ways, this does violence to the relationality inherent in nature as well as naturalising ecologically inert artifacts, camouflaging the radical divergence between the two systems and assimilating one to the other.

Particularly in the natural world, distinctions are often the basis of *relationships*, as the colloquial French phrase 'Vive la différence' suggests. Relationship is only possible between two entities which are also distinct; and this distinction can allow a complementary unity at the next higher level of the hierarchy. For example, the relationship between dairy ants and aphids, which benefits both, forms a mutualistic *system*; and this dialectic or 'dance' between separateness and alliance is particularly characteristic of the natural world. Likewise, the separateness of the queen and worker bees allows connectivity at another level, both roles being necessary for the survival of the hive. Similarly, predator and prey are part of the same ecosystem, and although they seem to be opposed at one level, ultimately the survival of each depends on the flourishing of the other. Life, then depends on the dialectic between separateness and connectivity – a dialectic that is often denied by our systems of categorisation, which tend to emphasise *difference* while downplaying *relationality*. Tim Ingold points out that in biological taxonomies, "every creature is specified in its essential nature through the bestowal of attributes passed down along lines of descent, independently and in advance of its placement in the world." Consequently, "difference is rendered as diversity. Thus living things are classified and compared … in terms of intrinsic properties that they are deemed to possess by virtue of genealogical connection, *irrespective of their positioning in relation to one another in an environment.*"[28]

This is not to suggest that such taxonomies do not capture an important aspect of biological organisation; only that they also *omit* the at least as important *ecological* aspects of life. Both scientific and folk biological knowledge are anchored to reality by their subject matter, as is indicated by the convergence between the biological classifications of widely separated societies. To take one of many possible examples, Scott Atran found high correlations among folkbiological classifications in rural Michigan, those of Itzaj forest-dwellers of Guatemala, and the corresponding scientific taxonomies. Specifically, the correlations between the scientific taxonomy and the Itzaj taxonomy ($r = .81$), and between the scientific taxonomy and the folk taxonomy of Michigan students ($r = .75$) were both highly significant.[29] Such 'triangulations' strongly suggest that both scientific and folk classifications are rooted in biological realities that are not 'culturally constructed' in arbitrary or culturally specific ways, but reflect realities that are common to both. According to Atran, such understandings of nature "appear to be naturally selected features of human cognitive architecture, which are designed to spontaneously and effortlessly map those regularities in the natural world that crucially influenced the course of hominid evolution."[30] Cultural theorists would do well to take to heart his further comment that

> Ignorance or disregard of our evolutionary heritage and of the basic cognitive domains upon which much in everyday human life and thought depend can lead to speculative philosophies and empirical programmes that misconstrue the natural scope and limits of the species-specific core of our common sense. This, in turn, incites misappreciation of the cognitive foundations of culture and cosmology, science and society. The intellectual and moral consequences of this misconstrual have varying significance, both for ourselves and for others, for example, in the ways relativism informs currently popular notions of separate but equal cultural worlds whose peoples are in some sense incommensurably different from ourselves and from one another.[31]

However, although these biological and cognitive commonalities may extend across cultures, there are also important differences. For example, whereas the Itzaj view individual characteristics as embedded within ecological relationships, biology frequently decontextualises them, viewing relation as an afterthought to an assumed individuality.[32] But within a *system*, relations are not established through the sharing of common properties, which is a simplistic conceptual algorithm, but rather through a *functional complementarity* which relies on their *differences*. For example, while the honeybee may be related conceptually to other members of the *apidae* family, *ecologically* they are related to the plants they pollinate and from which they gather nectar. Within industrial society, then, ecological understanding becomes something of an ephemeral 'optional extra', a footnote to an assumed individualism; and classificatory systems possess an ideological slant, portraying each species as part of a *cognitive* system – as a 'pest', as an 'invertebrate', as 'endangered', and so on – which bears little relevance to its *ecological* character. Categorisation therefore often serves as a tool of translation whose purpose is to relocate fragments of ecosystemic wholes within an industrialist discourse.

The priorities assigned to the artifactual and natural realms shows a good deal of cultural variation. One study, for example, compared the cognitive preferences of 221 children from rural Taiwan with those of 316 children from rural Indiana. When presented with pictures representing everyday items, American youngsters (but not the Taiwanese) tended to group the images taxonomically; for example, the saw goes with the axe 'because these are both things to cut' – an orientation which "may inhibit the development of a tendency to perceive objects in the environmental context in terms of relationships or interdependence." In contrast, the Taiwanese children (but not the Americans) tended to group them in a 'relational-contextual' manner – for example, human figures are grouped together 'because the mother takes care of the baby', illustrating a "tendency to perceive objects in the environmental context in terms of mutual dependence or relationships."[33] In other words, the American children showed their allegiance to a *cognitive* structure, a taxonomy, so prioritising cognition over embodied relations; whereas the Taiwanese were more attuned to an *ecological* structure.

Language plays an important role here, since it isolates purposes and actions from their contexts. If I say that I am 'growing beans', that is not quite accurate: I am also – to take examples from Thoreau's celebrated beanfield – feeding the woodchucks, rediscovering the "small implements of war and hunting" of "unchronicled nations", planting the seeds of "sincerity, truth, simplicity, faith, innocence, and the like", and disturbing the "cinqfoil, blackberries, johnswort", aided by the "dews and rains which water this dry soil". Thoreau was not, in other words, just growing beans; and he criticises farming which "is pursued with irreverent haste and heedlessness … our object being to have large farms and large crops merely."[34] Furthermore, to say that I am growing beans overstates my input; for after I have planted the seeds, the beans are pretty good at looking after themselves. With rare exceptions such as *Walden*, language usually fails to capture the full range of overtones of an activity; and as principles, laws, and textbooks are written and read, our experience is narrowed, like a symphony reduced to single notes on a piano.

Although the natural and industrial systems share a common elemental base, the divergence between them increases as we move up the hierarchy of organisation. At the most basic material levels, the natural and industrial orders are virtually indistinguishable. The elements that each uses – carbon, copper, oxygen, and so on – are identical, except, sometimes, where processes involving nuclear fission are concerned, in which case elements that are rare or unknown in the earth's crust occur as a result of nuclear power generation. Even at the next stage in the hierarchy – that of molecular organisation – the two systems have much in common: many of the chemicals used in industry – water, ethanol, calcium carbonate, and so on – occur naturally and are constituents of natural processes. However, there are also notable divergences at this level. Many organic compounds used in industry are rare or unknown in the natural world: polymers such as polythene or polyvinylchloride come into this category, as do most complex organic herbicides or pesticides, toxic byproducts such as PCBs, refrigerants such as organic fluorocarbons, and – at least on the surface of the earth – organic solvents such as benzene or carbon tetrachloride. Also, metals such as copper, aluminium, and iron occur only rarely in their elemental forms within the Earth's crust, although they may be commonly used in this form within industry. Extraction of the elemental metal from their ores generally uses processes that are unknown in nature. Similarly, physical laws such as the equation distance = speed $\times$ time represent the 'purified', 'ideal' forms of much more complex natural situations such as the swooping flight of a bird. Beginning with such simple elements, abstracted from nature, industrialism builds a quite different edifice.

At higher levels of organisation than the molecular, the two systems diverge more drastically, and are more overtly opposed. Metals and plastics, for example, are used to construct a host of machines such as supertankers, cars, sewing machines, and mobile phones that are completely alien to the natural world. The vitality of life is replaced by mechanical, electronic, and chemical interactions, generating a realm which is in some respects alien to our embodied selves. Industrialism therefore recognises natural realities at a basic, elemental level; but it denies and dismantles higher orders of natural organisation. This legitimises its view of the world as simply a source of 'raw materials', and justifies the imposition of its own structures, in much the same way as the European colonialists saw the lands they were colonising as 'empty' and the peoples as possessing no cultures of their own. Science is generally consistent with this emphasis on the lower levels of the material hierarchy, giving superbly accurate accounts of interaction at the molecular level, but – as we will see later – much less adequate accounts of more complex, emergent aspects of materiality.

We can therefore see that a systemic understanding that attends to the higher-order levels of organisation is essential to grasping the differences between the natural and the industrial systems. Andrew Collier succinctly summarises the significance of a systemic viewpoint as follows:

> there are hierarchies of structures in the world, e.g. molecules are composed of atoms, cells of molecules, organisms of cells, societies of people – and in no case are these 'wholes' reducible to their parts, or the parts to their wholes. There are irreducible mechanisms existing at each level, which could not for the most part be predicted from knowledge of the higher- or lower-level mechanisms. This view

contrasts with a number of one-level ontologies, which claim either that parts are mere aspects of some whole, so that ultimately there is only the Absolute, of which everything is an aspect; or that wholes are mere collections of parts, understood only when broken down into their components, which alone are ultimately real; or that some intermediate level of entity (e.g. 'selves') are the only reality, their parts being mere aspects, and the larger entities which they make up mere collections.[35]

If we ignore systemic functioning, higher orders of organisation are denied through their recategorisation in terms of the properties of individual 'things'. As Michael Taussig puts it, if "attention is focused on a single thing ... then that thing is seen as containing its relational network and surrounding context within itself; the thing is a system of relationships".[36] For example, the complex social arrangements of creatures such as the wolf are redefined in terms of individual 'instincts', thus denying the emergent properties of the pack as a whole. This denial of higher-order emergent properties allows us to treat nature as an amorphous collection of animals, plants, and physical characteristics that are ripe for exploitation by human ingenuity, rather than as a highly evolved *system*.

Equally, and equally disastrously, we deny the emergent systemic properties of industrialism, convincing ourselves that we are somehow 'in control', that industrialism exists 'for our benefit', and that it has no *telos* of its own.[37] Although the industrial world presents itself in terms of individual fragments, concepts, and processes, it possesses a systemic character that is cognitively opaque. The ways we are taught to think and feel, the legal system, scientific knowledge, and economic structure are integrated, often appearing as separate realms of knowledge and practice which through some more or less magical relation to 'reality' operate consistently with each other. As Jean Lave points out, it "is not so much that arrangements of knowledge in the head correspond in a complicated way to the social world outside the head, but that they are socially organised in such a fashion as to be indivisible."[38] Likewise, as Lave adds, "relations between science and the world it purports to investigate are mutually entailed in one another".[39] This occurs mostly through a process of *selection*; for science focuses on what is stable and separable from context, sidelining what is intrinsically systemic or fluid. As Lave continues, "science studies just that portion of the complex everyday world that we think we can know."[40] Such fragmentary thinking allows us to claim that the environmental crisis can be solved if we all recycle our bottles, drive hybrids, and grow our own vegetables: in other words, we are blind to our own engulfment by systemic forces we don't recognise and have little power over. This type of denial is in a sense 'built into' the industrialist system due to the degree of specialisation it depends on, each of us being a specialist in our own tiny pixel of the whole system.

Both the natural order and industrialism can in some respects be understood as incorporating 'dissipative systems' – that is, they maintain their structure by importing energy from and exporting disorder to their environments. In the natural world, various phenomena such as hurricanes, living creatures, cells, and termite nests can themselves be understood as dissipative systems; and as James Lovelock famously suggested, the biosphere as a whole can be understood in this way, regulating itself so as to maintain conditions suitable for life.[41] The industrial system even more obviously behaves as a dissipative system, assimilating energy and material for its own purposes, and exporting entropy and 'waste' to less industrialised areas. Individual parts of the industrial system also behave in this way: both corporations and individuals import materials and export waste in order to maintain their own more or less stable health. Such components can often be thought of as 'autopoietic'[42] – that is, rather than behaving as passive and determined only by their roles within the larger system, they have a measure of self-determination, acting to enhance their own vitality as well as that of the system. In Katherine Hayles' colourful description, "a dissipative structure is a raft which floats inexplicably but definitely upstream, against the current, gathering flotsam and organising it into its flotilla with some sort of autonomous force or direction."[43]

Dissipative systems develop through time, showing capacities for self-organisation; and in the case of industrialism this embodies a movement away from its residual connections with the natural order and the quintessentially human, so that nature's structure becomes industrialism's raw material. This also involves an ideological reformulation, as concepts such as value are disengaged from their original material and experiential sources and redefined in purely industrialist ways. For example, while in Marx's time the value of a commodity could be understood as related to the amount of human labour which produced it, today value is generally understood as *market* value – that is, defined in purely monetary terms or by comparison to other commodities. Similarly, since the abandonment of the gold standard, currency values have not been anchored to the physical availability of gold – a natural resource. Rather than being rooted in material realities, industrialism has become a self-referring system which then redefines materiality in its own terms. It will not do, then, to claim that the "solution to the problem of whether technology lies on the side of nature or human society is simply to dispense with the dichotomy, and with it the concept of technology that is predicated on this dichotomy."[44] While the boundaries between the technological and natural realms are complex, this doesn't imply that these realms are not fundamentally different. The boundary between day and night is also indistinct; but day is nevertheless not the same as night. It is clear that industrial society has diverged dramatically from the natural order; and that subsuming the spectrum of humanity from hunter-gatherers to those working within advanced technological organisations under the heading of 'tool users' glosses over essential differences. Industrial society is not just tribal society that happens to be more technologically advanced: it possesses emergent properties that make it *qualitatively different* to other societies, completely upending the original priority of nature over technique. It is, I think, grievously misleading to argue, as does Stuart Kauffman in his otherwise superb book *Reinventing the Sacred*, that the economy can be understood simply as exemplifying "the ceaseless creativity of the universe" – in other words, as an extension of *natural* process – or that we could "learn to harness this structure"[45] – implying that human cognition can control and direct economic development. Whatever else the economy may be, it is neither natural nor sacred – nor even, in some respects, human.

The rather tasteless analogy between industrialism and malignant growth has been made by many writers,[46] and seems so apt as to be inescapable. Industrialism plunders its host – the natural world – draining resources, converting them into commodities, and destroying the habitats of wild creatures. Its waste products pollute and damage ecosystems. It pretends to surpass its host, wild nature; yet it ultimately depends for its existence on wild nature's adequate functioning. It co-opts resources and mechanisms that previously were part of natural processes, using them purely for its own growth rather than as part of broader systemic interests. It replaces the balanced interaction of many factors by a single-minded striving for capital growth, sacrificing all else to this aim. Its colonial style of expansion establishes metastases in distant lands, which it then exploits for its own purposes regardless of indigenous social or ecosystemic patterns.

The problem, then, is how to distinguish between the cancer and the host – and to excise the former. Since both industrialism and cancer may best be understood as *systemic* disorders rather than as reducible to specific causes or genetic factors,[47] this is no simple matter, requiring a grasp of the system's dynamics.

Recognising the 'natural'

Given these systemic complexities, the distinction between what is natural and what is artifactual is neither clear nor obvious. If a natural area is industrialised, at what point does it cease to be natural? The Colorado River, for example, has been extensively dammed, and much of the previously flowing volume of water has been diverted for irrigation or other purposes; so is it still a natural feature? Is the intensively cultivated greenhouse tomato, pollinated by specially bred species of bumblebee, more natural than it is unnatural? And is the honeybee – a comparatively recent arrival in the New World, introduced by European colonists almost four centuries ago – now to be considered natural? The honeybee has out-competed many species of bee native to the Americas,[48] and this, in turn, has had repercussions on ecosystems; for many plants are dependent on certain species for their pollination and hence their survival. Can the honeybee, or the changed landscape it generates, be considered 'natural'? Or is it better regarded as a camp-follower of human colonists, helping to shape the world firstly towards human domination, before humans, too, succumb to whatever the next stage of industrialism brings?

Industrial development changes not only the species composition of ecosystems, but also the *forms of relation* between species. Some of the qualities we perceive in nature may have been introduced or exaggerated by industrial activities, sometimes in quite subtle ways. For example, Buchmann and Nabhan point out that "some plants may invest a considerable portion of their annual energy budget in producing flowers of the appropriate colour, size, shape, and odour to attract pollinators." Pollinators are often both critical to the survival of a species and in short supply: for example, only "2 percent of the pink lady slippers [in the eastern Massachusetts forests] are visited by bumblebees in a way that results in pollination and fruit set", because "there is an apparent scarcity of bumblebees … perhaps due to past pesticide spraying or other human activity."[49] Similarly, the introduction of the European honeybee to Australia has increased competition between nectar-feeding birds such as the New Holland Honeyeater.[50] This, say Buchmann and Nabhan, is "a much more insidious kind of extinction: the extinction of ecological interactions."[51] Industrial activity, then, may increase competition between species and reduce the extent of ecological connectivity, so that the view of nature as a competitive scramble for survival ironically becomes more accurate than would be the case in untouched landscapes. Likewise, just as cultural loss corrodes character and meaning, shrinking us towards an impoverished individualism, so the loss of ecosystemic interactions invisibly degrades the creatures that were once part of them; and as this process occurs our taxonomies of unconnected, separate species begin to seem more accurate, and well integrated ecosystems begin to approximate the 'cornucopia of raw materials' perceived by commercial interests. Industrialism is not only changing ecological makeup, but also effecting a convergence between shattered ecologies and the reductionist frameworks we use to assess and understand them, imprisoning us within a fragmented world that is both symbolic and physical.

Given our hubristic reluctance to recognise that natural situations may be beyond our powers of understanding, we prefer to categorise them as random, confusing such systems with mere collections of species. Ecosystemic order is not self-evident: it may be complex,[52] reflect multiple nonlinearities, involve periodicities that exceed the length of the individual research grant – or even the individual life-span;[53] and may be chaotic, so revealing itself only when plotted in phase-space.[54] Similarly, emergent properties tend to be only very indirectly related to the behaviour of individual species and populations, and so are generally difficult to predict; and so some authors dismiss them as unreal figments of our imagination. Stephen Budiansky, for example, suggests that if ecosystems have properties that are more than the sum of their parts, then "the study of these systems should be carried out by theologians rather than scientists",[55] thereby outlawing emergent properties as a fit subject for scientific study. It is unlikely, however, that Budiansky would insist that human beings be studied simply as a mixture of carbon, hydrogen, and other elements, or that the meaning of his writings can be located entirely within the words he uses. If reductionism can be perceived as absurd at one level, it is not obvious that it is any more acceptable at another just because our cognitive limitations cannot perceive higher-order wholes.

Because of this systemic character of industrialism, what is 'natural' depends as much on the context in which an activity occurs as on the character of the activity itself. Felling a tree, for example, might be considered consistent with the local ecology when carried out by an individual or family for their own needs, and when other trees in the area are left standing; but when the tree is felled as part of a massive clear-cut for purely commercial reasons, few would argue that it could still be considered 'natural'. Likewise, planting trees might be generally considered an ecologically benign activity; but less so if the trees were designed to 'prettify' a corporate headquarters, or were a commercially grown timber crop. Likewise, hunting may be justifiable as part of a long-term cultural relation between humans and nature, necessary for survival, but not as a leisure activity for affluent urban-dwellers equipped with high-tech weaponry. We can only assess whether an activity is 'natural', then, by considering its effects within the entire context – natural, social, and political – of which it is a part. The activity itself, like the material elements that are involved, is often ethically ambiguous, just as the meaning of a particular sentence depends on whether it is part of a conversation, a political statement, or a quote that we are critiquing. It is generally the higher orders of the hierarchy that determine ethical status; and as cognitive opacity is greater the higher the level, it is at these levels that ideology operates most effectively, largely hidden from critical scrutiny.

The ethical status of an act, then, depends largely on the context within which it is carried out. For example, in Toni Morrison's novel *Beloved*, Sethe kills her child, so freeing her from the suffering of slavery – a morally complex act the ramifications of which reverberate throughout the novel. Death, we believe, is to be avoided; and defending life is right. And so it often is; but reifying such principles results in a deeper form of death. Behaviour that is ethical in a healthy world may not be so in a degraded one; and our actions need to embody a wider recognition of their likely consequences and effectiveness. In this vein, Tim Luke argues that green consumerism and recycling,

> rather than leading to the elimination of massive consumption and material waste, instead revalorises the basic premises of material consumption and material waste ... [and by] providing the symbolic and substantive means to rationalise resource use and cloak consumption in the appearance of ecological activism ... [these activities remain] structurally invested in thoroughly consumerist forms of economy and culture.[56]

Simply to refer to such actions as 'ineffective' misses the point: by assuming that they are 'good' whatever their context, we ignore their assimilation to the industrialist system within which they occur, so ultimately reinforcing that system and undermining the challenge to it. Similarly, the more efficient use of petroleum is usually promoted within a context that assumes the survival of massively energy-dependent lifestyles into the indefinite future – in which case, 'efficient' use of energy becomes a way of perpetuating and legitimating industrialism.

While our difficulties in grasping the higher-order functioning of either the natural or the industrial system are largely due to cognition's inherent limitations in dealing with such complex and elusive phenomena, we should also note that systemic functioning is *inherently* indeterminate and unpredictable. Uncertainty and probability do not *only* reflect our lack of knowledge; they may also, as Robert Ulanowicz has pointed out, "reflect an indeterminacy in the process itself"[57] – which is an inescapable aspect of nonlinear systems. The misleading belief that the world is *in principle* entirely understandable derives from a desire for certainty, reflecting a sort of cognitive wish-fulfilment. As Ulanowicz summarises Karl Popper's viewpoint,

> the deterministic realm where forces and laws prevail is but a small, almost vanishing subset of all real phenomena, [which] are suffused with indeterminacies that confound efforts at deterministic prediction. Popper's is not the schizoid world of strict forces and stochastic probabilities, but rather a more encompassing one of conditioned probabilities, or deep-seated *propensities* that are always influenced by their context or environment.[58]

The natural order is indeed a tangled thicket of such 'propensities', inherently resistant to conceptual penetration, suggesting the need for a certain humility in the face of a reality that is, in Donald Worster's words, "more complex than we ever imagined, [and] some would add, ever can imagine".[59] Nor is the behaviour of the industrial system any easier to predict. The emergent behaviour of any complex system involving nonlinear variables will magnify some tiny perturbations while ignoring others, confounding any attempts at long- or medium-term prediction. This is not the same as saying that such systems are essentially random and disordered: rather, their order is not of the type that human cognition can cope with. Like my attempts at skiing, we can exert some influence on our progress while never being in control or being able to say where we will end up.

Given these indeterminacies, it is hardly surprising that environmental action has mostly been at a level that is reasonable, well-meaning, and largely ineffective, boosting our psychological comfort while having little discernable effect on the health of the biosphere. To the extent that our actions – often unconsciously – exist *within* industrialist 'rationality', we are largely impotent; and this applies to the majority of theory, writing, and activism. In order to be effective, we have to be standing on something that is *outside* the system – namely, our own embodiment and the natural world that is continuous with it.

Nominalism and the reduction of meaning

Our use of symbols has increasingly been absorbed into industrialism and used to lend it a superficially natural appearance, camouflaging its hostility to nature. Language, which originally evolved to represent and connect us to the physical realities we evolved within, is today as often used to conceal and replace these realities. In the past, for example, place names tended to reflect original geographical features: Forest Road, Well Lane, Beacon Avenue are my local examples. A mile away, however, more recently built roads are named at random from natural features of the Lake District National Park: Loweswater Drive, Langdale Avenue, Thirlmere Drive. The house of a neighbour has a plaque announcing that it is 'Rose Cottage', even though it is not a cottage and has no roses. Such labels cosmetically decorate areas that are bleakly functional, creating a comforting aura of groundedness within a system that is ungrounded.

This empty nominalism exemplifies the increasing detachment of the symbolic world from ecological realities, recreating the natural world as a fantasy rather than as a lived reality. The meaning of nature becomes restricted to words and their resonances, often recreating old memories rather than reflecting current experience, so that we retreat from unnatural actualities into a world of linguistically evoked meaning in much the same way as we escape into a romantic novel. Language, like all culture, was originally part of ecological structure in the wider sense, since it was a means of expressing and communicating vital relationships with the world; and so it is no surprise that the disconnection of meaning from context is paralleled by *functional* disconnections. The 'beacon' and the 'well' have disappeared, and the 'forest' is now a patchwork of domesticated woodlands.

But while loss of contact with nature can be demoralising, too much exposure to wildness can also be frightening, especially for those who have become used to urbanity; and naming is often the first stage in the assimilation of wildness into a safely domesticated realm over which we have more control. As the conquistadors knew, naming is the prelude to possession; and Patricia Seed notes that

> Beginning with a small strip of land renamed San Salvador, Columbus claimed to have named six hundred islands on his first voyage, leaving three thousand islands unnamed and so unpossessed ... Columbus's practice of naming – or more accurately renaming – rivers, capes, and

islands as part of the ceremony of taking possession was repeated throughout the Spanish conquest of the New World.[60]

Today, naming comes so naturally that we are unaware of it as a process of translation from one order to another. A recent experience of taking the shuttle bus in Zion National Park provides an example: the commentator announces that the cliff beyond the road "is the Court of the Patriarchs ... On the left is Abraham, in the middle is Isaac, and on the right is Jacob." This tell us nothing about the natural forms being referred to. On the contrary, it reduces them to bite-sized cognitive chunks, elements of a pre-existing human realm, so that tourists can relax, comfortable in the knowledge that they will not have to grapple with anything alien, and that the landscape has been safely integrated into the predictable realm of the cognitively predigested.

Compare these examples with place names from a culture more embedded in their land – the Western Apache society located around Cibecue, Arizona: 'Widows Pause for Breath'; 'She Carries Her Brother on her Back'; 'Bitter Agave Plain'; 'Scattered Rocks Stand Erect'. Keith Basso shows how these names are a living part of Apache culture, still central to the stories through which morality and conduct are learnt:

> For Indian men and women, the past lies embedded in features of the earth – in canyons and lakes, mountains and arroyos, rocks and vacant fields – which together endow their lands with multiple forms of significance that reach into their lives and shape the ways they think. Knowledge of places is therefore closely linked to knowledge of self, to grasping one's position in the larger scheme of things, including one's own community.[61]

Unfortunately Basso, like many anthropologists, suggests that the meaning of places is a purely human, imposed meaning, thus perpetuating the intelligent-thinker-in-a-passive-world viewpoint, and safely re-enclosing potentially wild features of the landscape in a domesticated frame:

> [P]laces may seem to speak. But ... such voices as places possess should not be mistaken for their own. Animated by the thoughts and feelings of persons who attend to them, places express only what their animators enable them to say; like the thirsty sponges to which the philosopher alludes, they yield to consciousness only what consciousness has given them to absorb. ... Human constructions par excellence, places consist in what gets made of them ... and their

disembodied voices, immanent though inaudible, are merely those of people speaking silently to themselves.[62]

For Basso, then, places are merely pegs on which to hang human meanings; and having briefly glimpsed a natural realm that transcends the human, we are reimprisoned within the manufactured world, so that what we hear is merely the echo of our own voice. Others such as Ann Spirn, however, are more willing to venture outside this constructed world of human meanings:

> The meanings landscapes hold are not just metaphorical and metaphysical, but real, their messages practical; understanding may spell survival or extinction. Losing, or failing to hear and read, the language of landscape threatens body and spirit ... Rivers reflect, clouds portend.[63]

While symbolic classification in the industrial world often separates natural entities from their contexts, this is not inevitable or culturally universal. Daniel Nettle and Suzanne Romaine tell us that in "the native American language Micmac, trees are named for the sound the wind makes when it blows through them during Autumn, about an hour after sunset when the wind always comes from a certain direction".[64] In the industrialised world, however, such ecologically anchored forms of wording – for example, describing ourselves as 'rooted in' a place, needing 'breathing space', or 'out of touch' with a situation – tend to have a vestigial quality to them.

Words may be used to reach out into the world, embedding us within it and bringing it alive for us; or they can be used to shrink the world to fit our cognitive requirements. Words are also used to cosmetically reframe less concrete aspects of the social environment so that these aspects are made to appear consistent with dominant ideologies. Terms such as 'freedom', 'community', and 'democracy', for example, are often used to reframe social realities in ways preferred by powerful interests, reminding us that a conscious emphasis on certain notions may serve to cover up the absence of the corresponding realities. To say that the world is 'linguistically constructed' is to attempt to make such deceptive portrayals of reality into an ideological principle.

In much the same way as words are often used to summon up earlier memories of ecological embeddedness, imagery is also often employed by advertisers to convey the aura of 'naturalness', power, or sensuality. Popular music, by incorporating a range of borrowed styles, melodies, and phrases, allied to politically-resonant imagery – the pop star as soldier, waif, addict, or blue-collar worker – summons up memories of authentic experience; but on closer examination, these are generally the fragmented icons of older traditions, stitched together to construct an illusion of meaning and integrity. Behind the images of trendy ethnicity, gritty working-class reality, or impoverished minstreldom are the carefully researched calculations of marketers. Thus is the glossy pseudo-authenticity of the capitalist world presented to its affluent 'consumers'.

Quantification and the monetary reduction of nature

The elimination of natural meaning is expressed in its starkest form in the drive to quantify, abstracting the dimension of a situation that can be most easily expressed mechanically or digitally, and excluding all other qualities. Quantification needs to be understood in the context of the Cartesian concentration of meaning within the mind: the world is describable simply as substance, of certain types, in certain positions, and present in a certain quantities – each of these characteristics being mathematically describable. For example, the elements can be characterised in terms of their atomic numbers, defining the number of protons present in the nucleus of an atom; and our position on the surface of the Earth can be specified in terms of GPS coordinates, ignoring the characteristics of the local landscape. By thus reducing complex entities to their elements, we avoid the cognitive difficulties inherent in having to deal with their myriad emergent properties, falsely implying that emergence is merely a cosmetic addition rather than an essential constituent of the world. Meaning is thus squeezed out of the world, becoming an *idea* rather than something we *experience*.

To describe a 'crowd of 10,000 people', or 'an area of 20,000 acres', or '20,000 board-feet of timber', tell us little about the entities concerned except their quantity. This maximises their use-value to industrialism, since it jettisons all 'irrelevant' properties – the political opinions of the people, or the lives of the creatures that inhabit the 'area', or the ecological relations previously served by the 'timber'. Such skeletal description serves an analogous function to that of the bulldozers that level a piece of land in preparation for building work: it removes all characteristics and structures that get in the way of of its industrialist reinvention. Small wonder, then, that quantification is often experienced by tribal peoples as alien and threatening. As one ethnographer reported of his attempts to conduct a census among the Konds of central India in the mid 19th century, "I tried hard to establish a registry of the men, their wives and children, but was compelled to abandon the attempt. On discovering my intention the people fled in great alarm, asserting they were sure to die if I persisted in my design of numbering them".[65] As Felix Padel notes, "the contrast with tribal forms of understanding could not be more extreme ... maybe the villagers ... were sensing accurately something that was essentially alien and hostile to their culture: the collecting of a kind of knowledge that was geared toward controlling them."[66]

Such reductions-to-quantity are examples of the way that industrialism relies on essentially *reductive* processes to break down complex substances and assemblages into simpler elements. Because this is generally a much more rapid process than the evolution of new forms, industrialism can grow explosively faster than evolutionary processes, so that thousands or millions of years of evolution can be undone in a few years. For example, agriculture reduces the complexity of original ecosystems and replaces it with monocultures achieved by killing 'unwanted' species – a process which can take place virtually overnight; and globalisation has introduced an enormous cultural and linguistic simplification as indigenous groups have become assimilated to economic forces and the demands of 'labour'. Industrialism dissolves any structure that impedes the drive towards market efficiency, so that trade laws, for example, involve sweeping away local customs and traditions, efforts to protect local farmers, and small producers' ability to compete with corporations.

Quantification is essential to this drive towards simplification, as the value of everything and anything becomes *monetary* value, allowing us to calculate and compare the economic worth of such apparently uncommodifiable entities as whooping cranes,[67] the view from the the Blue Ridge Parkway,[68] and the relative cost of pollution in Third World and 'advanced' countries.[69] As Jack Manno notes, things that are not commodifiable are invisible to economic indicators: "when grandparents provide voluntary babysitting, nothing registers in the GNP ... Barbie adds to GNP, snow angels simply melt away."[70] What is more, many of the ecological benefits we enjoy are not legally 'ours', and so have no legal protection: think of scenery, snow, sunshine, countryside. Consequently, if they are taken away, we cannot be said to have experienced any legally recognisable 'loss'. Whatever cannot be ascribed a monetary value either disappears or is pushed to the periphery of life.

Aspects of living that were previously embedded in culture are now services to be bought and sold. For example, loss and grief have traditionally been expressed through ritual and family life, but are now the source of income for an army of counsellors and therapists. Sex, too, has become part of an array of commercial activities from subtly suggestive advertisements to computer dating. Particular human characteristics become distorted in order that they fit an economic scheme: Native American tribes are now often reliant on "gambling (a commoditized response to hope), gasoline (a commoditized form of energy), and nicotine ... a commoditized form of longing."[71] In David Smail's terms,

> by far the most important reason for the disappearance of what one might call a 'public ethical code' is that its continuation would place unacceptable constraints on commercial exploitation. However brutal the mediaeval feudal lords may have been, and however corrupt the Church, at least, officially, they were not supposed to be. We have now reached a stage when the precepts of exploitation and self-interest are being slid into the place of moral values which can scarcely any longer be remembered.[72]

Roy Rappaport, drawing on Simmel, notes the crucial symbolic inversion which will ultimately destroy industrial society. Money, he says,

> has a most peculiar and interesting ability: it annihilates quality. That is, it dissolves the distinctions between qualitatively unlike things, reducing those distinctions to mere quantitative differences by providing a common metric in terms of which all things can be assigned values directly comparable to the values of other things. But the world on which this metric is imposed is not as simple as the metric itself. Living systems ... all require a wide variety of qualitatively distinct materials to survive ... The imposition of the more-less logic of money on systems – organisms and ecosystems most obviously – that do not operate in terms of a logic of more-less ... is bound to be destructive ...
>   If money becomes the standard by which all value is assigned and compared, then it itself becomes ipso facto the highest of all values ... [However] Economic values, indeed all economic systems, are entirely contingent upon the prior existence of biological systems, both organisms and ecosystems. The reverse is not the case. Biological systems are not contingent upon economic systems and had been in existence for 3.5 billion years or more before anything that could plausibly be called an economic system emerged.[73]

Consequently, Rappaport argues, when "money itself becomes the ultimate standard of value, and when such an epistemology is embedded in societies whose ultimate goals and values are represented by dicta like Coolidge's famous 'The business of America is business', such an epistemology [is] an instance of that form of falsehood ... in which ... the value of the fundamental [is] subordinated to the status of contingency, and as such 'relativised'". Thus "monetary accounts of organic, ecological, and social systems misrepresent the natures of that which they purport to describe so falsely as to distort or mislead our comprehension of the world as a whole."[74] Fundamentals such as the health of the natural world are relegated to the status of preferences, to be protected only as long as they do not impede economic growth; while derivative abstractions such as economic value are falsely seen as fundamental. As Anne Pomeroy remarks, capitalism can be viewed as "a form of lived misplaced concreteness in which human beings are literally ruled by abstractions, particularly the abstraction of value".[75] Money connects things that are ecologically separate, and isolates things that are ecologically related by assigning them to separate categories, thereby replacing ecological structure by economic structure.

Some of the bluebells in my local market, for example, are likely have been illegally taken from nearby woods. As they sit in neat rows on the shelves, their 'value' displayed on price tags, their existence has become anchored to a commercial rather than an ecological system. Both of these systems are, of course, invisible; and because our *understanding* of 'bluebells' is *already* of a separate entity, the replacement of the natural system by a commercial one passes unnoticed. The 'bluebells', it seems to us, survive just as well when transplanted to our gardens, even if the ecology has been invisibly scrambled. Imperceptibly but profoundly, they have been changed through their assimilation into a system quite different from and hostile to the ecological system they were taken from. Similarly, as Vandana Shiva has noted, rare species existing in wilderness areas are sought out by corporations and patented, transplanting them from the natural order to the economic system.[76] In Frederick Buell's terms, "no longer was wildness, as Thoreau wrote, 'the preservation of the world'. Now wildness was the patentable instrument of a specific set of new industries".[77] As Zbigniew Herbert notes, the "order of the stock market was introduced into the order of nature. The tulip began to lose the properties and charms of a flower: it grew pale, lost its colours and shapes, became an abstraction, a name, a symbol interchangeable with a certain amount of money".[78]

As this relocation occurs, values lose their rootedness in ecological and ethical systems and instead become concretised abstractions within a *commercial* system, so that ethics becomes a matter of *negotiation* and *exchange* rather than of preordained, absolute values. Thus Colin Blakemore, defending the use of animals in experiments, claims that "two mice and half a rat per person is a fair price to pay for medical purposes".[79] But human meanings need to be based on more solid foundations than 'market values'; and to use a parallel example, Warwick Fox has pointed out that "it is always wrong to torture another person. It is no defense to say that you took their desire not to be tortured 'into account' but nevertheless reached a decision that, 'on balance', your desire to torture them … outweighed this 'other factor'".[80] Torture, like natural value, is not something that can be negotiated; and the attempt to relativise fundamental values leads to a society that has lost its moorings in the real world.

Appearance and reality

An implication of the previous section is that many of the changes associated with the induction of an entity into the industrial system may be in realms that are imperceptible to consciousness. In the past, the purpose of a device such as a plough, a knife, or a building could not be easily separated from its appearance; and functionality was strongly associated with aesthetic qualities. This is not to say that artifacts were not decorated; but such decorations were additions to or enhancements of the basic form of the object, and were not intended to *disguise* this form. By the 1830's, however, style and image began to be separated from function, buildings often being constructed around 'balloon frames', and walls becoming decorative structures rather than structural necessities.[81] Part of the inauthenticity attributed by critics to this dissociation of function and appearance was due to the scrambling of *temporal* structures, as historical forms became particular *styles* within a present that could abstract appearance from its original historical epoch. Richard Terdiman's description of the Paris of the 1850's, partly reconstructed by its prefect, Baron Haussmann, can more generally apply to an underlying ideological shift: "'reality' was itself increasingly tenuous, was slipping away under the pressure of appearance, image, and manipulated belief."[82] The *relatively* direct awareness of reality was being replaced by the interposition of a veil of manufactured facades – a formula that would be repeated in many spheres of industrialised life, and which reaches its peak in the computer simulation of experiences and images. Haussmann's many critics poured scorn on the superficiality of this reconstruction. "As this contemptuous rhetoric evolved", says Terdiman, "three linked metaphors came to dominate its corrosive imagery: *merchandising, theatre,* and *prostitution.*"[83]

A world that could be intelligently sensed was giving way to one in which sensation was decoupled from reality, one in which cosmetically-enhanced facades provided a psychologically comforting but delusory habitat behind which the original ecological fundamentals of life were being silently supplanted by commercial mechanisms. While the sensed world became an increasingly 'virtual', contrived one designed to humour our preferences and expectations, the industrial mechanisms behind this façade embodied the uncontrolled materialisation of a specific genre of symbolic possibilities.

Artifacts generally lack all the ecological, microcellular, and emergent 'depth' of the natural things they replace, resembling them only in their *perceptible* properties. Simulated 'log fires', stone-cladded concrete walls, 'wood finish' plastic kitchen work-surfaces – such 'skeuomorphic'[84] features convey the reassuring illusion that we are still grounded in previous, more natural forms of existence while disguising the radical discontinuities that accompany the severing of human life from nature. Similar points could be made about social and political traditions such as the pomp and ceremony associated with the British parliament, evoking historical connections that camouflage the degree to which British 'democracy' is now influenced by concealed financial and political interests.[85] From the individual choices of the family home to the great institutions of state, a façade of solid cultural tradition disguises the much more pragmatic grinding of economic wheels.

Advertising increasingly relates to this 'world' of appearance rather than to functionality. Fifty years ago, goods were advertised according to their practical merits: 'buy Sunlight soap because it washes better'. Today, we have moved out of the practical realm into one of symbolism and status: we buy jeans, cars, cosmetics in order to assert our status, or individuality, or success; or to comfort ourselves. The gulf between the embodied realities of our lives and the airbrushed, Photoshop-enhanced images of perfection that inhabit our TV screens is not generally taken to indicate the unrealities of images, but rather as a reminder of our own imperfections, and, therefore, our 'need' for the product advertised. In other words, we have learnt to regard the image-realm as our *primary* reality, and our embodied material existence as a *secondary*, inadequate approximation to this reality. The loss of self-esteem which results from this process of comparison is one of the driving forces of capitalism,[86] so that feelings of inadequacy are not merely unfortunate 'collateral damage', but are essential to the functioning of the industrial system. Hair dyes, cosmetics, and shopfronts are the material expressions of denial – denial of mortality, ageing, and the by now obvious unsustainability of the lifestyles they are part of, pandering to a narcissism that is assumed as a basis of individuality ('Because I'm worth it'), and fostering 'incongruence' as the self-concept diverges from organic experience. Terms such as 'honour', 'grace', and 'integrity' have almost disappeared from use, and others such as 'envy' and 'avarice' have been replaced by the laundered terms 'self-interest' and 'utility maximisation'. 'Honesty' and 'sincerity' are endangered or – in mouths of politicians – used to camouflage their absence. As Elizabeth Farrelly relates, socialisation into consumerism incorporates a forgetting of past values and their relevance to the present, requiring

> a highly-developed tolerance of the inauthentic, and a number of equally sophisticated denial strategies: denial that the truth was ever accessible, or ever existed, or was ever worth pursuing in the first place. The whole of postmodern relativism is, arguably, such a strategy[87]

Today, deception has become multi-layered and more difficult for consciousness to penetrate. If we have been seduced by dishonesty, fashion, and kitsch, as Farrelly argues, then we need to rediscover the connection between beauty and functionality. There is something beautiful – and authentic – she suggests, about Aussie shacks, African Queen type houseboats, Chinese abacuses, Persian weighing machines and Portugese sextants, because they are functional and therefore honest. "Work, being necessary, feels like a kind of truth, and truth is the medium in which we humans dwell."[88] Likewise, a bicycle, a garden hoe, or a pair of walking boots facilitate our connection to the earth; but a DVD player, a Barbie doll, or a Hummer generally do not, since these are players within a *symbolic* realm of status, fantasy, and an escape from embodied realities. Conversely – as Farrelly relates – there is an inherent dishonesty in modern suburban homes, with their garden gnomes, Aphrodite-shaped pepper grinders, and high-tech kitchens. "With their glam and polish, their air-con and their mod-cons, they whisper the same false promise that high office whispers to politicians – exemption from nature and necessity. They lie, and we love them for it."[89] In Michael Steinberg's terms, "[c]ommercial media and consumption are only broader, more vulgar and more resource-intensive realisations of the modernist ambition to replace life with ideas."[90]

Invisible forms and self-organisation

If the higher levels of either the natural or the industrial order are largely imperceptible to us, physics has nevertheless provided some tantalising glimpses of the sorts of order that govern our world. I am not referring to the often-quoted subatomic uncertainties discovered by nuclear physicists and frequently extended with dubious validity to the macroscopic world, but rather to the complexity and unpredictability that are essential aspects of the self-organisation of systems,[91] as every weather forecaster knows. Self-organisation is a pervasive property of the world, underlying phenomena as basic as the behaviour of boiling liquids and dripping taps or as complex as ecosystems or the stock market. Indeed, since Stanley Miller's classic experiments, which I referred to in the previous chapter, demonstrated that a simulated early Earth atmosphere generated significant quantities of amino acids, it has been clear that the organisational and life-generating capacities of the biosphere do not imply any external intelligence or divine influence, but are intrinsic to matter itself. The significance of these developments cannot be overestimated; for they undermine the entire Cartesian separation of a symbolic realm of order and intelligence from a base world of mere matter – a founding assumption of industrial society. Miller's early work has been extended by others such as Stuart Kauffman and Robert Ulanowicz who have demonstrated that the statistical implausibility of life's self-generational capacity is overcome through the influence of autocatalytic loops which are intrinsic to systems development.[92] As a result of this work, autopoiesis has become not so much a fanciful hypothesis as an established fact.

While the processes through which systems change have become better recognised over the past few decades, *predicting* systems behaviour is another story entirely. Natural processes are generally nonlinear; and this leads to types of complexity that are recognisable only in their simplest forms. The fact that natural systems often change unpredictably, and do not necessarily stabilise around a 'climax' state, has been used to shore up the argument that nature is *random*; but as the discovery of strange attractors demonstrates, unpredictability can equally well imply that the order of nature is beyond cognition's power to recognise. William Schaffer and others have shown that the behaviour of such variables is not random, but simply unpredictable. Although, for example, the population of a species such as the Canadian Lynx may vary greatly and *apparently* randomly, if we plot population on a three-dimensional 'phase diagram', representing the relations between population levels at certain time intervals, then an unsuspected order sometimes emerges in the form of a 'strange attractor'.[93] What such phenomena indicate is that the replacement of the wild world by industrialist monocultures involves the disappearance not only of species, but also of nested temporal patterns without which ecosystems would be mere collections of creatures. There is, in other words, an unsuspected temporal order, often somewhat misleadingly referred as 'chaos', concealed within the apparently disordered variation of the lynx population. Such complexity is fairly typical of even simple natural systems; so how much more complex – and therefore difficult to recognise – must be processes involving large numbers of species and relations? The assumption that such systems behave in 'random' or 'disordered' ways is therefore a defence against recognising the limits of our own conceptual capacities.

In contrast, machines – think of a car production line, for example – tend to be highly predictable, centrally controlled, and intrinsically consistent with the sort of causal hierarchy cognition feels comfortable with. In other words, human design, mediated by computer systems, controls the speed of the production line, which itself determines the rate at which other processes occur. Thus there is a simple unidirectionality in the system, making it enormously simpler than an ecosystem, which embodies various feedback loops and autocatalytic circuits. At a higher-order level, however, unsuspected forms of organisation seep even into humanly constructed systems such as the stock market, making industrialism far from predictable, and upsetting the myth that civilisation is a bridgehead of self-conscious order pushing into an essentially disordered universe. Although we can clear spaces such as assembly lines where the world behaves according to our simplified cognitive models, the larger world that surrounds these spaces contains forms of order that are beyond our comprehension.

This illustrates a fundamental difference between the natural and industrial systems. As Emily Martin points out, in "the machine world, regularity is a sign of health; irregularity a sign of disease or impending death. In the chaos model, it is just the opposite."[94] Healthy natural systems will often be unpredictable; that is, *cognition cannot predict them.* Chaos is a pointer to the world beyond cognition: we can recognise its existence, but seldom its structure. The epistemological significance of chaos theory lies in its meta-scientific implications – that the order of the world stretches far beyond our ability to understand it, reminding us of our often forgotten place in the world as lowly participants rather than as controlling masters. This contradicts the assumption that complex natural realities are just 'stuff', waiting to be arranged into commodities, in rather the same way that indigenous populations were viewed by nineteenth century missionaries as cultureless savages waiting to be filled with the Holy Spirit. This colonialist stance emphasises our role in constructing our environment, while drawing a veil over the destruction that precedes it; so that as we destroy what we do not recognise or understand, nature increasingly comes to resemble the sort of amorphous rubble that we thought was there in the first place.

Part of what we mean when we talk of 'wildness' is the complexity we dimly perceive just before we sweep it aside, imposing a techno-cratic system that is not only simpler than but also hostile to wild nature. This process is equally a destruction of humanity; for humanity evolved to roam freely within a world that was largely mysterious and unpredictable, and the focus on cognition and the marginalisation of non-cognitive faculties is an annihilation of awareness and a reduction of human imagination to mechanism. As the world itself is correspond-ingly reduced, becoming consistent with this reduced awareness, so we become sealed within a self-consistent realm of thought and physical 'reality'.

Life in fragments

Industrialism, then, operates through the fragmentation of previously whole structures, whether natural or cultural, rather as our digestive systems break down the structures of foods. While the fragmentation of organic substances is easily recognisable, the dissolution of our own subjectivity into a jostling collection of 'needs' and impulses is seldom acknowledged. Culture is understood as constructed from 'lifestyle choices' selected from commercially available possibilities rather than something that grows informally, organically from embodied relations with others. Spiritual awareness is translated into religious dogma, becoming integrated within industrialist consciousness rather than offering an alternative to it. In our working lives, increasing specialisa-tion isolates what we do from broader social and personal engagement, so that our ability to see and understand the whole system is reduced.

These subjective dissociations have profound implications for democ-racy, which is based on the assumption of the self as an integrated and coherent entity, capable of making intelligent decisions and aware of the interests and powers it relates to. As Stuart Ewen remarks, "it is not that the people are in charge, the people's *desires* are in charge";[95] and it follows that there exists a dictatorship of those individuals or organi-sations that manipulate our desires. Broadly, the fragmentation of the self reflects and reproduces a *social* and *political* fragmentation; and the resulting structural vacuum allows our manipulation by powers we cannot recognize, and which reconstitute the fragments as part of the industrialist machinery.

Since meaning cannot generally be focused within specific, commodifiable aspects of life, but is rather an emergent property of multiple interactions, meaning becomes an accidental by-product of life rather than an essential ingredient. Growing vegetables may help to pay the bills; but it also fosters the reintegration of a range of physical and mental capacities in ways that we sense but find difficult to articulate, contextualising otherwise compartmentalised processes of living and dying. Conversely, buying vegetables from the supermarket offers none of these important but conceptually elusive benefits. Subjectivity is reduced to its fragmentary elements – 'I feel bored'; 'I'm worried about the exam'; 'I want a cheeseburger' – which often float free from any broader scheme of meaning.

This fragmentation of the life-world generally results in a cognitive 'condensation' of properties within 'things' and 'individuals', and the abstraction of these qualities from other less obvious aspects. For example, we can deny the existence of any social qualities outside the individual – a denial eloquently expressed in Margaret Thatcher's infamous statement that "there is no such thing as society", so that qualities such as 'warmth', 'intelligence', or 'conscience' are assumed to be solely *individual* qualities rather than qualities of a society. We are losing our intuitive abilities to sense the emergent properties of wholes; and just as Maupertuis presumed that there is within each atomic particle "some principle of intelligence, something akin to what we call desire, aversion, memory",[96] we view the difficulties caused by the absence of meaning within our fragmented social structures as *individual* problems.

All too often, science commits a similar error. For example, Margaret Boden argues that the mating strategy of the hoverfly is determined by a very simple and inflexible rule. This rule, which is apparently hard-wired into the insect's brain, transforms a specific visual signal into a specific muscular response.

> This evidence must dampen the enthusiasm of anyone who had marvelled at the similarity between the hoverfly's behaviour and the ability of human beings to intercept their friends. The hoverfly's intelligence has been demystified with a vengeance, and it no longer seems worthy of much respect.[97]

Certainly, the behaviour of the individual hoverfly is simple compared with that of an individual human. But in context, the hoverfly is part of a larger system which is highly intelligent: it is adapted to its environment via camouflage, it reproduces successfully, it is an important pollinator, and so on. The unspoken assumption in Boden's remarks is that intelligence is always *individual* intelligence floating in an amorphous sea; but if we are prepared to accept that the systems which individuals inhabit can also be intelligent, then the picture changes. The hoverfly is superbly adapted as part of a system involving preda- tion, pollination, and reproduction; whereas the industrial human is often related to our environment in a somewhat pathological way, destroying it and extinguishing other species.

As agents of the industrialist symbolic order, we unthinkingly impose this isolation of organism from context, unaware of its destructive and arbitrary character. The caging of animals, like the ideological isola- tion of 'the individual', obliterates the (invisible) higher-order levels of the creature's being – those that integrate it within the natural order, such as its social interactions and ecological roles – so forcing an alien individualism on it. This sort of reduction is mostly imperceptible to industrialised humans, since we have learned to accept our ecological isolation simply as 'the normal state of affairs' from infancy onwards.

In a healthy natural system, no one level has autocratic power over any other. Large carnivores may appear to dominate their part of the animal world, but they are also ruled by lower level entities such as parasites, availability of prey, and so on, as well as by larger ones such as climatic and ecosystemic forms. It is characteristic of the emergence of industrialisation that the complex dependencies between levels are replaced – at least initially – by the more uni-directional exercise of power. For example, the media and advertising tend to manipulate needs while claiming to articulate individual freedom, so distorting the character of individuality and concentrating power in boardrooms. In this way the ideology of individual freedom is used to *reduce* individual freedom.

The fragmentation of time

Time, like other aspects of ecological structure, is fragmented by indus- trialism. Today, time is understood as an objective, regular cognitive lattice, abstracted from some simple mechanical or nuclear sequence, and draped over the world. To be of use in industrial processes, time has to be 'emptied',[98] abstracted from its embeddedness in cultural and astronomical contexts in a similar way to that in which 'timber' is abstracted from its previous incarnation as part of a forest. For nature, time is inseparable from life, not an abstraction from it. For example, many species embody the diurnal pattern of day and night which itself varies with season. For cicadas, a significant natural pattern involving

17-year sequences governs their emergence. And Holling and Sanderson point out that in a typical boreal forest

> fresh needles cycle yearly, the crown of foliage cycles with a decadal period, and trees, gaps, and stands cycle at close to a century or longer periods. The result is an ecosystem hierarchy, in which each level has its own distinct spatial and temporal attributes. ... The cycles are all operating concurrently, influencing one another. They are rhythms within rhythms, providing not the static structures of a well-oiled machine shop ... but rather those of a jazz band, building rhythms and riffs around each other, coalescing into both short and long rhythmic structures.[99]

Many creatures embody the lunar cycle. In the Palau area of Micronesia, for example, the spawning behaviour of fish is intricately related to the phases of the moon, and an understanding of the lunar calendar therefore becomes essential for survival in this fish-dependent society.[100] If the Palauans instead used the Western calendar, the behaviour of fish would appear to vary unpredictably from year to year, and fishing would become more of a hit-or-miss affair. Likewise, as Stephen Mithen notes, "modern hunters in glaciated environments keep a very close lookout for the annual arrival and departure of ... birds, since such information gives a clue as to when the big freeze of winter, or the spring thaw, will happen."[101] In such cases, the calendar is *in the landscape*; in other words, it has not yet become part of a separate *symbolic* sphere which is detached from natural realities.

Relying on clock-based time, we lose sight of the relations between natural processes such as the cycles of the moon and spawning. While the fishermen of Palau, like the fish they catch, are embedded within ordered natural processes, we often distance ourselves from these processes, from the other creatures that embody and express them, and from our own archetypal knowledge. For example, comparing the ethologist Tinbergen's fragmented understanding of stickleback behaviour with Darwin's more empathic description, Eileen Crist remarks that these two accounts embody

> different conceptions of the experience of time. Darwin's male stickleback lives in a continuous stream of time ... in which actions merge seamlessly into one another. Within the stream of time no expression is isolated: each moment of action is meaningful in virtue of being part of the larger pattern and of a single feeling.

The understanding of the cohesiveness and continuity of time *for* the stickleback allows the stickleback to emerge as an inhabitant of a meaningful world – a world in which fish can be mad with delight. With Tinbergen's sticklebacks each set of the chain reaction (a set being composed of one male behavioural pattern plus one female behavioural pattern) is complete as a stimulus-response unit, but discontinuous from the previous set and the next set. This discontinuity is equivalent to breaking the stream of time of each fish into separate, isolated segments. The sticklebacks, then, figure as inhabitants of a fragmented world.[102]

Crist notes how the destruction of meaning that results from breaking up the continuity of animal life into brief fragments permits the assimilation of behaviour to an economically-inspired paradigm, allowing "the elaboration of a nexus of interconnected economic terms [such as] monopoly, advertising, budgets, efficiency, investment, value, costs, benefits, maximising, minimising, winning, losing".[103] This *symbolic* re-ordering is continuous with a *material* one: we scavenge energy from the past, using the remains of life-forms as fossil fuel; and steal from the quality of life of future generations, not only by using up nonrenewable resources, but by leaving our pollution and nuclear waste for them to deal with. Economists frequently use the current interest rate as a basis for discounting, so that a human life in around two generations is worth a small fraction of a current life. Such mechanistic calculations disguise what is really an *ethical* issue. As James Buchan suggests,

> The delusion lies in the conception of time. The great stock-market bull seeks to condense the future into a few days, to discount the long march of history, and capture the present value of all future riches. It is in his strident demand for everything right now – to own the future in money right now – that cannot tolerate even the notion of futurity – that dissolves the speculator into the psychopath."[104]

Just as industrialism assimilates, redefines, and restructures everything natural as it transforms the world, so it sucks the future and the past into an intensely destructive present, giving rise to a sense, as Katherine Hayles puts it, that "the future is already used up before it arrives."[105] The image of the 'dissipative system' is compelling here, illustrating the way industrialism sucks up order, energy, and meaning before spitting it out as entropy and waste. The complex continuities between past, present, and future are destroyed and replaced by simpler relations: any attempt to relate the present to the past is viewed as 'nostalgia', and the future becomes merely an extrapolation of the present – capitalism with space ships and teleportation.

Actualities and realities

At this point it may be useful to outline Roy Bhaskar's distinctions among the actual, the empirical, and the real,[106] which have been implicit in some of what I have said so far. Firstly, the 'actual' refers to a specific situation or structure – what has actually occurred out of the myriad of possibilities, physical laws, and predispositions that constitute the 'real'. Out of all the possible forms of social structure that could exist, we inhabit one particular form; and while nature may take a bewilderingly diverse range of forms, only one of these actually exists at the present moment. Furthermore, we cannot be aware of every aspect of these social and ecological actualities; so there is a subset of the actual which we refer to as the 'empirical', containing those aspects of the actual which we can recognise. By examining a range of actualities we may be able to infer the various underlying biological and ecological processes and laws which contribute to the 'real', such as the Second Law of Thermodynamics, or the generality of autocatalytic processes in biological emergence,[107] or the tendency of social, biological, or ecological processes to self-organise.

For example, humans have the cognitive and motivational capacity to trade – a capacity that is expressed to the full in industrial society, shaping the way we live and possibly discouraging other capacities such as empathy, spirituality, and altruism. Now, if we adopt a naively empiricist psychology, then we will draw conclusions from the observable aspects of the actualities around us, concluding that humans are intrinsically rational, economic creatures. But this is misleading, since the economic and social structures we are embedded within represent very specific, historically dependent conditions that draw out certain human characteristics while discouraging others. Current human character has been channelled by the nature of industrial society towards an economically focused 'actual'; but it could just as easily be channelled in other directions if circumstances were to change. As Philip Cushman has argued, the dominant form of selfhood in industrial society is the 'empty self'[108] – a needy, acquisitive self that is constructed around the requirements of consumer capitalism – but as he would be the first to recognise, quite different configurations of selfhood would occur in alternative forms of social organisation. The potentials and capacities within human being – which are in the domain of the real – contain far more than the characteristics we can currently display. Given the extent to which our symbolic capacities allow us to be colonised by a potentially almost unlimited variety of ideologies, this distinction between underlying biological realities and the partly ideologically contructed actualities to which they currently give rise has never been more marked; and this introduces certain vulnerabilities that form the subject-matter of this book. In general, the actualities we live among – whether physical objects, forms of personhood, or social conditions – are the end-products of processes; and the processes are therefore more fundamental than the 'things' they give rise to at any one time. As Robert Ulanowicz points out, researchers have often got the cart before the horse in studying actualities as the starting-point for theory, regarding the processes that gave rise to them as too complex and ephemeral to be studied.[109] Thus current economic, psychological, or environmental conditions are generally the starting-points of theory; while the more fundamental propensities and processes that gave rise to these, and might also give rise to vastly different consequences, are pushed into the background.

For example, intensive agriculture may be essential for the survival of current populations; but this actuality is not a basic law of survival, but rather the result of a specific and historically unique situation which arguably distorts the relationships between ourselves and many non-human species. Our current actualities, while they give some of us lives of relative affluence, have a curiously provisional taint, reflecting the colonisation of the world by a system that is likely to represent a cul de sac in the evolution of life; and as Katherine Hayles notes, "time splits into a false future in which we all live and a true future that by virtue of being true does not have us in it."[110]

Just as humans have the potential to develop many different characteristics, so do ecosystems, which can vary enormously depending on the conditions prevailing during their emergence. As Daniel Botkin notes, the "biosphere has had a history and what it will be tomorrow depends not only on what it is today, but also on what it was yesterday. Like an organism, the biosphere proceeds through its existence in a one-way direction, passing from stage to stage, each of which cannot be revisited."[111] Thus if 'ghost' species which are essential to the emergence of the ecosystem are extinct – even if these are not part of the intended system – reassembly will be impossible – a phenomenon that Stuart Pimm has referred to as the 'Humpty Dumpty Effect'. Furthermore, even if all necessary species *are* available, the *sequence* in which they are introduced can drastically affect the final outcome.[112] In other words, the *history* of an ecosystem is not irrelevant to its present state, but survives in an implicit way, embodied within the structure of the system. When industrial development destroys ecosystems, this is not a simple, reversible change; and it is an industrialist conceit that damaged ecosystems can be reconstructed simply by adding the ingredients and stirring.

The specific form that an ecosystem takes, therefore, is the endpoint of a very specific journey; so it is a grievous mistake to conclude that the current form of the system is an adequate or complete presentation of all its potentials. This is particularly the case in the contemporary world, where technologically powerful interventions have forced discontinuities onto ecosystems, violently disrupting their relation to their histories. Because the various mechanisms and processes that underly the system – in the domain of the real – are capable of giving rise to many different ecosystemic actualities with many different empirically observable characteristics, changing conditions may give rise to dramatic changes in ecosystems. For example, a desert may appear lifeless; but it may contain the seeds of species such as the shrimp-like *Triops*, which can exist in anhydrobiotic form for decades – until rain occurs, when desert pools will suddenly be filled with *Triops*. Likewise, one *Canna Compacta* seed was found to be around 600 years old. It germinated to become a six-foot plant.[113] Thus 'nature' is best understood as a reservoir of underlying predispositions and potentials rather than as any current state or states.

Given that the domain of the real can give rise to a diversity of actualities, a common error is to mistake the actual for the real. This occurs, for example, when we examine various landscapes that have been affected by human activity and then conclude that 'nature is culturally constructed', forgetting that in the distant past most landscapes were affected minimally or not at all by human activity, and that they will exist long after humans have disappeared from the earth. For example, one contemporary social theorist expresses the view that when

> appeals are made to 'what nature requires' or assertions of knowledge made regarding nature's true 'essence' or 'telos', all that happens ... is that particular socially mediated conceptions get projected onto a supposedly pre-social world and then illegitimately claimed to have been grounded there.[114]

Here, 'nature' is identified with its current, domesticated form rather than with the real mechanisms, laws, and predispositions that underly this actuality as well as the indefinitely large number of other possible actualities. This leads on to a further erroneous conclusion that embodies a sort of born-again Cartesianism: that nature, being devoid of real predispositions, laws, and elemental realities, is endlessly transformable, a sort of 'raw material' waiting to be defined by human vision and activity. But while we have some control over nature in specific and localised situations, the higher-order aspects, often of global reach, are generally beyond our control and, often, our understanding. For example, James Hansen and his colleagues have shown that climatic conditions on earth can 'flip' – and have 'flipped' – between different states, with potentially catastrophic implications for the earth's human inhabitants.[115] Nevertheless, industrial society generally assumes its own immortal reality, denying the deeper realities of the natural order, so that as Bhaskar points out, superficial levels of reality actually dominate as well as occlude the level of reality on which they are parasitic.[116] Thus although culture and technology are very recent outgrowths from the natural order, the view of nature as 'culturally constructed' inverts this order, suggesting that nature is *secondary* to culture.

These errors are given a further idealist twist when it is suggested not only that the natural world is the passive recipient of our technological manipulations, but also that its very physical reality is a phantom devoid of real physical properties and tendencies. For example, William Cronon claims that it "hardly needs saying that nothing in physical nature can help us adjudicate amongst ... different visions [of nature], for in all cases nature merely serves as the mirror onto which societies project the ideal reflections they wish to see".[117] The 'real', then, is not just viewed as infinitely malleable; it disappears entirely, and the 'actual' is a product not of technological power but of something more ethereal – the 'projection' of our wishes. Here, the impatience of the symbolic to displace ecological structure short-circuits technological transformations and becomes a sort of magic wish-fulfillment.

This idolatry of the actual reflects a colonisation of the mind which is ultimately untenable; and therein lies the cause of many of our current problems. There are (at least) two levels of pathology here: firstly, mistaking a specific actual for the real, and secondly, the peculiarly destructive qualities of the particular actual we have constructed. Unfortunately for us, the assumptions which form part of our own actual – that economic growth can continue indefinitely, that other species exist for us to use, that technology will always find solutions to environmental problems – seem increasingly to be contradicted by those characteristics of nature that are most deep-rooted and therefore most 'real'. Viewing future states of nature simply as extrapolations of current ones is therefore misleading. For example, Botkin, assuming precisely what needs to be questioned, states that "nature in the twenty-first century will be a nature that we make".[118] Here, industrialism is taken as the primary reality which *produces* nature, ignoring the fact that industrialism relies on and takes for granted an enormous range of 'ecosystem services' such as the hydrologic cycle and photosynthesis without which life on earth would collapse. At the time I am writing this, the human capacity to control nature is looking decidedly shaky, with climate change and imminent peak oil hinting that natural limits and the consequences of human blindness may have unforeseeable consequences.

Despite this, Bruce Braun and Noel Castree assert that

> 'first nature' is replaced by an entirely different produced 'natural' landscape. The competitive and accumulative practices of capitalism bring all manner of natural environments and concrete labour processes upon them together in an abstract framework of market exchange which, literally, produces natures(s) anew.[119]

Similarly, Neil Smith enthusiastically advocates the capitalist transformation of nature:

> Much as a tree in growth adds a new ring each year, the social concept of nature has accumulated innumerable layers of meaning in the course of history. Just as felling the tree exposes these rings – before the timber is sent to the sawmill for fashioning into a human artifact – industrial capitalism has cut into the accumulated meanings of nature so they can be shaped and fashioned into concepts of nature appropriate for the present era.[120]

What is repressed here is the embedding of the present 'actual' within the more basic 'real' context of nature, so that what we aim to replace are in fact the irreplaceable foundations of life on earth. A new, capitalist nature, supposedly, simply supersedes an old one; and the very word 'past' has taken on the evaluative overtones of obsolescence and outdatedness, as if present structures not only *need* have no connection with past structures, but in fact *should* distance themselves from the past. According to Braun and Castree, capitalism has replaced ecology in our understanding of nature, since the latter "has offered only weak understandings of the nature and materiality of transformed environments."[121] This is like trying to build the spire of a cathedral by taking stone from the foundations, and reflects an anthropocentric epistemology that has entirely forgotten its anchorage in the real world.

Violence and forgetting

Just as the view that animals are mechanisms devoid of feeling has been used to justify their experimental abuse, so more generally the idea that nature is essentially passive and without its own structure or tendencies allows us to distance ourselves from the violence through which industrialism has been established. As I argued in *Nature and Psyche*, there are striking parallels between our current attitudes towards nature and those of the early colonists of the New World towards the wilderness and its indigenous inhabitants; and the notions of an 'empty' landscape and 'primitive' natives served much the same function then as the supposedly 'culturally constructed' character of nature does today. A difference, however, is that the latter term has more of the quality of a post hoc rationalisation, naturalising nature's current, domesticated form and therefore drawing the curtain over the possibility that in other circumstances it would have had its own distinctive forms of flourishing and diversity.

This naturalisation of industrialised landscapes sweeps aside awareness of the industrial activities that were necessary to convert the landscape to its current, 'natural', form. The view that nature is culturally constructed requires an intense social and psychological effort of forgetting – both of the way nature was before it was transformed and of the effort needed to transform it. This is why 'nostalgia' for past 'Edenic' forms of nature – unspoiled landscapes, old-growth forests, the endless prairie grasslands – has been denigrated as a 'romantic fantasy' in recent times: it is the embodied memory of the nature we evolved to inhabit and expect, the painful recollection of the way things were once and might be again. Such embodied memories are suspect, taboo, in a world in which we are encouraged to be 'forward-looking' and 'realistic', and in which success is defined as economic success.

In the same vein, Paul Connerton has suggested that "the modern world is the product of a gigantic process of labour, and the first thing to be forgotten is the labour process itself."[122] Commodities, he points out, take on what Marx called a 'phantom-like objectivity' when viewed in the shop window; and he refers to Lukács' argument that "the capitalist process of production was constituted by the loss of its memory of the very process through which it is produced."[123] Looking through a mail-order catalogue, for example, there is no need

> for the customer to wonder where [the product] came from, or how
> it had been created, or by whom, or with what implications for the

place where it had been made. ... One could buy merchandise ... without troubling to reflect on a web of economic and ecological linkages that stretched out in all directions.[124]

Even when the industrial world remains utterly parasitic on the natural world, these connections are often obscured. Referring to Cronon's history of Chicago, Connerton argues that

> Chicago both fostered an ever closer connection between city and country, and concealed its debt to the natural system that made it possible. ... The field was separated from the grain, the forest from the lumber, the rangeland from the meat. ... The easier it became to obscure the connections between Chicago's trade and its earthly roots, the more casually one could forget that the city drew its life from the natural world around it.[125]

Such forgetting is today a taken-for-granted aspect of contemporary subjectivity, as we are socialised into focusing on present and future. As Connerton tells us, most "young men and women [today] grow up in a sort of permanent present lacking any organic relation to the public past of the times they live in."[126]

This is not so much an *individual* process of forgetting as "something societies, or indeed civilisations, might do."[127] Memory, as we saw above, is *an essential element of ecological structure*; and as such it becomes essential that it be pushed aside in order to allow industrialism's imposition of alternative forms. Such acts of forgetting are intrinsic to contemporary language, as in terms such as 'natural resources', 'raw materials', or 'development', so closing any ideological gap between our transformation of the world and the ways we describe it.

It is not only labour that has been forgotten as the contemporary world is naturalised as 'just the way things are'. Violence is also intrinsic to the process – violence against both ecosystems and the humans who lived among them. The current tranquillity of the Scottish highlands, for example, draws a deceptive veil over the clearances of the eighteenth and nineteenth centuries, in which thousands of crofters were thrown off their land by large landowners. Similarly, the history that lies hidden behind the ordered rows of crops on a modern farm is one of the eradication of the natural grassland or forest, and of the fauna that inhabited it. The extreme case is the total annihilation of life that might result from a nuclear exchange: the devastated landscape that would result would be 'peaceful', since there would be no life, no predation, indeed scarcely any movement: but its historical basis would be one of extreme violence. This latent violence, then, is violence that is sedimented into history as the basis of a present which denies its violent past, presenting an appearance of peacefulness and naturalness. Despite all the noise and rush, the numb emptiness of latent violence pervades industrial society; indeed, the noise and rush are often unwitting attempts to fill the vacuum of life, to conceal the emptiness disguised by the functioning of mechanism. The latent violence of contemporary life embodies past oppressions, reflecting the continuing suppression of alternatives, the maintenance of a disparity between what is and what could be, the mistaking of the actual for the real. It is therefore often invisible and unsayable: it cannot be exemplified by what is, only by what might be but isn't. Like the peace maintained by an occupying army, the silence that haunts the ruins of a bombed city, the uniformity of a field of cabbages, the serenity of a corpse, the sullenness of a caged lion, the conformity that stems from fear, or the bleakness of a clear-cut, it is the absence of something rather than its presence that is oppressive. It is not just the aftermath of violence: it concerns lifestyles and structures that *depend* on violence, perpetuating the oppression that began with violence, ingraining 'incongruence' into our character, and implanting a half-conscious guilt within us.

Latent violence exists plentifully in most manufactured artifacts: the engine block, derived from ore dragged from the earth; the garden furniture, made from redwoods once growing on the northern California coast; the smooth coatings and sauces in *Elle*, designed to "disguise the primary nature of foodstuffs, the brutality of meat".[128] It is present not only in artifacts, but also in the lives of humans and nonhuman creatures. For example, Stjepan Mestrovic has referred to 'post-emotional society' as one in which people are not overtly oppressed, but are distinguished by their incapacity for emotional commitment and empathy.[129] Likewise, Richard Silverstein has referred to 'pre-emptive censorship'[130] – that is, the phenomenon whereby issues or events are stifled due to the fear that they *may* cause controversy. This is only one step removed from the complete repression of a possibility, the removal even of current events from public awareness. For example, in January 2009 the BBC refused to transmit a humanitarian appeal on behalf of the recently bombed, shelled, and otherwise brutalised people of Gaza, claiming that showing the appeal would 'compromise its impartiality'.[131] What links these situations is that whole areas of debate or emotional reality are damped down and denied social expression. Thus the full spectrum of subjective feeling and awareness is shrunk to fit pre-ordained political frameworks.

Something similar happens in the ecological realm. If forest is clear-cut, then that is an act of violence that is easily recognisable; and, given enough time, it is partially reversible. But if the once-forested area becomes an industrial estate, a more fundamental disappearance has occurred: the area that was once forest has been inducted into another realm entirely, that of the industrial world; and while the new area may appear peaceful and even attractively landscaped, no trace will remain of the violence of its birth. The death of the forest, like its life, is forgotten, becoming a 'romantic myth'; and the industrial estate seems immaculately conceived, born out of nothing. To talk of the 'violence' inherent in the building of the estate seems as inappropriate as talking of the violence associated with my armchair: these entities are made to appear as if they belong to an entirely separate realm from the natural, so that the death throes of the natural entities involved in their production are dissociated from the realities we live amongst.

Affluent areas of the modern world often seem non-violent, even peaceful. But if something should contravene the industrialist pattern, it will quickly be snuffed out, revealing the violence that lies beneath the surface of a smoothly functioning social order as well as the normally hidden mechanisms that enforce it. Two gear wheels, for example, mesh smoothly so long as nothing gets between them – in which case, it is crushed. A field of corn may seem peaceful; but if any 'weeds' appear, they will quickly be doused in chemicals. I am 'free' – just so long as I follow the rules and regulations; so that conformity is engineered into my character, reducing the need for coercion. A seamless integration of law, 'common sense', and technology is harmoniously allied with the built environment: notions of private property neatly complement the technologies which map borders and produce fences, and the idea that the new has to be constantly updated is embraced by industries which obligingly churn out the latest models and fashions. Violence is *unnecessary* in a society in which the repressions and controls are already incorporated *within* people, institutions, and infrastructure, so that as Morris Berman points out, censorship becomes unnecessary when nobody reads books anyway.[132] The face of violence in the modern world is as often the Eichmann-like figure of the bureaucrat in his office as it is the tank in the market place. By accepting a situation as no more than it appears in the present, by our blindness to its history, we conceal unpalatable aspects of the way we live, and deny the disparity between what is and what could be. The past is no longer seen as a foundation for what lies ahead, but rather as something that has been rubbed out and *replaced by* the present. Thus an organic metaphor of growth and connectedness is replaced by an industrialist one of obsolescence, in which 'progress' forgets its origins.

As 'human nature' is reconstructed, the acute pain of loss experienced by recently industrialised tribal peoples is replaced by chronic amnesia and emptiness. Peter Kahn has referred to this as 'environmental generational amnesia'[133] – the tendency for each new generation to take as the norm whatever environmental situation they are born into. The same applies to the experience of selfhood: we not only identify mainly with our symbolic capacities and ignore our embodied ones; we also identify with *particular* ideologically generated forms of the symbolic – fashion, political quiescence, material status-anxiety, and so on. As reduced beings, we can only become aware of our reduction if our subjectivity is expanded to include the past and the wild world – in which case, a door is opened that we previously never knew existed. The crushing of alternatives generates a sort of emptiness that translates into mild chronic depression or anxiety; and it is not surprising that both these disorders seem to be becoming much more widespread in the industrialised world.[134] Human nature is thus chronically reduced to fit industrialism, and the harmony between these two reduced forms of being converts the violence of oppression into the sort of vacuity which is the absence of structure and vitality. Academia often reflects rather than critiques these developments, replacing critical theory with fashionably relativistic forms of intellectual amnesia, celebrating the frivolous diversities of individual consumer choice, and so colluding in the replacement of the natural by the industrial.

Despite this well-orchestrated ideological assault, however, the real remains more fundamental and more enduring than any socially conjured, technologically aided actual. Wild nature is distantly echoed by countryside and gardens in the same way that the possibility of wild rivers is implied by the still waters of reservoirs held back by dams – a potential that is always present. Similarly, every child born – at least until the advent of industrial-scale genetic manipulation of human life – embodies the memories and potentials of natural being. The replacement of these embodied memories and potentials by those that are better fitted to life in the industrial era therefore requires a sophisticated process of socialisation into current 'realities' – a process we will examine in the next chapter.

3
Growing Out of the World

Empathising with the world

Each of us comes into the world as a result of the natural processes of conception, gestation, and birth, the products of an evolutionary process spanning hundreds of millions of years – enormously exceeding the time language has existed. Our embodiment, including our neurological capacities, reflects this evolutionary heritage, imposing constraints on the sort of world we can adapt to flourish in. We are both natural and symbolically capable beings, well equipped to function in a world that is physically and intellectually demanding, but still different in important respects from the industrial world we are born into. In Edward O. Wilson's words, "the brain evolved in a biocentric world, not a machine-regulated world. It would therefore be quite extraordinary to find that all learning relating to that world had been erased in a few thousand years".[1]

Children are unsurprisingly predisposed to interact with the natural world in ways that reflect this evolved complementarity. As Peter Kahn and Batya Friedman found in a study of inner-city black children in the Houston area,[2] even "the serious constraints of living in an economically impoverished urban community cannot easily squelch ... children's diverse and rich appreciation for nature, and moral responsiveness to its preservation."[3] For example, children typically believed that it was wrong to pollute a bayou – often for anthropocentric reasons involving the effects on other people, but also, sometimes, for biocentric reasons involving the effects on wildlife and the intrinsic value of nature.

Scott Atran and Douglas Medin found a similar anthropocentric bias among urban majority culture American children; and they argue that this may be due to "lack of intimate contact with plants and animals". These children readily recognised biological affinity between humans and other animals, but also tended to justify "a failure to extend a property from an animal to humans on the grounds that 'humans are not animals'". Comparing these results to those of similar-aged Menominee children from Wisconsin, Atran and Medin found that "even the youngest Menominee often reasoned in terms of ecological relations … and showed no reliable human-animal asymmetries." They point out that the "Menominee origin myth has people coming from the bear, and even the youngest children are familiar with the animal-based clan system. In short, there is cultural support for a symmetrical relation between humans and other animals."[4]

Atran and Medin therefore suggest that the "anthropocentric pattern of reasoning in young urban children is an acquired cultural model. That is, 3-year-old urban children do not show the anthropocentrism that we observe in 4- to 5-year-old urban children."[5] This explanation is supported by their finding that Menominee children's performance on biology undergoes a precipitous decline in successive grades. Menominee children, they found,

> have a precocious understanding of biology. Indeed, on standardised tests, fourth-grade Menominee score above the national average in science and it is their best subject. Strikingly, however, by eighth grade, science is their very worst subject and they score below the national average. [This may be because] there is a mismatch between science as it is taught and Menominee culture on at least three levels; (1) specific facts, (2) knowledge organisation, and (3) cultural values and practices.[6]

Atran and Medin point out that "in Menominee culture (and in many cultures around the world) all of nature is alive, including not only plants and animals but also rocks and water". This suggests a conflict between Menominee understandings of 'aliveness' and those suggested by biology textbooks, in which ecology was not given prominence. Furthermore, "both the texts and the majority-culture parents tend to imply that nature is an externality to be exploited, cared for, learned about, and so on [whereas] Menominee parents and urban Indian parents … tend to emphasise that we are part of nature and that nature is not an externality." Atran and Medin conclude that some aspects of science education may be "alien to Native Americans in the same way that an economic approach to a family (e.g. How many dollars is a daughter worth?) is repugnant to most people."[7]

While I would not want to argue that ecological knowledge is 'innate' in young children, it certainly seems plausible that it is at least consistent with the expectations and predispositions implanted by our evolutionary past, and that we have a readiness to become attuned to the natural environment. Children need little encouragement to view nature as alive and intrinsically valuable, and to form relationships with animals: as Stephen Mithen suggests, "Give a child a kitten and she will believe it has a mind like her own."[8] Gene Myers, in his important study of children's interactions with animals, confirms this view, showing that despite predominant cultural assumptions to the contrary, children invariably perceive other creatures as purposeful social beings like themselves and other humans.[9] Likewise, children tend to prefer to interact with animals as social equals rather than from the perspective of a conceptually or physically imposed distance: for example, Gary Nabhan reports that when he took his 2- and 4-year-olds to a zoo, they preferred to spend their time feeding the ground squirrels that had "broke into the zoo" rather than looking at the other creatures that were separated from them by ten-foot wide moats.[10] Only reluctantly are we socialised into believing that there is an impregnable barrier between ourselves and the rest of the natural world and that we should interact with creatures as 'things' rather than as sentient, intelligent entities like ourselves. In this respect, 'transitional objects' such as teddy bears may unwittingly enable a different kind of 'transition' than that which Winnicott had in mind[11] – the transition from relating to other creatures as intelligent, sentient beings to the 'adult realisation' that they are merely 'things'.

This taught injunction to interact with and understand other creatures only in the terms laid down by our anthropocentric conceptual system is part of an even broader principle: the replacement of embodiment by symbolism. As Susan Buck-Morss puts it,

> the first great cognitive leap is the prototypical experience of alienation. It is the ability of the child to divorce subject from object, hence to grasp the building block of ... industrial production. ... With the attainment of object permanency, the idea of an object ... becomes a substitute for the thing itself, indeed ... is granted greater cognitive value than the material object, and the child is capable through symbolic play of leaving reality unchanged.[12]

Conceptual functioning and alienation

Has this gulf between our evolutionarily based expectations and the knowledge system we are socialised into been a constant feature of Euro-American societies; or is it a relatively recent feature of modernity? Atran and Medin find evidence for the latter hypothesis, focusing on references to trees in different historical periods. Using the *Oxford English Dictionary*, and assessing the number of tree-related quotations relative to the total number of quotations for a given historical period, as well as the relative number of sources from which the quotations were drawn, they found a marked decline in references to trees over past centuries. They note that "the start of the decline corresponds closely to the start of the industrial revolution."[13]

The period during which this change has occurred is, of course, insignificant compared with evolutionary timescales; and the implication may be that the environment children today are growing up in diverges significantly from their embodied expectations. While this would not be simple to confirm in any empirical way, it is a plausible hypothesis, and one which has enormous implications for child development, mental health, and indeed the future of industrial society. It is also a hypothesis which has been largely ignored – partly, perhaps, because of empirical difficulties, but also due to the pervasive splintering of research into the 'hard' sciences such as biology and those disciplines which are preoccupied with a cultural focus. As Stephen Kellert remarks, one "wonders if the relative absence of published material ... may be indicative of a society so estranged from its natural origins it has failed to recognise our species' basic dependence on nature as a condition of growth and development".[14] Similarly, John Schumaker has suggested that modern society is "characterised by a collective dissociative amnesia that involves a complete forgetting of the human-nature relationship".[15]

This rupture between ourselves and the world also has *moral* implications. As Myers points out, although contemporary theory overwhelmingly represents moral development in terms of the 'social construction of morality', there is plentiful evidence that evolutionarily-derived, embodied predispositions are also involved. For example, a teacher explains to five-year-old Joe that

> Some big animals have only one baby, and they're the kind of animal that take care of their young, but when an animal has many, many, many babies like a mosquito or like an ant or like a bee, they don't

take care of them, and some of them are meant to be food for other animals. Wow, that's wonderful to know.[16]

At first, Joe seems to accept this explanation; but later, he has the following conversation with his mother:

MOTHER: You told me that some of the babies, that there were some other babies but they died?
JOE: Don't – I don't want to talk, about that.
MOTHER: You don't want to talk about it? … How come?
JOE: 'Cause it makes me sad.
[Omitted text]
MOTHER: I see. Okay, well, let's not talk about it then.[17]

This extract suggests that Joe is being taught to replace a *felt* relation to the turtles by an *instrumental* one; and more generally, to develop a detachment from natural entities. David Eaton has examined Myers' findings in detail; and he comments that

[i]t seems very clear here that Joe is experiencing feelings that are totally incompatible with the repeated attempts of the adults concerned to socially construct the predation of the turtles in an emotionally neutral, or even positive, way. Myers is therefore almost certainly correct to conclude that Joe's feelings here are predominantly independent of social processes – or at least of discursive and exclusively human ones.[18]

As Myers points out, his results do not fit the belief of many cultural theorists that the natural world is assimilated by human cultures through 'arbitrary codes': that is, natural features and entities are co-opted as symbols within cultural and linguistic forms which owe little to natural structure. Rather, Myers shows that symbolic understandings can grow out of interactional experience with other creatures such as turtles:

MR LLOYD: "How do you think it feels, what does it feel like to be a turtle?"
SOLLY: "Safe … *safe*."
MR LLOYD: "You think it feels safe, why?"
SOLLY: "Because you have a shell."

Later, the class discusses how the turtle reacts to the appearance of a shark:

> BILLY: "When a shark comes ... He just puts his whole body in his shell" ... As he says this, Billy pulls his arms in tightly towards his sides.[19]

Myers comments that for Billy, "the turtle symbolises not only safety and coherence but also the whole affective experience of surviving an imagined life-threatening situation. Notably, Billy's symbolisation took the embodied form first of a tightly closed-off protective posture and then of an expansive, mobile, and agentic one – conveying affective qualities that would be hard to represent verbally."[20]

Myers suggests that concern about the fate of animals may often be lost as the child is exposed to the abstractions of school learning, noting that "children's books idealise farm life",[21] and that "the larger society ... uses distancing mechanisms to reduce the conflict. ... These misrepresent exploitation, thus relieving the discomfort we might otherwise feel."[22] In other words, children's empathy with and concern for other creatures is overlain by a symbolic web that protects us from kinds of embodied knowledge that are uncomfortable and politically inconvenient.

Developmental psychology, however, has generally framed moral development in terms of cognitive *abilities* rather than embodied *feelings*, so that in Lawrence Kohlberg's sequence of stages of moral development, for example, each stage reflects greater cognitive sophistication than the previous one.[23] However, the ability to *reason* morally is not necessarily reflected in moral *behaviour*,[24] since behaviour reflects a range of influences and motivational factors as well as the individual's perceived control over a situation. The person inhabits a world that is not only reasoned about, but also experienced through sensation and feeling; and trying to explain moral behaviour purely in terms of abilities, attitudes, and hypothetical scenarios fails to bridge the gap between the the symbolic and the embodied aspects of our existence. For example, an 18-year-old student interviewed by Cynthia Thomashow traces the beginnings of her environmental activism to the sale of a piece of land:

> My parents betrayed me by selling this piece of land. I used to wander there, spend time thinking and sorting out things. I loved playing back there. As the bulldozers moved in, I remember feeling like my arms and legs were being torn from my body, I felt it deep

down inside myself, like I lost a part of myself. It was awful and sad and unreconcilable. ... I will never forgive them.[25]

Morality may require cognitive sophistication, but it is also a matter of felt experience, of empathy, loss, recognition, identification, pain, and delight. As William Barrett suggests, we can intuit the falseness of purely theoretical, cerebral forms of morality that have no roots in embodiment:

> Some dumb inarticulated understanding, some sense of truth planted, as it were, in the marrow of my bones, makes me know that what I am hearing is not true. Whence comes this understanding? It is the understanding that I have by virtue of being rooted in exist-ence. ... We become rootless intellectually to the degree that we lose our hold upon this primary form of understanding.[26]

To study and understand morality primarily as a *cognitive ability* is to perpetuate rather than expose the fundamental moral error of industrial society – the isolation of mind from embodiment and world. If their felt ethical sense is disregarded, then children will be socialised into a sort of institutionalised incongruence – a situation that is likely to have both social and psychological repercussions.

The world as parent

In keeping with its decontextualised focus, the emphasis in devel-opmental research is generally on the 'mother-child dyad'. But the mother and child do not exist and communicate within a self-sufficient bubble; rather, they are nested within wider contexts – the family, the community, the society, and ultimately, the ecological totality of the Earth itself. The relation between the child and the natural world, however, is not *only* an indirect one, mediated by the parent: even young children quickly learn to interact with the world and its various living and non-living constituents; and as they do so, their subjective reach widens and becomes more inclusive. In Peter Marin's words, "the natural direction of human ripening is from the smaller to the larger world",[27] from egocentrism to socio-centrism to ecocentrism; and we can see this movement clearly in tribal societies, for whom the land typically plays a parental as well as an instrumental role. Tim Ingold points to the analogy between "the most intimate relations of human kinship and the equally intimate relations between human persons and the non-human environment" in groups such as the Mbuti, the Batek Negritos of Malaysia, and the Nayaka of southern India.[28] Ingold and other anthropologists such as Laura Rival emphasise that natural entities and processes are not *metaphors* for *social* processes, because social processes themselves are understood as a subset of the *natural* realm. As Rival explains, for example, the Huaorani of Amazonian Ecuador do not so much assimilate the growth rates of the peach palm trees as metaphors for human life: rather, human life is viewed as *part of* the larger natural realm which *also* includes the trees around them, so that "it is a non-mediated perceptual knowledge which orders social relations between people, and between people and other living organisms".[29] Thus the division which we tend to make between human and nonhuman does not apply, since the social is, at root, natural. In contrast to some anthropological conceptions, notes Ingold,

> there are not two worlds, of nature and society, but just one, saturated with personal powers, and embracing both humans, the animals and plants on which they depend, and the features of the landscape in which they live and move. Within this one world, humans figure not as composites of body and mind but as undivided beings, 'organism-persons', relating as such both to other humans and to non-human agencies and entities.[30]

This perspective profoundly changes our understanding of the role of nature in child development. Rather than the parent being viewed as a uniquely important channel of cultural knowledge, direct sensory and practical experience of nature can be seen as underpinning much of the basic knowledge that may later be symbolised in specific cultural forms; and the parent's role becomes less one of direct instruction, and more one of orienting the child's attention in particular directions – as we saw in Chapter 1 in this case of Tim Ingold's father. Arne Naess described the parental role of the mountain after the death of his own father when Naess was an infant; and while rare in industrialised countries, this understanding of the world as playing a 'meta-parental' role is common theme elsewhere. Likewise, the psychoanalyst Phyllis Greenacre found that children destined to become artists were protected from poor parenting by what she called a 'love affair with the world'.[31] But we can turn this idea around: suppose the 'love affair with the world' might be a *normal* childhood experience, and its absence – and the absence of a world we *could* love – might itself be a form of deprivation, albeit one that is unrecognised because it is so widespread? And in this case, if the family, or even the single parent, is forced to take responsibility for providing the experiences previously intrinsic to childhood interaction with the natural world, is this responsibility realistic? Viewed from this perspective, many apparently social or individual problems might more accurately be understood as deriving from our lack of early engagement with a healthy natural world.

Socialisation to an unwild world

Education in industrial society often seems designed to alienate young people from the natural world, focusing as it does on abstractions, concepts, principles, equations, and language rather than on embodied experience. On the campus where I teach, all the recently constructed lecture theatres are windowless dungeons without natural lighting or ventilation. The implication is clear: to learn about the world, you must first cut yourself off from it, and from your own embodied experience of it. Although Descartes' dualistic philosophy is ritually denounced in academia, it seems that his ghost lives on, an unsuspected presence in our lecture theatres and classrooms. There are two fundamental and related lessons being taught here; firstly, that the world should be *thought about* rather than *sensed* or *empathised* with, and complementarily, that we are *thinking* rather than *feeling* beings. This inevitably builds up conflict between embodiment and mind, leading to the sort of situation described by Jerome Bernstein, who reports that one of his clients, while learning about nature at school, had been encouraged to bring insects into class so that they could be put in the 'sleeping jar' before being mounted on a cork board. This girl's response, however, was not what the teacher was expecting:

> I could hear the bugs dying. Some were very quiet about it; most made gasping and moaning sounds as the air in the jar was replaced with unbreathable fumes. The butterflies screamed. It was a high pitched staccato sound. I could not stand it, and ... I asked to be excused, went into the bathroom, turned on the water, to drown out the screams, put my hands over my ears.[32]

Clearly, this girl had failed to learn the Cartesian dogma that to know the world, we have to reject our empathic capacities. As Myers points out, an emphasis on language is often used as a means of separating mind from embodiment and humans from other creatures, replacing the Cartesian soul in, for example, George Herbert Mead's influential account of the origins of selfhood. As Daniel Stern remarks, the child's indoctrination into language

> is a double-edged sword. It … makes some parts of our experience less shareable with ourselves and with others. It drives a wedge between two simultaneous forms of interpersonal experience: as it is lived and as it is verbally represented. Experience in the domains of emergent, core- and intersubjective relatedness, which continue irrespective of language, can be embraced only very partially in the domain of verbal relatedness. And to the extent that events in the domain of verbal relatedness are held to be what has really happened, experiences in these other domains suffer an alienation. … Language, then, causes a split in the experience of the self. It also moves relatedness onto the impersonal, abstract level intrinsic to language and away from the personal, immediate level intrinsic to the other domains of relatedness.[33]

Likewise Eric Berne suggests a scenario that many of us can relate to:

> A little boy sees and hears birds with delight. Then the 'good father' comes along and feels he should 'share' the experience and help his son 'develop'. He says: "That's a jay, and this is a sparrow." The moment the little boy is concerned with which is a jay and which is a sparrow, he can no longer see the birds or hear them sing. He has to see and hear them the way his father wants him to.[34]

Our indoctrination into language, despite its unquestionable powers and advantages, may leave us scrabbling to recover from the resulting alienation from the world and the associated loss of our own experience, undermining our ontological security and fatefully locating us within a detached symbolic realm. There is also a more specific alienation: if subjectivity is made dependent on language, then clearly animals cannot be subjective beings, and as Mead puts it, 'man' is "a more advanced product of evolutionary development than are the lower animals."[35] Consequently, we cannot have relationships with animals in the same way as we can with humans; nor can they play a significant role in our development as persons.

But the notion that subjectivity depends on language is contradicted by the results of Myers' thorough exploration of young children's interactions with animals. Myers found that contrary to Mead, subjectivity was not linguistically based: rather, the sense of self was formed as children interacted and empathised with others, including non-human others. Furthermore, children treated other animals as fully subjective beings, recognising four key characteristics of subjectivity which Myers labels

> *agency* (the animal moves on its own and can do things like bite, crawl, look around, and so on), *coherence* (the animal is easily experienced as an organised whole), *affectivity* (the animal shows emotions …), and *continuity* (with repeated experiences the animal becomes a familiar individual). … crucially, [these] are qualities that form the child's sense of self and other, beginning at a very early age.[36]

According to Myers, the recognition of these qualities and the subsequent acceptance of nonhuman animals as subjective beings precedes any relevant linguistic or cultural input. In fact, Myers found that even young children were often reluctant to accept adults' constraints on animals' autonomy and freedom. For example, he describes the scenario that unfolds when a spider monkey, its tail firmly held by its owner, is brought into the classroom:

> The monkey strains briefly to get crumbs on the floor. Drew: "Why don't you let go of his tail?" Ms Dean: "He's got my hand. I'd let it go in a second if he's let go of my hand." Actually, her hand is tight around the monkey's tail but its prehensile tip is visibly relaxed in her hand. Drew points to this, gesturing with one hand on the other: "He's let go'd of your hand … He let go of your hand." Ms Dean holds her hand behind her leg, out of Drew's sight.[37]

If such experiences are important formative agents in the growth of empathic subjectivity, then it is worrying that embodied experience of the natural world is becoming rarer, for a variety of reasons such as the destruction of wild areas for agriculture, mining, and urban construction, the absorption of people into cities from the countryside, and the availability of television, computer games, and the internet. Gary Nabhan reports that "children today spend more time in classrooms and in front of the television than they do interacting with their natural surroundings. … a significant portion of kids today have never gone off alone, away from human habitations, to spend more than half an hour by themselves in a 'natural' setting. None of the six Yaqui children responded that they had; nor had 58% of the O'odham, 53% of the Anglos, and 71% of the Mexican children". Furthermore, "77% of the Mexican children, 61% of the Anglo children, 60% of the

Yaqui children, and 35% of the O'odham children ... had seen more animals on television and in the movies than they had personally seen in the wild".[38]

I remember visiting Land's End – the extreme westernmost tip of Cornwall – as a child, and savouring the wildness of the place as I stared across the Atlantic towards America. Such situations don't tell us what to feel or experience; rather, they invite reflection and personal connection. That evocative landscape has now been erased by the conversion of Land's End into a theme park, complete with aquarium, film shows and a 'multi-sensory audio-visual extravaganza entitled *The Last Labyrinth*', signalling its assimilation as part of the commercial world. What is destroyed by such changes is not just a place, nor just the particular subjective feelings evoked by the place; but the union of feeling and ecological realities within a single coherent world. If I returned to Land's End, I would be an observer, merely a witness to the wreckage of structures that were as vulnerable to capitalism as was the Great Auk to the clubs of sailors. Today, wildlife cameramen trawl for images from the last few wild places, just as fishermen drag their huge nets across the sea bottom to scrape up any remaining life. Increasingly, 'nature' is experienced through nature documentaries, pretty adornments bought from the Nature Company, or visits to Sea World. It is more and more difficult to find nature except within the context of these industrial aliases.

Mainstream developmental theory depicts the child as moving ever more adeptly within a symbolic universe in which humans are portrayed as unquestionably and rightly dominant, and interacting with the material world in selective, and symbolically and technologically mediated ways. Indeed, it is an article of faith among many theorists that all experience beyond infancy is necessarily symbolically and culturally mediated, and that direct experience of the world is simply impossible. 'Growing up' is therefore a movement away from interdependency with nonhuman others and a repudiation of felt links with the world as the person develops an 'internal locus of control'; and in Piaget's terms, "thought becomes free from the real world."[39] As Harold Searles pointed out, many writers have emphasised the child's progressive *differentiation* from the world. This, as Searles argued, leads to a "sense of inner conflict"[40] as the child's naturally developing *empathy with* the world is challenged and eventually overwhelmed by socialisation into a separate, symbolic realm of principles and ideology.

This widely accepted story of individual development simply as a growing autonomy ignores the possibility that development, ideally, *also* involves a *growing together* of the individual and the world. As David Smail puts it:

> Even more fundamental than our relations within society ... is a rootedness in the physical environment encountered by us as infants as we taste, smell, feel, hear, and see ourselves into existence. ... the rock-bottom basis of a secure subject might be traced to his or her passionate embrace, from the very first moment of existence, of the world as physical environment. It is as if the infant is presented with a choice: to accept the evidence of its senses, or to bend to the demands of ideological power.[41]

The irony of this process of growing 'autonomy' is that we are being assimilated into a system which saps autonomy far more completely than any immersion in the natural world.

Growing into – what?

There is now plentiful research showing that experience of the natural world has positive effects on psychological and physical health. For example, Howard Frumkin has reviewed evidence that contact with animals, plants, and wilderness all benefit health; and Rachel and Stephen Kaplan's *The Experience of Nature* makes a similar point.[42] This evidence has to date been mainly used to justify wilderness therapy[43] – which is rather like saying that vitamin C is a useful medication in cases of scurvy. Could it be that in a healthy society, direct experience of wild nature should be a part of *all* our lives rather than an 'optional extra'? If so, then it is necessary that we inhabit a world that can be *grown into* – in other words, one that resembles our 'environment of evolutionary adaptation' more closely than the contemporary world. Today, the cognitively elusive but emotionally necessary qualities of wild nature have been trampled in the drive for economic growth; and the result-ing symptoms are often treated as *individual* problems to be addressed by social workers, the law, psychotherapy, and an educational system increasingly aimed at shaping children to fit economic demands rather than at articulating their humanity.

Paul Goodman referred to this covert environmental poverty when he stated a rather obvious truth that is nevertheless widely ignored: "growth, like any ongoing function, requires adequate objects in the environment to meet the needs and capacities of the growing child, boy, youth, and young man"[44] – and, one might add, young woman. In the absence of these conditions, we will be prone to psychopathology, unhappiness, or some other, probably undesirable, form of deviance. Goodman's words, written in the 1950's, apply with as much force today, demonstrating how little social progress we have made since then:

> Thwarted, starved, in the important objects proper to young capacities, [young people] naturally find or invent deviant objects for themselves: this is the beautiful shaping power of our human nature. Their choices and inventions are rarely charming, usually stupid, and often disastrous; we cannot expect average kids to deviate with genius. But on the other hand, [those] who conform to the dominant society become for the most part apathetic, disappointed, cynical, and wasted.[45]

Given that 'economic rationality' represents a considerable narrowing of human potential, it is hardly surprising that what remains unexpressed and unarticulated through the lack of appropriate cultural patterns and natural landscapes may emerge in a variety of unwelcome forms. As Robert Bly puts it, "When a young man in our culture arrives at the end of adolescence, the river of secularity typically carries him over the waterfall and he's out in the big world. The speakers at his high school graduation will say, 'The future belongs to you.' But the speaker does not mention to whom the student belongs. He belongs to nothing ... he belongs to light beer, and sitcoms about bars, and forgetting".[46] Adrift in this vacuum of meaning, the 'empty self'[47] is by default dependent on a constant 'life-support system' of television, music, and social interaction, so that as Richard DeGrandpre has argued, "stimulation has become a substitute structure for the child."[48] Such an arid cultural landscape predisposes young people to a range of 'psychological' problems such as anxiety and depression, and not least, prejudice; for if we are predisposed to seek structure and meaning, then in the absence of appropriate cultural structure we will inevitably grasp at distinctions such as race, accent, colour, and so on.

Thus socialised, miseducated, and entrained to fit a technologically advanced but morally groundless world, our embodiment is largely forgotten except inasmuch as bits of it are hooked by advertising which appeals to our insecurities, greed, need for status, sexual desire, and so on. In DeGrandpre's terms, we are torn between two 'selves' –

> the technological self in which the interface with technology synchronises our conscious minds to the rhythm of the microchip, and gives us increasingly realistic virtual worlds in which to live

and dream ... [and] the social self, which moves within human relationships that operate within relatively slower speeds and with old-fashioned rules of engagement.[49]

DeGrandpre's 'technological' and 'social' selves seem strongly related, respectively, to the symbolically and ecologically embedded aspects of selfhood. To the extent that the former aspects dominate industrial society as we become governed by clock-time, deadlines, and the rapidly shifting imagery of television and computer, so, he argues, "the external structures that now govern behaviour are no longer of the sort that can be internalised by the child to produce self-governance". This situation contrasts with that experienced by children in indigenous societies such as the Huaorani studied by Rival, in which individual development is part of natural patterns and processes within the forests they inhabit.[50] It is therefore unsurprising that, cast adrift from the natural flows and rhythms that our embodiment resonates with, we suffer a range of anxieties, conflicts, and deep uncertainties. Nancy Scheper-Hughes' comments on Japanese children could equally be applied almost anywhere in the industrialised world:

> children today are colonised subjects in homes and schools where they are coached to assume the normative adult role of 'ceaseless production' that is expected of all Japanese citizens ... Japanese children – denied free play and physical liberty – are increasingly manifesting the chronic stress illnesses of adults: hypertension, obesity, ulcers, and elevated cholesterol.[51]

Such symptoms are reminders that we are still embodied creatures, and that the attempt to educate children to suppress their embodiment while learning to inhabit a largely symbolic world can never be entirely successful. The child's predicament is reminiscent of the 'double-bind' situation identified by Gregory Bateson and his colleagues,[52] one form of which involves a conflict between the verbal and non-verbal components of a message. Constantly having to deny what her embodied senses are telling her, and instead learning to focus on detached symbolic understandings, the child learns that sanity and success depend on closing oneself off from the world and denying one's empathic capacities in order to cope with the contradictions and denials that are woven into the fabric of industrial 'culture'. This culturally acquired incoherence will continue so long as we remain both embodied and symbolic beings living within a society that focuses on one of these human dimensions while denigrating the other. As DeGrandpre notes:

> It thus seems that until the ... technological self, takes over the ... social self – a time when we will live only among virtual beings of our own design – the divided self will continue to be deeply conflicted.[53]

That conflict is the human manifestation of the fundamental conflict that defines our era – that between industrial symbolism and the natural order. Until we recognise and engage with this underlying conflict, solutions to the social, psychological, and environmental problems that plague us will remain elusive.

4
Lost in (Symbolic) Space

Embodiment and symbolism

One might reasonably assume that our physical embodiment in a physically real world would be the foundation from which our understanding would grow; and it is bizarre and tragic that much philosophy and social science, as well as some of the major religious traditions, begins from precisely the opposite assumption – namely, that our embodiment leads us away from truth and understanding. It is an indicator of the estrangement of academia from reality, as Paul Stoller tells us, that it has become necessary to defend the view that "the human body is not primarily textual ... rather, it is consumed by a world filled with smells, textures, sights, sounds, and tastes, all of which spark cultural memories".[1] Arthur Kleinman, too, has referred to the "tyranny of meaning that overvalues coherence and other intellectualist priorities", adding that

> sensory conditions of sounds, smells, tastes, feel, balance, sight, and more complex and subtle sensibilities (moral and aesthetic), as well as muscular agency and action also constitute everyday experience. So do social relations and social memories; fragmentary, contradictory, changing, unexpressed and inexpressible though they often are.[2]

In a similar vein, Robert Desjarlais has pointed out that even when the body is referred to, writers have a "tendency ... to privilege the linguistic, the discursive, and the cognised over the visceral and the tacit".[3] Language, for Desjarlais, is a means of expressing something beyond language: "We cannot forsake a study of discourse and meaning, but only because words and images are one of the key means through which people learn and express sensibilities."[4] Likewise, Paul Connerton complains that as consciousness and language have become emphasised as the defining features of humanity,

> the life of human beings, as a historical life, is undertaken as a life reported on and narrated, not life as physical existence. When the defining feature of the human species was seen as language, the body was 'readable' as a text or code, but the body is regarded as the arbitrary bearer of meanings; bodily practices are acknowledged, but in an etherealised form.[5]

Consequently, according to Connerton, embodied experience "has recently been subjected to a cognitive imperialism and interpreted in terms of a linguistic model of meaning",[6] so that what social theorists talk about is

> the symbolism of the body or attitudes toward the body or discourses about the body; not so much about how bodies are variously constituted and variously behave. It is asserted that the body is socially constituted; but the ambiguity in the term constituted tends to go unexamined. ... Practices and behaviour are constantly being assimilated to a cognitive model.[7]

Drew Leder sees this tacit takeover of embodiment by the symbolic as a pervasive problem within philosophy, noting that

> a certain telos toward disembodiment is an abiding strain of Western intellectual history. The Platonic emphasis on the purified soul, the Cartesian emphasis on the 'cogito' experience, pull us toward a vision of self within which an immaterial rationality is central.[8]

Thus in the contemporary repackaging of the Cartesian doctrine, the body itself is not seen as *embodying* meanings: rather, it supposedly *carries* meanings that exist in a separate, linguistic, domain that somehow hovers above physical existence. The problems with this approach have been summarised by Ingold, following Michael Jackson, emphasising that the

> 'subjugation of the bodily to the semantic' diminishes the body and its experience in two ways. First, body movements – postures and gestures – are reduced to the status of signs which direct the analyst in search of what they stand for, namely *extra*-somatic cultural

meanings. Secondly, the body is rendered passive and inert, while the active role of mobilising it, putting it to use and charging it with significance is delegated to a knowing subject which is both detached from the body and reified as 'society'.[9]

As David Smail summarises these problems,

> The great paradox of the 'linguistic turn' ... is that, at the same time as helping to construct a mythical, essentially interior world of 'discourse', it radically undermined our ability to talk about a real, exterior world.[10]

Given these difficulties, it is unsurprising that there has been a good deal of bet hedging about whether or not language creates or represents embodied realities. For example, Derek Edwards has his cake and eats it when he claims that "it is our texts, our discourses, our descriptive practices, that bring their objects into being. At least, they bring them into being as the objects of our understanding."[11] Similarly, Jonathan Potter manages to defend linguistic constructionism while simultaneously capitulating to realism, all in the same sentence, when he claims that "descriptions and accounts construct the world, or at least versions of the world."[12] As Andrew Collier ruefully remarks, "the kind of idealism that treats the world as dependent on our cognitive choices ... has really come into its own" in the past few decades.[13] So it is that the spirit of Descartes undergoes a reincarnation in theories that would claim to take us beyond dualism.

As Terrence Deacon notes, symbolic systems such as language enable us to share aspects of experience that might otherwise remain idiosyncratic.[14] The danger, however, is that by focusing on the communications themselves, we lose touch with the physical and experiential realities we originally intended to communicate, so that these realities begin to appear as if they were created *by* language. But although words can *sometimes* create subjective states, much subjectivity is independent of language. When we are stung by a wasp, or admire a beautiful view, or hear a clap of thunder, these experiences have nothing to do with language. As Lyng and Franks state:

> Words by themselves cannot conjure up the actual feeling of love or clinical depression. If one could express a particular quale in certain words, we could produce a particular sensation just by saying it: we could produce the taste of the best meal ever by just describing it, or

we could verbally drum up some other sexual experiences and be in eternal ecstasy.[15]

Furthermore, as Richard Shweder points out, "There are documented cases … of people who talk a lot about an emotion … yet rarely experience it, or experience an emotion … yet have no word for it and rarely speak of it. … A people's lexicon for emotions is a rather poor index of their emotional functioning." And he goes on to point out that "by 18 months of age, that is, prior to language learning, children experience and know the difference between anger, surprise, distress, interest, fear, and disgust".[16] What is more, the claim that "any knowledge of the things of the world is … articulated in language",[17] can be easily contradicted by obvious exceptions. We know how to walk, throw a cricket ball, judge our distance from another car, decide whether we like a particular wine or food, interpret another's glance, and turn in the direction of a sound – all without any necessary involvement of language. We share this sort of knowledge with many other creatures: a spider's web, for example, demonstrates superb understanding both of the materials involved and the spider's own physical capabilities – all without any linguistic involvement.

Nevertheless, we may be misled into believing that 'objectivity' is located in the shared *description* rather than in the reality – an error that is enshrined in social constructionist theory. As Timothy Reiss puts it: "Galileo's 'I' becomes Descartes' 'we' and the objectivity of 'common speech' from Locke to Lavoisier."[18] There is a parallel here with the way that linear perspective vision, with its built-in assumption of the detached individual gazing out over the world from a precisely defined location, unwittingly conveys a certain sort of world, so that in Robert Romanyshyn's terms, "a way of seeing has become for us a world that is seen."[19] Through such transformations, the mind, rather than being the means through which we recognise and align ourselves with the world, instead becomes an agent of colonisation. As Robert Nadeau summarises the history of this error,

> after the human mind emerged, the world that previous generations of hominids perceived as single and entire became two worlds – an inner world in which the self that is aware of its own awareness exists, and an outer world in which the self seeks to gratify its needs … One of the large compulsions in this linguistically constructed symbolic universe was to code and recode experience, to translate everything into representation.[20]

Symbolism reshapes the world

If this symbolic takeover flourished merely in certain academic departments, it would remain merely a cultural curiosity. But while it has achieved cult status in some disciplines, it is also pervasive throughout the industrialised world in ways that profoundly affect us all. For example, our cars are as much vehicles for symbolic meanings of power, speed, and virility as they are practical means of transport. Cocooned in an enlarged, air conditioned carapace of fast and luxurious movement, we do not so much inhabit the world as inhabit the car which shields us from the world-become-alien outside. Unwanted sounds are replaced by those generated by our in-car stereo systems as we float over smooth tarmac within a road network that materialises our dreams of effortless travel between destinations. Such materialised fantasies of power and comfort entwine themselves into almost all aspects of our lives, surrounding us with a manufactured environment that accords with our colonised understandings. In additional twists of the spiraling dominance of symbolism, manufactured situations are themselves re-presented in ways that further distort the embodied realities they involve, as when the South Manchester University Hospitals Trust was found to have successfully met government targets for reducing waiting times – by simply removing from the list those patients who had been waiting longest.[21]

Just as our physical abilities are being replaced by those of machines, so intelligence, as we saw earlier, has increasingly moved out of the individual into the industrial environment itself as we become reliant on systems such as computers, satnavs, and communication technologies. What began as the human ability to represent and envision something in symbolic terms has mutated into a shared symbolic system that provides templates around which we reconstruct the physical world. While we may originally have been the agents of colonisation, we are now subjected to it; and while idealist philosophies reconstruct the world conceptually to accord with what we understand as 'mind', industrialism does much the same materially. This diversion of subjectivity into symbolically ordered channels is essential to the functioning of industrial society, and is strongly defended by those who have an interest in 'business as usual'. For example, a British minister argues that those who are angry with the government's foreign policy should "put their point of view vigorously", but should not resort to other nonviolent forms of protest.[22] Unfortunately, and especially given the control of the media by large private interests, to follow Hazel Blears' advice condemns one to impotence. Similarly, the 'Have Your Say' section of the BBC's website seems to act primarily as a safety valve for feelings which might otherwise be expressed more effectively. The political systems in the USA and the UK, generally, foster an illusion of choice by allowing citizens to vote in elections; but as both main parties in each of these countries are subtly or overtly influenced by corporate interests, the political situation could more accurately be portrayed as a corporate dictatorship.[23]

What has happened here is that the symbolic realm has begun to orchestrate its own relations to the material, selecting the manner, extent, and subjective accessibility of particular forms of relation. In the cases just referred to, protest is safely contained within the symbolic realm, while industrialist processes continue unscathed: you can say whatever you like, so long as it doesn't make a difference. As Michael Steinberg puts it:

> Capitalism embraces discourse, but it does so only because what we say no longer points outside itself. In a world of totalising discourse, any glimpse of genuine life lies outside language and thus appears nonsensical or chaotic. … [Thus] Political and social debates … function not to resolve anything but to reinforce the unspoken presuppositions shared by both parties.[24]

Under these conditions, freedom of speech becomes part of a repressive dynamic. Industrialism operates in two complementary spheres, both of them symbolically ordered: firstly, a sphere of partly concealed material operations, and a second sphere of discourse that is separated from material activities. Increasingly, individual choices are limited to the second of these spheres; and as Steinberg notes, as "a replacement for living, thought is just as addictive as shopping".[25]

The diversion of subjectivity away from embodiment and action into the more easily manipulable symbolic realm also has its academic resonances, as occurs when the differences between speech and action are blurred by linguists' references to 'speech acts'. This notion originates in J. L. Austin's claim, in *How To Do Things With Words*, that by "saying something we do something".[26] As Mark Johnson puts it:

> Our twentieth-century obsession with language has led linguists, philosophers of language, and cognitive psychologists to focus almost exclusively on concepts, sentences, propositions, and other highly abstract structures in their accounts of meaning and reason … There is virtually no mention of structures of bodily orientation, manipulation,

and movement as crucial to semantics. And there is virtually no mention of how our cognitive and 'rational' acts are prefigured in patterns of bodily experience.[27]

This prioritisation of the symbolic means that ecological structures and relations come far down the list of priorities in policy making compared to economic outcomes, legal precedent, or commercial interests, illustrating what Mark Fettes has referred to as a preference "for conceptual coherence over organic coherence".[28] But human well-being is substantially *embodied* well-being, which in turn requires that decisions reflect the health of the physically real world we evolved to be part of: feet evolved to give us traction on hard surfaces, hands to grasp things, and lungs to transfer oxygen from air to blood. None of this has much to do with symbolism or language.

Homo sapiens, says Norman O. Brown, is "the 'animal symbolicum' ... which has lost the world and life, and which possesses in its symbol systems a map of the lost reality."[29] Increasingly, *statements about* the world take precedence over the material and ecological realities of the world itself. Furthermore, words may be used not only to *represent* reality, but also to *hide* it: forest destruction continues behind a smokescreen of wildlife programmes and weasel words such as 'thinning', and brutal colonialist occupation flourishes behind the pretence of the 'peace process'. Until the industrialist reconstruction of the world and of human character is complete, such propaganda will be necessary to symbolically veil, distort, and reinterpret material realities that are still chilling to the incarnated human being. The media complement this propaganda by providing a steady stream of emotionally engaging and vacuous daytime soaps, programmes about 'animal hospitals', celebrity gossip, and human interest stories which divert emotionality into harmless channels unconnected with significant situations and events.

Sometimes, of course, words *are* used to point to something important in the material world – something that is not just at an elemental, purely factual level, for this is unlikely to provide any challenge to industrialism – but at a relational, higher-order level which diverges more starkly from industrialism. In such cases, other defensive mechanisms are likely to swing into action to maintain the comfortable separation between the symbolic and embodied realms. For example, when whistle-blower Jeffrey Wigand revealed in an interview how cigarette company Brown and Williamson fostered the addiction of smokers, the in-house counsel to CBS News ordered that the interview by buried. Similarly,

ABC News apologised to Philip Morris – a major advertiser – for having accurately reported the company's manipulation of nicotine levels in its cigarettes.[30] And when Cherie Blair – the wife of then Prime Minister Tony Blair – was asked about the Palestine issue, and remarked that as "long as young people feel they have got no hope but to blow themselves up you are never going to make progress", Downing Street apologised.[31] Each of these vignettes conveys something important about how discourse is forced into consistency with the industrialist system while betraying embodied, experienced realities.

The enslavement of embodiment

We are living in two parallel realms. On the one hand, our embodied lives are governed by material processes we know little about – within our own bodies and in the remnants of the natural world around us. This realm is poorly articulated, mostly taken for granted, and permeated by intrusions from the second, symbolic realm. This second realm, while it articulates some aspect of the embodied world, mostly serves to colonise and restructure embodiment while distracting us from embodiment's original qualities. Recent social theory, especially that in the Foucauldian tradition, has overwhelmingly emphasised the 'construction' of the body by symbolism; and what is more, it treats this symbolic takeover as a normal, natural state of affairs – just the way 'things are' – so incorporating a pathological process of colonisation within our conception of the human, and abandoning any critical perspective.

As we noted above, the functioning of industrialism depends on the isolation of embodiment from the symbolic order, and the selective reintegration of these two spheres. Just as water is collected in reservoirs before being allowed to resume its natural downward course by being diverted through turbines, our emotions are selectively harnessed to drive consumption. For example, the 'soft and fluffy' images that inhabit television advertisements for 'farm eggs', appealing to our empathy and warmth towards other creatures, are not generally allowed to influence the ruthlessly economistic operation of poultry farms; nor is this operation accurately reflected in advertising material. As Humphrey McQueen records, a "leading practitioner of the 1920s warned advertising agents against visiting the factory where the product was made because ... 'when you know the truth about anything, the real, inner truth – it is very hard to write the surface fluff which sells it'".[32] Similarly, 'terrorists' are generally portrayed as anonymous symbols of evil devoid of human qualities; while at the other extreme, the death of

Princess Diana in 1997 generated a media-induced sanctification that opened the floodgates for an international outpouring of grief.

Thus the symbolic realm is a space of extensive manipulation within which our emotions can be stifled or amplified by different styles of presentation. Only occasionally is there 'leakage' from one realm to the other, as occurred in 1998 when two pigs escaped from an abattoir in Malmesbury. The exploits of the 'Tamworth Two', later personified as 'Butch Cassidy and the Sundance Kid', were given extensive press coverage as they roamed free in the Wiltshire countryside for a week before being recaptured. When the owner announced that he would still send them to slaughter, an international outcry resulted in their sale to the *Daily Mail* and their eventual retirement at the Rare Breeds Centre in Ashford, where it was recently announced that Butch had 'passed away'. Once the realm of personal feeling has leaked past the barriers that 'protect' us from the bleak industrial realities of 'bacon production', it is hard to put the genie back in the box again.

Although the various meanings involved here are in a sense 'symbolic', their adoption by the industrial economy is quite different to the unconscious free play of symbols that Freud had in mind. Symbols are systematically selected to accord with the diverse needs of capital in a way that is superficially acceptable to affluent consumers, so that the unconscious realm is also colonised by industrialism, winnowing down the meanings that are allowable. Jean-Joseph Goux refers to the "loss of any sacred dimension, of any properly symbolic value", and the reduction of the symbol to "a merely material and quantitative sign." He continues:

> Modern currency is governed no longer by spiritual authority, but only by economic and political power. This dwindling of the sacred, this reduction of value to the secular horizon, as in the function of the monetary instrument, exemplifies the 'reign of quantity' ... In contrast to social formations whose dominant ideological coherence is magico-mythological or mytho-cosmological – that is, entrenched in a 'true symbolism' ... capitalist society and the entire mode of symbolising which characterises it suffer from a deficiency of meaning.[33]

As Goux argues, there is a tendency toward intellectual abstraction, the "reduction of reality to a flat rationality. The symbol may lose its depth, its verticality: instead of being a sign of the unfathomable, it becomes a signifying articulation, a structure, a machine".[34]

Symbols are originally *felt*, as when we experience a mountain as symbolising the power of the divine; but through their induction into an abstract and autonomous 'symbolic' realm, they lose their connections to embodiment – which is itself *reinterpreted* by means of industrialist 'symbols'. For example, as Nancy Scheper-Hughes and Margaret Lock point out,

> We rely on the body-as-machine metaphor each time we describe our somatic or psychological states in mechanistic terms, saying that we are 'worn out' or 'wound up', or when we say that we are 'run down' and that our 'batteries need recharging'[35]

In this respect, Scheper-Hughes and Lock argue, those of us who inhabit the 'developed' world differ from non-industrialised peoples, who "think the world with their bodies":

> like Adam and Eve in the Garden they exercise their autonomy, their power, by naming the phenomena and creatures of the world in their own image and likeness. By contrast, we live in a world in which the human shape of things ... is in retreat. While the cosmologies of nonindustrialised people speak to a constant exchange of metaphors from body to nature and back again, our metaphors speak of machine to body symbolic equations.[36]

These authors are pointing to a reversal in the direction of symbolism. Rather than relying on an imaginative empathy between the body and the world, the body and the world are themselves redefined through metaphors derived from industrialist symbolic structures – a redefinition that is eventually realised in material form by the physical reconstruction of the world. As Goux suggests, industrialism instigates a symbolic monoculture which obliterates the diverse symbolic potential of natural entities: "to say that algebra or mathematics operates with symbols, to say that paper money is a symbol of economic value, to say again that computers or electronic calculators combine symbolic links from a list, is already to apply the term in a very different order from myth or dream or Egyptian art".[37] But Goux holds out the hope of throwing off the "domination by a single mode of symbolising", and replacing it by a fully felt, 'polysymbolic sociality'. This 'polysymbolic sociality' can be seen as one aspect of a 'polysymbolic ecology': just as industrialism reduces individuality to single roles, so within a healthy ecology, an entity may regain its multiplicity. A tree need not only be a source of timber: it may recover its multiple symbolic and ecological roles as a source of shade, fruit, shelter, inspiration, spiritual revelation, as well as all those roles that are beyond our powers to recognise.

What is banished from capitalist society, says Goux, is this 'surplus of meaning' that our symbolic capacities could make possible. Just as a landscape's ability to grow a diversity of species is the basis of industrial agriculture's imposition of a particular monoculture, so symbolism's multiple references and meanings are selected and channelled so that they articulate only the specific forms required by industrialist ideology and its materialisations. There is nothing inherently destructive about symbolic ability itself; it is its vulnerability to colonisation by reductive ideologies that is its Achilles Heel. Symbolism, like language – and like the evolutionary changes which made these possible – provides the basic soil out of which industrialism grows; but it does not make industrialism inevitable or 'natural'. In looking for the specific properties of the industrial system, we have to look beyond symbolism, and towards the elusive and rapaciously self-organising capacities of the system itself.

Language is not reality

Illustrating the 'preference for conceptual coherence over organic coherence' noted earlier, the 'linguistic turn' in the humanities has led to a type of environmental writing having a primary allegiance to textual rather than ecological rules. Meaning, we are told, resides in the text, not in any world of nature which the text represents. As one theorist claims, "meaning exists, if anywhere, only in the relationship speaker-language-hearer, not in any one of the three, and least of all in any connection with the extralinguistic universe."[38] David Miller and Greg Philo rightly criticise such discursive approaches, which "bracket off reality and focus largely on the analysis of texts, leading to an inability to discuss material determinants of psychological states".[39] Similarly, Eugene Gendlin notes the way that many philosophers lose sight of experience as they retreat into the more ordered world of language, moving "from experience to what they say, and then on and on. Their explications seem designed to replace experience, to make it no longer necessary".[40]

Such denials of reality are usually implicit; but occasionally there are disarmingly frank admissions that what is being written is side-lining crucial issues. In one paper that argues that "cities are part of nature", we read in a footnote that "We disregard … the complex and pervasive mechanisms that sustain urban living".[41] Another writer argues that environmental sociology should not involve "fleshing out a 'new human ecology'", but rather "understanding how claims about environmental conditions are assembled, presented and contested".[42] A third, after raising the question of whether we can "find ways of living rich, fulfilling, creative, complex, nonrepetitive lives without putting too much stress on nature", tells us that these "highly abstract and intractible questions" will not be directly addressed, since this "would be to shift out of an ecocritical mode into something more sociological or political".[43] If academics do not engage with these vital issues, which are not so much 'abstract' as urgent and practical, then their writing will be of interest only to other academics who are equally estranged from reality.

This preference for symbolic activity uncontaminated by material relevance is also popular outside academia, as in the carbon trading approach that involves industrialised nations buying permits to emit carbon dioxide from less industrialised nations. So far, this has had no perceptible impact on the CO_2 emitted; but it has the 'advantage' of convincing us that steps are being taken to reduce carbon emissions, reducing cognitive dissonance within a symbolic sphere that is only tenuously related to the material realities of CO_2 emission. If we were to be serious about reducing CO_2 emissions, then we would need a solution which exists within, or impacts upon, the material sphere, such as George Monbiot's apparently naïve but actually entirely realistic suggestion that we should leave fossil fuels in the ground rather than extracting them.[44] Clearly, once extracted, fuels are likely to be used – however they are traded or governed by permits. Such real-world solutions, however, tend to be unpopular, since they perceptibly impinge on industrial activity.

Discussing nature in abstract ways that ignore ecological implications is consistent with the industrialist objective of defining a 'human' realm that claims to be self-sufficient. As Louis Sass argues, the "old idea of relationship with the world or with others that is mediated *by* words is replaced by an isolating relationship with language alone – and with a language that in its emphatic self-sufficiency, has itself come to stand as an epitome of isolation and self-involvement".[45] Cognition tries to construct itself as autonomous, free from the material limitations inherent in the life of mortal creatures; and the Cartesian dream of an intellectual life detached from the uncertainties of embodiment is reborn as a social or discursive realm that claims to have liberated itself from the inconvenient restrictions of reality. In one book, for example, we read that viewing nature as something external to society is problematic because this "renders nonhuman objects and processes intractable barriers to which humans must at some point submit".[46] Recognising the constraints associated with our animality, our mortality, or the finite character of the earth is a bitter pill for some to swallow; but it is a recognition that is basic to any sustainable form of life. One of academia's honourable traditions is to comment critically on the directions taken by the world outside; and it is unfortunate that this tradition should be cast aside in favour of this sort of disembodied fantasy dressed up as serious intellectual debate.

Given this isolation of dominant forms of symbolism from the real world, it is not surprising that the conventions of discourse often take precedence over the requirement that beliefs be checked against any external criterion. One writer claims, for example, that "one should be 'symmetrical' ... [that is,] we must treat what have come to be accepted as 'facts' and what have come to be accepted as 'mistakes' in the same way".[47] But to do so is to ignore the very good reasons why 'facts' and 'mistakes' should be treated *differently*: namely, because facts generally express some valid characteristic of the world, whereas mistakes have no reliable relation to reality. Similarly, discursive approaches invariably ignore our long history of social development out of an originally asocial nature. 'Nature' and 'society' are not simply discursively invented categories of equivalent status; and when the same author complains that "we have routinely engaged in dualism ... we have represented nature as transcendent, while society is seen to be our free construction",[48] he loses sight of the fact that nature *is* transcendent – that is, it is ontologically, historically, and functionally prior to social life. To imply otherwise is to ignore the evolutionary record and the entire history of social life as a recent outgrowth of, and as dependent on, the natural order. Conversely, (and un-symmetrically), nature is seldom dependent on social life: if humans vanished from the earth overnight, nature would flourish in most respects.

The intellectual bid for freedom from mortality and animality does not remain an individual one; but becomes 'built into' our institutions through the systemic qualities of industrial society, generating the principle of social organisation that Bauman, building on Tonnies' notion of 'Gessellschaft',[49] refers to as 'legislative reason'.[50] The 'linguistic turn' is therefore a move towards self-sufficiency that applies only in the industrialised world, rather than being a more widely applicable attempt to understand how language functions across a range of cultures. Clearly, only the affluent can background, export, and take for granted the material and ecological costs involved in maintaining such delusions; and in the poorer areas of the world, those whose lives are at every moment governed by material and ecological circumstances would greet any claim for the causal priority of language with bewilderment and derision. Such claims sideline our immediate dependency on nature and so are part of an ecologically destructive apparatus, as Jeffrey Wollock points out:

> Obsessed with the conception of language as a purely arbitrary system of signs, modern linguistic theory cannot accept the idea that there are really existing systems outside of language and outside of humans, to which humans, and therefore language, bear a definite relation. ... Because language is the main guide to action, a theory of linguistics that regards social constructionism as the exclusive source of meaning is itself, I believe, a threat to biodiversity.[51]

There are other dangers, too. If nature is 'socially constructed', then of course we can repress politically awkward ecological realities such as population growth; and such taboos can be understood as "attempts to combat a malign reality by denying that it can be spoken of".[52] Similarly, any recognition of the opposition between nature and industrialism is often claimed to be 'dualistic' or 'polarised': in other words, the conflict between them is abolished by dissolving it discursively while pretending that it does not exist physically and systemically.

Forms of world denial: Individualism and its kin

The social sciences and humanities are ostensibly divided between the individualist paradigm which has dominated since the Enlightenment, and the reaction to this paradigm which has emerged over the past several decades, dissolving individuality into social, cultural, and linguistic frameworks. Both these ways of framing human life, however, are reductive simplifications which deny the interplay between the various organisational levels of human ecology. Thus the debate between social or linguistic determinism, on the one hand, and individualism, on the other, serves an analogous function to the political 'opposition' between the Labour and Conservative parties (in the UK) or the Democrats and the Republicans (in the USA), focusing on minor differences to divert attention from the important issues on which both sides tacitly agree. I will deal firstly with the individualist version of this ideology before passing on to its social and linguistic variants.

Individualism plays out the Cartesian scenario of an active, intelligent mind within a passive, meaningless world. This paradigm will be familiar to many readers, so I will deal with it briefly. An example is Freud's well-known reversal of his initial hypothesis that neurosis was the result of abuse and trauma, so that psychoanalysis, rather than challenging the prevailing social denials and repressions, took refuge within a safely autonomous mental realm.[53] In the Dora case,[54] Freud reveals the extent of his world denial, arguing that Dora's discomfort at being molested by her father's friend was in fact due to her repressed *wishes* for sexual contact with him. Thus attention is switched away from Dora's real-world situation to supposedly 'internal' factors. Dora is here being inducted into an area of the symbolic domain labelled 'psychoanalysis', and a real-world situation of sexual bullying is translated into technical concepts and symbolic relations that are only distantly related to what happens outside the consulting room. In the same vein, the psycho-analytic term 'object' does not generally refer to something 'objective' in the outside world, but to a *mental* construct; and this dissociation of psychology from the real world has continued and grown more marked since Freud's times. For example, as Sarah Hrdy relates,

> When John Bowlby saw an emotionally disturbed and hyperactive boy and his even more emotionally disturbed mother, he felt certain that the boy's difficulties had something to do with their relation-ship; but talking to (let alone attempting to treat) the mother was off-limits. For Bowlby, ambivalent, remote, negligent, or abusive parents were not necessarily imagined; some were all too real. Yet as a psychoanalyst, his job was to work through the child's fantasies.[55]

Bowlby described how his views were rejected by other psychoanalysts:

> I held the view that real-life events – the way parents treat a child – is of key importance in determining development, and Melanie Klein would have none of it. The object relations she was talking about were entirely internal relationships.[56]

Anna Freud, too, rejected the possibility that events in the real world affected emotional development, commenting that as "analysts, we do not deal with happenings in the external world as such but with their repercussions in the mind".[57] But these repercussions are not so much in the mind of the *patient* as in the mind of the *analyst*. Other psycho-analysts such as Ferenczi and Reich who argued that both symptoms and their mainstream psychoanalytic theorisation needed to be under-stood *politically* were sidelined; and the taboo these theorists violated, which was policed by Freud and the Vienna Psychoanalytic Society, was hostile not so much to the reformulation of theory *per se* but rather to the escape of explanation from the discursive realm of mind and theory into the material world. And while Winnicott's insistence that "the true self must never be affected by external reality"[58] is quite valid as a statement about the withdrawal of the 'real self' from an uncongenial world outside, it *also* fails to recognise that a landscape more consistent with the evolved needs of the self would invite a quite different theory – one that rather than reflecting the *opposition* between the 'real self' and the world, would instead reflect their complementarity. In other words, Winnicott's statement refers to current actualities; and existing theory, being itself an adaptation to these same actualities, cannot envision a healthier situation beyond them.

Thus psychoanalysis, as Norman O. Brown cogently argues, bears only "a fugitive relation to reality":

> If ultimate cure consists in finding real objects in external reality ... then what psychoanalysis does is to effect an immense withdrawal of libido from the macrocosm of the external world, which Freud (and modern man in general) rightly could not love, and to direct it to the microcosm of the internal world. ... In so far as this libidinal posture goes with the recognition of the unloveliness of the external world, it represents an advance in reality thinking. But in so far as psychoanalysis deflects attention from a further advance, to make external reality such that it can be loved, it can be an obstacle in the way of a final attainment of truth.[59]

According to Brown, "orthodox psychoanalytic therapy, with its emphasis on the role of verbalisation in consciousness and its de-emphasis of the relation of consciousness to external reality, cultivates word-consciousness and calls it true consciousness".[60] As Nigel Mackay has argued, psychotherapy has in recent years continued this movement towards viewing the psychic world as independent of physical realities, increasingly claiming that meaning, "not being a part of objective objects or things, must then be autonomously ... created by the person, or, in social constructionism, in the discursive practices of persons".[61] I am not arguing, of course, that meanings should be regarded as rigidly *determined* by physical realities; only that such realities allow us a certain, variably limited, amount of freedom in the meanings we can discover in them. Meaning is to be found *in the world* as much as *within the person*; and damaged, reduced environments can be lacking in meaning. In other words, lack of meaning is an *environmental* problem at least as much as it is a psychological one. As Raymond Rogers remarks, "meaning is not always internal and repressed, but may be external and gone".[62]

The trend towards isolation from the real world has widely infected many areas of psychotherapy. James Hillman gives a good example of the problem:

> I'm outraged after having driven to my analyst on the freeway. The trucks almost ran me off the road. I'm terrified. I'm in my little car, and I get to my therapist's and I'm shaking. My therapist says, "We've gotta talk about this."
>
> So we begin to talk about it. And we discover that my father was a brute and this whole truck thing reminds me of him. Or we discover that I've always felt frail and vulnerable, so this car that I'm in is a typical example of my thin skin and my frailty and vulnerability. Or we talk about my power drive, that I really wish to be a truck driver. We convert my fear into anxiety – an inner state. Anxiety is fear whose source we do not understand. Individualism prevents us from understanding systemic sources of danger – e.g. environmental threats. We convert the present into the past, into a discussion of my father and my childhood. This requires only cognitive action, as we cannot alter the past. Present threats, in contrast, can be resolved through *action*. And we convert my outrage – at the pollution or the chaos or whatever my outrage is about – into rage and hostility. Again, an internal condition, whereas it starts in *out*rage, an emotion. Emotions are mainly social. The word comes from the Latin *ex movere*, to move out. Emotions connect to the world. Therapy introverts the emotions, calls fear 'anxiety'. You take it back, and you work on it inside yourself. You don't work psychologically on what the outrage is telling you about potholes, about trucks, about burning up oil, about energy policies, nuclear waste, that homeless woman over there with the sores on her feet – the whole thing.[63]

Clearly illustrated in much psychotherapeutic theory and practice, the individualistic understanding that Hillman criticises is more generally part of a society-wide mechanism that encourages us to fit our experience to existing social and environmental conditions rather than recognising the often external and material sources of our psychological distress. As David Smail succinctly puts it,

> people are injured, psychologically as [well as] physically, not essentially by errors of their own judgement, the vagaries of their

consciousness, lack of insight into their own motives or failures of their will, but by the operation of basically material powers and influences in the world around them. What really upsets the apple-cart and buggers up people's lives and relationships is threatening their livelihoods, throwing them out of work, stripping them of social meaning, depriving them of health and education, pillaging and destroying the environment, and so on.[64]

Symbolic Intoxication Disorder: A new DSM category??

In recent decades, the reaction against individualism has swept areas such as cultural studies, linguistics, and psychoanalysis while leaving other fields such as experimental psychology almost untouched. However, what has often been portrayed as a necessary corrective to individualism is not what it seems, and can better be regarded as a *generalisation* of the problem which underlies individualism. That is, if individualism is understood as involving the abstraction of the person from context, then the reactions to this often move the focus away from a decontextualised individual to a reified realm of symbolic interaction that is *itself* decontextualised. Thus in many areas of the social sciences and humanities, the focus on the allegedly autonomous psyche has transmuted itself into an equally exclusive focus on *language*. Far from being a corrective to decontextualisation, this movement introduces it in another form, alienating us further from the natural world, and so can most accurately be regarded as a continuation of the trend towards an autonomous symbolism.

Stephen Mitchell, for example, argues that "both preverbal and nonverbal dimensions of experience can be retrieved, experienced, and expressed only within a socially shaped system of linguistic meanings";[65] and quoting Terry Eagleton, he suggests that "we can only have ... meanings and experiences ... because we have a language to have them in."[66] Like many theorists, Mitchell does not assess or critique this movement towards a linguistic emphasis: he simply suggests that psychoanalysis should 'be informed by it'. If, however, this movement is an explicit shift away from the material realities of the world towards a purely symbolic realm which denies our embodiment, then this simply illustrates Michael Steinberg's observation that through our fear "of change and death, we are persistently drawn away from the ceaseless transformations of bodily life into the delusory shelter of discourse".[67]

In Mitchell's writings, we are no longer the biologically driven creatures originally envisaged by Freud, torn between instinct and socialisation. The basic unit of study, Mitchell tells us,

> is not the individual as a separate entity whose desires clash with an external reality, but an interactional field within which the individual arises and struggles to make contact and to articulate himself. Desire is experienced always in the context of relatedness, and it is that context which defines its meaning.[68]

In Freud's work, the conflict between the embodied individual and social realities remains, even if Freud's allegiance is ultimately to the latter. But in Mitchell's writings, the conquest of nature by civilisation has moved on a stage, as the very existence of a biologically grounded individual is denied, and the person becomes an epiphenomenon of social relation. The struggle between nature and industrialism is over: and nature is tacitly assumed to have been eliminated. If, as Mitchell argues, we are symbolically constituted by *social relations*, then no part of us can be in conflict with the social context. Thus in "more radical statements of the relational position, the very notion of a single mind [Note: 'single mind', not 'single person'] as a meaningful unit for study is called into question".[69] No longer are we the creatures physically striving to survive and reproduce in the natural world, driven partly by our evolved instincts and preferences, guided by ecological relations that have developed over millennia; rather, we are the ethereal creations of a symbolic – largely linguistic – universe. This latter system does not define us as organisms which are at one level identifiable as individuals, but simply absorbs individuality into language, so that there is a reductive flattening of the hierarchy of relationships. While ecosystems relate entities at many different levels, so that cells, organisms, and ecosystems form interrelated levels of structure, in Mitchell's approach the individual is *dissolved* by language rather than located within it.

But Mitchell's portrayal of the person is not entirely misleading, since it is accurate enough as a prophesy about where we are heading – towards the loss of any groundedness in the wild world, towards the manipulation and overwriting of evolved characteristics by genetic engineering, and ultimately towards the determination of human character and social realities by a tightly-integrated industrial system. As is the case in much theory influenced by postmodernism, Mitchell's account intuits a pathological social trend and repackages it as a new vision of the person.

Mitchell quite reasonably argues that Freud's renunciation of the seduction theory, because it turned the focus onto internal psychodynamics, individualised psychoanalysis. But this is only a transient effect if the individual psyche thus removed from its real-world context is immediately incorporated within a larger system – the symbolic system for which psychoanalytic theory serves as a portal. As is the case with the contemporary inhabitant of industrialism, a superficial individualism conceals our more profound incorporation within a larger order. Freud was no ethnographer, entering the patient's world and taking part in their work and leisure; and neither was he a client-centered therapist, walking in the client's world and sharing their existential angst. Rather, he remained largely emotionally detached as the patient was inducted into the arcane symbolic world of psychoanalysis; and his understanding was of a more technical nature. Freud was essentially the gatekeeper of a symbolic order; and he encountered the patient only inasmuch she became a character and a process within this symbolic mapping.

When Mitchell claims that his own, 'relational' approach restores the lost context, then, this is somewhat misleading; for psychoanalytic theory serves much the same function for Freud as language does for Mitchell, even if somewhat less explicitly. Both theories contextualise individuality; but in neither case is the context that of embodied experience. Although Mitchell suggests that "the drive and relational models … are based on very different fundamental presuppositions about the generation of experience and meaning", both theories fundamentally operate as techniques of translation from the natural order to the symbolic order. Thus the context that has been lost from *both* theories is the entire *ecology* distantly nodded at by the concept of 'drive' – the context of physical relations and emotional attachments more adequately recognised by the early object-relations theorists such as Fairbairn and Guntrip.[70] To accuse Freud of individualism is only superficially accurate, for his 'individualism' is merely a first step towards our reincarnation as symbolic beings. Freud plucks us from our ecological context in the same way that a bluebell is plucked from its woodland habitat; and while his quasi-biological approach seems at first glance to embed psychoanalysis in our animal nature, as Frank Sulloway notes, "once he had finally achieved his revolutionary synthesis of psychology and biology, Freud actively sought to camouflage the biological side of this creative union".[71] Freud's dictum that "where Id was there shall Ego be" seems to refer not only to human development, but also to the emergence of psychoanalysis as a purely symbolic practice. From this perspective, Mitchell's move towards a linguistically focused theory of self seems more like a logical next step than a radical alternative to Freudian psychoanalysis. Eliminating our ecological context and then 'restoring relationality' by symbolically reincarnating us within language is like catching butterflies and ordering them in glass-fronted boxes in a neatly ordered collection. The function of both theories is to transform us into denizens of a symbolic world.

The supposed 'freedom' from embodied constraints this relocation allows has introduced a certain postmodern flippancy into some areas of therapy; and as David Smail parodies the common overemphasis on language in this field, "the world is made of words, and if the story you find yourself in causes you distress, tell yourself another one."[72] For a profession that claims to align its clients with reality, this is a disadvantage, as Smail suggests:

> That such a preposterous notion could be seriously put forward and maintained by people considered to be social scientists is inexplicable unless one introduces into the explanatory framework the notion of *interest*. In other words, it cannot be that the proposition in question is *true*; it can only be that it is *useful*, i.e., that it suits the interests both of those who assert it and those who assent to it.[73]

It is useful, in other words, in the sense that both therapist and client share and collude in the denial of embodied realities, and maintain the assumption that psychological distress is necessarily a 'mental' problem to be addressed solely within the symbolic sphere of talk and ideas. As Nigel Mackay puts it, the "focus of psychotherapy has moved from how objective conditions may control and be manipulated to alleviate symptoms, to the ways in which persons' perceptions of their world generate, and may be adjusted to alleviate, pathology."[74] In situations where these 'objective conditions' are unhealthy or potentially dangerous, this detachment from reality has three undesirable consequences: firstly, the therapy does nothing to change the unhealthy material realities in the client's situation; secondly, clients are misled as to the causes of their distress; and thirdly, the long-term effect of therapy is to reinforce the pervasive culture of incongruence between symbolic and embodied functioning.

Most approaches to psychological healing have continued this trend, acting as techniques through which the patient is re-socialised into current symbolic 'realities', rather than engaging with their lived experience. Arthur Kleinman, considering the cross-cultural diversity of approaches to healing, has observed that in industrial societies the capacity to enter the emotional world of the patient, which he sees as a "core task of doctoring", has "atrophied in biomedical training"[75] to the extent that it focuses on *technique*. For Kleinman, "what is necessary for healing to occur is that both parties to the therapeutic transaction are committed to the shared symbolic order".[76] But while the symbolic order may indeed be shared by both parties in the *indigenous* societies he refers to, this is unlikely to be the case within industrialised ones. Consider an example documented by him later in the same book, comparing the patient's and, firstly, the psychiatrist's accounts of the same interview:

> 40-year-old white male physician with several months of depression, hopelessness, helplessness, anhedonia, irritability, guilt, insomnia, and energy and concentration disturbances. Associated anxiety. Not acutely suicidal but has some suicidal thoughts. No plans. No delusions or hallucinations. No family history. No prior episodes. No mania. Sinus headaches, chronic, amplified by depression. No other medical problems. No alcohol or drug abuse.
>
> Bereavement 13 months ago following death of father. Grieving with ambivalence and modeling of father's symptoms seems to have extended into depression. Depression deepening over past few months, with greatest increase over past month since anniversary of father's death. Worse past month. Work, marital, family problems contributory.

> *Impression*
> Axis 1: Major Depressive Disorder secondary to bereavement (prolonged and pathological).
> Axis II: No personality disorder.
> Axis III: Chronic sinus headaches.
> Axis IV: Severe bereavement reaction, 4/5.
> Axis V: Reasonably good level of functioning, some work-related problems and chronic mental tensions, worsening. But able to cope.

> *Plan*
> 1. Doxepin, begin at 50mg, bring up to 150 mg qhs over course of 1 week.
> 2. A few psychotherapy sessions to do grief work.
> 3. ENT consult to rule out serious sinusitis and ? CNS effects.
> 4. See in 1 week.

Now compare this with the patient's account, as recorded in his diary:

> I don't think he heard me. I wanted him to listen to me not for the diagnosis but for the story, my story. I know I'm depressed. But I wanted him to hear what is wrong. Depression may be the disease, but it is not the problem. The problem is my life. 'The centre doesn't hold. Things fall apart.' It's falling apart. My marriage. My relationship with my kids. My confidence in my research. My sense of purpose. My dreams. Is this the depression? Maybe it caused the depression. Maybe the depression makes it worse; or seem worse. But these problems also have their own legitimate reality. This is my life, no matter if I am depressed or not. And that is what I want to talk about, to complain about, to make sense of, to get help to put back together again. I want this depression treated, all right. There is something more I want, however. I want to tell this story, my story. I want someone trained to hear me. I thought that was what psychiatrists do. Someone ought to do it, ought to help me tell what has happened. But all he seemed interested in was the diagnosis and my dad's death. I'm sure that is part of it, but so much else is going on. I need to talk to someone about my whole world not just one part of it.[77]

There is little here in the way of a 'shared symbolic order'. The psychiatrist's symbolic order is 'objective' and unemotional, with its own technical concepts, vocabulary, and taxonomy of disorders. The patient, on the other hand, is concerned with meaning, continuity, and the emotional wholeness of his world, including his work and relationships with family members. He will be 'cured' to the extent that he leaves this context and operates within the symbolic landscape of psychiatry, behaving as the sort of biomedical entity created by that profession and so tacitly renouncing his membership within the troublesome realm of embodied creatures. As in the case of carbon trading, the problem is not so much solved as relocated to a domain wherein it becomes irrelevant.

Generally, the psycho-professions have embraced this retreat behind technical and professional barriers; but one who mounted spirited resistance to it was R. D. Laing – which is perhaps one of the reasons for his current unpopularity in professional circles. Whatever his excesses and vulnerabilities, Laing's lasting contribution to psychology and psychiatry is his recognition of the reality and validity of the patient's *experience*; and in so doing he challenged the development of psychiatric jargon and clinical detachment that is so characteristic of the profession. In *Sanity, Madness, and the Family*,[78] for example, Laing and Aaron Esterson show with devastating clarity how 'meaningless' and 'bizarre' symptoms become entirely comprehensible when viewed in their relational context, and complementarily, how the psychiatric judgement of these symptoms as 'mad' depends on their *de*contextualisation. In thus challenging the culture-wide dissociation between lived experience and technical symbolism, Laing was not only picking a fight with the psychiatric establishment; he was also swimming against the tide of industrial society as a whole as it becomes engulfed by scientistic and bureaucratic symbolism.

Psychoanalytically influenced theory has continued this movement away from the body, and the results are often bizarre and of little relevance to everyday life. If Mitchell has overplayed the importance of disembodied symbolism in understanding psychological life, Lacan has in certain respects parted company with reality entirely, demonstrating not so much a redefinition of reality as a blithe disregard for it. Furthermore, Lacan seems to defend his writings from the harsh corrective light of reality by means of deliberate impenetrability and mystification. "Whereas Freud set himself the moral duty of persuading the reader through clear argument and telling examples", Michael Billig comments, "Lacan is too haughty to explain what he means", adding that "difficulty is not to be confused with depth".[79]

But Lacan's writings are not just difficult; they are in places the product of an idiosyncratic imagination that has become detached from the world. To take just one example, Sokal and Bricmont have focused on his use of mathematics, showing that his 'analogies' between mathematics and psychological states are fanciful at best. Take the following sentence, which demonstrates both Lacan's fascination with topology (an area of mathematics which focuses on the properties which are preserved when surfaces are deformed) and his failure to understand it:

> In this space of *jouissance*, to take something that is bounded, closed [*borné, fermé*] constitutes a locus [*lieu*], and to speak of it constitutes a topology.[80]

Sokal and Bricmont comment that in this sentence,

> Lacan has used four technical terms from mathematical analysis (space, bounded, closed, topology) but without paying attention to their *meanings*; the sentence is meaningless from a mathematical point

of view. Furthermore – and most importantly – Lacan never explains the relevance of these mathematical concepts for psychoanalysis. Even if the concept of *'jouissance'* had a clear and precise meaning, Lacan provides no reason whatever to think that *jouissance* can be considered a 'space' in the technical sense of this word in topology.[81]

I refer readers to the appropriate chapter in Sokal and Bricmont's book for many more examples of this flamboyant use of mathematics. For our purposes we simply need to note that a hubristic intoxication with symbols combined with a disregard for their meanings has led Lacan to float entirely free from the real world. If much post-Kantian philosophy has tried to sideline the real world in much the same way that mediaeval priests tried to sideline sexual urges – and with much the same degree of success – then Lacan exemplifies social science's vulnerability to the same error. But we should recognise, in passing, that there is no need for abstract thinking to lose touch with reality in this way: as Anthony Storr remarks, for example, the power of Einstein's theorising was based a "capacity for extreme abstraction combined with a retention of contact with reality".[82] His theories are not only empirically testable, they are rooted in embodied feelings and images which anchored him to the world, rather than originating in any symbolic 'free play'.[83]

In contrast, Derrida engaged in a long struggle to collapse the world into discourse, arguing that "[w]hat I call 'text' implies all the structures called 'real', 'economic', 'historical', socio-institutional, in short, all possible referents".[84] A case in point is his study of drawings exhibited in the Louvre. Published as *Memoirs of the Blind*, Derrida argues that "a drawing of the *blind* is a drawing *of* the blind"[85] – that is, both the artist and the model are blind. Derrida's account of drawing is a sort of science fiction fantasy: like science fiction, it refers to real things in the world – blindness, drawing, the experience of vision – but weaves a fantasy text around them to 'prove' that the artist cannot see what he is drawing.

There are, Derrida argues, three reasons for this, which I will all-too-briefly summarise, inasmuch as I can grasp Derrida's somewhat obscure reasoning. Firstly, in the moment between looking at the model and turning his attention to the drawing there is a gap filled by memory; so that drawing requires us to forget the corporeal reality that stands before us as we concentrate on the act of drawing from memory. This is reminiscent of Descartes' view, in Meditation V, that he cannot be certain of anything experienced through the senses because "I cannot fix my mental vision continually on the same thing", and that memory is an unreliable guide.[86] Secondly, the lines we have drawn are not seen for their own sake: rather, they fade as we focus on the spaces which are separated by the lines and on the *whole* drawing, so that the image depends on our not seeing the lines as such. Thirdly, Derrida argues that words are always hanging in the background of drawing: the request to make the drawing, the description of the drawing, the discussion of the drawing, and so on, are all unavoidably verbal, and so drawing can be only a temporary retreat from language. Consequently, Derrida would have us believe, drawing is secondary to an inescapable realm of language; and so he replays the drama which is reproduced within industrial society in a myriad of ways – the subordination of the real and the sensory to the symbolic.

However, Derrida's rather counter-intuitive – and, some would say, tortuously derived – conclusion seems to have been motivated by very real emotional factors rooted in his own doubts and anxieties. Here is how he describes his reaction to being invited to write about the Louvre drawings:

> I was more than just honoured by the invitation that was extended to me; I was intimidated, deeply worried, even, by it. And I still am, no doubt well beyond what is reasonable. The anxiety was, of course, mixed with an obscure jubilation. For I have always experienced drawing as an infirmity, even worse, as a culpable infirmity, dare I say, an obscure punishment. A double infirmity: to this day, I think that I will never know *either* how to draw *or* how to look at a drawing. In truth, I feel myself incapable of following with my hand the prescription of a model: it is as if, just as I was about to draw, I no longer *saw* the thing. For it immediately flees, drops out of sight, and almost nothing of it remains; it disappears before my eyes, which, in truth, no longer perceive anything but the mocking disappearance of this disappearing apparition. ... The child within me wonders: how can one claim to look at both a model and the lines *[traits]* that one jealously dedicates with one's own hand to the thing itself? Doesn't one have to be blind to one or the other?[87]

Derrida goes on to explain, with considerable honesty, the origin of these feelings: his "wounded jealousy" in the face of an artistically gifted elder brother whose drawings decorated the family home, bringing out a "fratricidal desire" and a decision to strategically outflank his brother:

> in place of drawing, which the blind man in me had renounced for life, I was called by another *trait*, this graphics of invisible words ... I am

speaking of a calculation as much as a vocation, and the stratagem was almost deliberate, by design. Stratagem, strategy – this meant war. ... I draw nets of language about drawing.[88]

Derrida's determination to reduce everything to language, then, is traceable to his own insecurities, rather as Descartes' insecurities motivated the efflorescence of an entire philosophical system. In both cases, we see the construction of a symbolic refuge: in Derrida's case, a flight from family tensions and artistic deficiencies into a haven of language; in Descartes', an escape from an unpredictable world into the certainties of 'objective' thought. But Descartes and Derrida are far from alone in this retreat; and as Noam Chomsky has noted, if "it's too hard to deal with real problems", some academics tend to "go off on wild goose chases that don't matter . . . [or] get involved in academic cults that are very divorced from any reality and that provide a defense against dealing with the world as it actually is".[89] Derrida's book is indeed based on blindness; but it is not the ethereal forms of blindness he claims to be intrinsic to drawing, but rather his own half-admitted blindness to the reasons which underly his obsession with language, which we all, to varying extents and in varying ways, share. If he cannot accept the world as it is, Derrida seems to be saying, then he will invent an alternative world – which, in microcosm, exemplifies the entire project of Western industrial civilisation. "I find myself writing without seeing", says Derrida; and this is a particularly clear example of the drive to cut out sensory awarenesses in order to focus on the symbolic. The brilliance of Derrida's writing cannot disguise that fact that it inhabits a purely discursive realm, and that like Descartes', it rejects the anchors of sensation and embodiment that would hold it to the real world. As Jason Powell has noted in his biography of Derrida, "Being blind to empirical things, the artist or believer is free to impose meaning on them."[90]

The cutting out of sensory awareness in order to impose symbolic meanings, however, can take many forms, some less subtle than those of Derrida and Descartes. An example is Neil Evernden's description of the work of the famous French physiologist Claude Bernard. Bernard's work involved the vivisection of dogs; and not wishing to hear the screams of the animal as it was dismembered, Bernard would cut its vocal cords. As Evernden points out, this is both a denial and a simultaneous recognition that the nonhuman world has a voice, and that we can hear it.[91] For Bernard as for many others, the concern is not to understand and articulate this voice, but to suppress it, and so evade its protest against the replacement of the real by a symbolic simulacrum.

Each of these theorists is, in their unique way, brilliant; but it is a brilliance that employed precisely to *avoid* any connection to the real world or to the theorist's own life. As Bernard Williams remarked, the impression that the humanities have fled from reality into denial is reinforced when they "adopt a rhetoric of political urgency which represents only the café politics of émigrés from the world of real power, the Secret Agents of literature departments".[92] In cultural theory there is endless talk about 'liberation', 'freedom', 'resistance', and so on, seeming to function only to cover up the absence of these qualities in lived reality. As Todd Gitlin puts it,

> Resistance, meaning, all sorts of grumbling, multiple interpretation, semiological inversion, pleasure, rage, friction, numbness, what have you – 'resistance' is accorded dignity, even glory, by stamping these not-so-great refusals with a vocabulary derived from life-threatening work against fascism – as if the same concept should serve for the Chinese student uprising and cable TV grazing.[93]

Words such as 'resistance' and 'liberation', then, are sometimes used to cover up the absence of their referents in real-world practice, in much the same way that leafy rural images are used as fig-leaves by multinationals to greenwash their destructive activities. Likewise, Nancy Scheper-Hughes and Carolyn Sargent point to "the Paris intellectual elite, led by Foucault, Deleuze, and others who, in the wake of the 'revolution' of '68, published a special issue of the leading Parisian journal *Recherches* extolling cross-generational sexual encounters between greying French intellectuals and nubile Arab street boys and prostitutes".[94] In such cases, the inconsistencies between 'liberatory' theory and embodied practice are clearly apparent. But such is the giddy intoxication implicit in much cultural theory that critiquing the "postcoherent thinkers"[95] seems almost unsporting; and my primary task in this book is not so much to point to the efflorescence of symbolic indulgence which characterises certain corners of academia as to lay bare the more widespread loss of contact with reality which infuses the entire industrialist realm.

The supposed autonomy of the symbolic has been elevated almost to the status of an article of faith across the entire cultural spectrum. For example, in a book entitled *Cultural Pessimism*, the author earnestly explores the psychological reasons for despondency without ever seriously entertaining the possibility that such 'pessimism' might be related to our global situation.[96] Likewise, in geography, history, and nature writing, information collected in the field[97] is often subordinated to the language used to describe it, so that 'knowledge' becomes based more on what authors write than on physical and behavioural realities. As Barnes and Duncan enthusiastically proclaim, for example,

> texts draw upon other texts, that are themselves based on yet different texts ... there is only intertextuality, defined as 'the process whereby meaning is produced from text to text rather than, as it were, between text and world' ... what is true is made inside texts, not outside them.[98]

This lack of groundedness has extended outwards from academia to infect other areas of art and literature, as the playwright David Hare complains:

> A majority of films and books could have been conceived at any time in the past 30 years because they are effectively reactions not to life itself but to other imitations of life. The deadly question 'Who are your influences?' presupposes of any writer that the primary source of inspiration will not be what is happening now on the street but what has already happened between the covers of other books.[99]

Barnes and Duncan fall into the same trap as Mitchell – mistaking a pathological trend for a universal necessity. One might guess that in real life they do not take seriously what they themselves write, and are quite capable of distinguishing menus from meals. But when viewed in its wider context, this sort of reality-denial should be taken seriously, for it amounts to a dereliction of duty of staggering proportions, as academics who should be assessing and critiquing the current pillaging and exploitation of the world and its peoples instead retreat into vacuous theories which ignore or even collude with brutal material and political realities.

Any adequate grasp of our situation must recognise the fundamental roles of embodiment and experience. As Merlin Donald puts it, "the surface form of language is subsidiary to a deeper process, which cannot be, in its origins, linguistic or even symbolic. ... words and sentences define and clarify knowledge that resides elsewhere, in foundational semantic processes that we share with other primates".[100] As Donald argues, it is the metaphoric character of language which indicates that it is based on somatic experience, so that if we are not to get the linguistic cart before the embodied horse, it is essential, as George Orwell observed, "to let the meaning choose the word and not the other way round."[101]

Rediscovering the world

The pawprint, the smoke, the dark cloud, and the roar of an approaching flash flood are signs that do not depend on any specifically linguistic or human interpretation: their meaning is already 'out there' in the world, to be recognised by any creature that has the necessary faculties. In Kathleen Raine's words, they form a "language far more powerful and no less exact; A language in which bird and beast and tree were themselves the words. Full of otherwise inexpressible meaning".[102] One of the curiosities of the recent history of the industrialised world is that although technology has enabled us to become enormously more able to understand the world in some ways, in other respects technology has weakened our senses' capacity to 'read' the world directly. Instead, we read texts *about* the world. While some non-human creatures can quickly learn to sense the earth's magnetic field or the differences between nutrient-deficient and healthy diets,[103] industrialised humans mostly lack such abilities. These issues would not arise were it not for our inward focus on symbolism as our major source of meaning. Language, as Steinberg remarks, "always conceals the wordless thought from which it begins, thought that belies language claims. Beneath its spurious universality ... lies our common grounding in the life of the body".[104]

All too often, anthropologists have interpreted embodied knowledge and practices simply in terms of their symbolism and religious significance, as Laura Rival points out:

> For Levi-Strauss, there is no doubt that people's interest in plant and animal species ... stems from an intellectual concern with difference and analogy, that is, with the codification of discontinuities. Therefore, people's concrete knowledge of the world they live in is meaningless, unless transposed to an abstract level where it can be used to classify and order the social.[105]

According to such views, Rival comments, "the practical experience of, or communication with, plants and animals is irrelevant" for understanding people's belief systems, "for these should be seen as a linguistic code to think of the world of nature only *in so far as it can be contrasted with the cultural world of human beings*". However, when "the relationship of people to nature is re-considered in terms of engagement, practical experience, and perceptual knowledge, nature ceases to be a mere reflection of society".[106] Other writers, too, have criticised the view that embodied practice simply reflects symbolic realities. Robert Williamson, for example, recognises that Inuktitut is not just a language but also a metaphysical system which expresses the union of symbolism and the world – one which in Arctic latitudes is essential to survival.[107] Likewise, Morris Berman points out that for the Basseri – a nomadic group of Iranian herders who migrate to their summer pastures in the mountains each spring, meaning is embedded in *practice*. In contrast to some anthropologists, who have debated at length on the symbolism underlying the migration, Berman considers a heretical possibility. "Suppose", he suggests, "the meaning of the migration were ... the migration?"[108]

The differences between an attitude that deals directly with the world and one that tries to assimilate it into a symbolic system is poignantly illustrated by Berman's experience of attending a service

> at a very small church, in the Pacific Northwest, to honour the completion of an altar engraving that had been done, on commission, by a Native American artist. The engraving showed two fish in a circular pattern, beautifully carved in wood. Everybody, the artist included, sat for forty-five minutes as the pastor went on and on about the symbolism of the fish, their Jungian and Christian associations, the balance of yin and yang, the integration of male and female principles, etc etc. When he concluded his sermon, exhausted but flushed with pride from a job well done, he turned to the artist – I'll call him 'No Bull' – and asked him to come up to the front and explain his engraving. No Bull did as requested, put his hand on one of the fish, and turned to his audience.
>
> "This here," said No Bull, pointing to the eye of the fish, "is the eye." "And this, this here," he said, indicating the fin, "is the fin." "And this right here," he concluded, tapping the tail, "is the tail." Whereupon No Bull returned to his seat and sat down. The silence in the chapel, as you might imagine, was deafening.[109]

This attachment to physical realities is one of the deep and sometimes unacknowledged taproots of the environmental movement, surfacing in writers such as Edward Abbey, who rejects abstract "patterns of unifying relationships" in favour of physical immediacy. "I am pleased enough with surfaces", he declares. "Such things as the grasp of a child's hand in your own, the flavour of an apple, the embrace of friend or lover, the silk of a girl's thigh, the sunlight on rock and leaves, the feel of music, the bark of a tree, the abrasion of granite and sand, the plunge of clear water into a pool, the face of the wind – what else is there? What else do we need?"[110] Environmentalism, thus understood, expresses a profound distrust of the technologically materialised pseudo-world into which symbolism has led us, and an equally profound preference for a fully embodied existence in the natural order.

Abbey's words express the notion – no, the *experience* – that meaning exists within the world, rather than floating above it in an ethereal, cognitive, realm. This experience is one that seems to be shared by every non-industrialised society I have either read about or personally encountered. This may be difficult to understand if one lives in an industrial society, where for many centuries we have separated meaning from embodiment, aided by the major religious and philosophical traditions. When we 'understand' something, we do not extend it, but rather create a symbolic simulacrum of it; and consequently, as Michael Steinberg puts it, "we live under a kind of linguistic totalitarianism".[111]

Unfortunately, unlike 'important' human activities such as industrial production, share dealing, or sporting performances, natural processes and features are just an unnoticed part of the context, and so apparently need little attention or care, tending to enter the symbolic realm as important only when they become endangered.[112] And even when this happens, the symbolic realm often functions as a substitute for the embodied realities rather than as a means of protecting them, and the material realities are allowed to slip away under cover of their symbolic representations. Thus biodiversity only became an issue when monocultures were becoming widespread; and individual freedom became a fashionable topic for discussion around the time trading in slaves reached its peak.[113] The accelerating devastation of remaining wilderness areas is accompanied by ever-glossier, Photoshop-enhanced calendar images and wildlife documentaries; and Albert Borgmann tells us that when "the communal context for the realisation of Bach's cantatas had evaporated", it was suggested that "we entrust the cantatas to the hyperreal perfection of compact discs and give them the imperishable existence of an insect in amber".[114]

Replacing the real by the symbolic

The redefinition of reality as primarily discursive may be partly motivated by the fear of being deterministically anchored to the world, dragged down into a realm of iron necessity. There is good reason for this fear: when European life was dominated by famine, disease, lack of reproductive control, climatic variation, and so on, the vision of a heaven in which one could be free of all bodily ailments and needs was irresistible. Since those times, technology and medical science have given the affluent inhabitants of the world a clearing within which we can be relatively free of these mortal fears. Unfortunately, many contemporary theorists have tried to take this freedom a step further, generalising the notion of social and political freedom to reproduce in another form the religious aspiration of freedom from the flesh and a miraculous independence from the laws of physics. Thus Lee Braver, summarising Foucault's project of throwing off domination, claims that "minimal domination would require a regime to renounce realist claims to the one, absolute truth which naturally freezes relations and restricts options, in order to allow people to modify their selves as works of art".[115] But the notion that a realist view involves 'one absolute truth' and 'freezes relations' is a travesty. True, certain options are restricted, such as those involving immortality, time-travel, invisibility, or guaranteed lottery wins; but within a physical world that has certain essential qualities, there is nonetheless a vast array of options. In fact, it is precisely the immutable physical realities of the earth and of organic life that *enable* certain options. For example, aircraft can fly *because* air, steel, aluminium, and kerosene have certain reliable and measurable properties; and sex is fulfilling *because* we have certain physical and psychological characteristics. While the ability to imaginatively or technologically extend such physical and biological realities is an important human attribute, losing the thread which connects our symbolic formulations *to* these realities produces a state that is delusory and potentially psychotic.

To the extent that the fears and fantasies which motivate this flight from reality are centred around our mortal bodies, meaning has to be detached from embodiment; and the body, like nature, becomes the mule which carries symbolic meanings. In Foucault's work, for example, confession is viewed as a hermeneutics of the self because it does not *express* an inner, embodied self: rather, it *creates* the self in the moment of confession. But this discursive self is, I think, built on a denial of the organic self and its ecological embeddedness. Although Foucault's work on sexuality involves a critique of the 'repressive hypothesis', it is important to recognise what both Freud and Foucault have in common: the aim of transforming embodied realities into discourse. This is why psychoanalysis is *about* sex but does not *express* it, and why reading Freud is relentlessly unerotic. As Jeremy Carrette points out, the confessional – whether it occurs in a psychoanalytic or a religious context – inflicts a double repression on the body; for not only is confession an admission of the sinfulness of the flesh, but also the very act of translating bodily desires into words is a neutering of these desires:

> What is secret and silent becomes sinful; silence becomes a resistance to authority, but silence also becomes the very act of rendering to authority. The ability to utter and extinguish is the mark of disposing of evil thoughts. However, in the very act of articulation there is a silencing, a rejection, of flesh. ... The Christian demand to 'dissociate', 'disinvolve', 'renounce', and 'sacrifice' the flesh is to silence the body in the very act of verbalising its reality.[116]

Within a discursively defined world, the body is silenced, whether it is silent or not; and "Christianity paradoxically constructs a self in the very sacrificing or silencing of the embodied self."[117] More generally, while the symbolic realm pretends to provide an adequate representation of embodiment, it necessarily represses those aspects which cannot be represented symbolically. To the extent that we are forced to translate experience into digital, and especially verbal, form, and then to identify with this form, we become strangers to ourselves. Embodiment is driven inwards, separated from the world, and forced to lie low.

The reasons for this denial of embodiment are not difficult to perceive. As Whitehead put it, "the body is a part of the external world, continuous with it. In fact, it is just as much a part of nature as anything else there – a river, a mountain, or a cloud".[118] To acknowledge that the body is an essential aspect of selfhood would, then, conflict with industrialism's denial of natural structure and challenge its ability to symbolically redefine and materially transform the world. In keeping with this project of transformation from embodiment to discourse, history is pre-eminently *cultural* rather than *natural* history, and life becomes the exclusively *human* life of the *disembodied* mind and its consequences. As Raymond Rogers remarks, it

> is the anthropocentric perspective which assumes that all meaning is socially created by humans, and that humans brought nothing with them from their previous relationships with nature. ... [This] dismisses the manifold ways in which human identity is rooted in natural being.[119]

In keeping with this transformation, as Margaret Archer has pointed out, there is a tendency among some theorists to absorb the sense into the concept, which often loses its anchorage in lived realities and instead floats free within the currents of symbolism.[120] But if the problem we are addressing is precisely the disembodiment of ourselves and the world, and this problem is effected *through* discourse, then clearly solutions cannot be found *within* the realm of discourse; and as Alf Hornborg points out, the "impact of the disembedded language of modernity upon the material world is such as to facilitate ecological destruction".[121] This is what occurs in cultural constructionism; for only if physical and biological realities are denied can they be absorbed into a self-sufficient world of symbols, untroubled by material limitations. Thus, as Ian Hacking remarks, "when we read of the social construction of X, it is very commonly the idea of X ... that is meant". Thus using the example of women refugees, "X does not refer to individual women refugees. No, the X refers first of all to the woman refugee as a kind of person, the classification itself, and the matrix within which the classification works".[122] What is happening here, then, is that an impoverished conceptual *representation* of the person is replacing the *person-as-embodied-being*, and writ large, the symbolic realm is replacing the natural world. These replacements are seldom explicit; rather, they are slid into place as if the concept, the symbol, were identical to the reality. This blurring between the way we think and what we think *about* pervades not just the humanities but also everyday life, as we assimilate people and other entities to our concepts and stereotypes.

This pervasive infusion by symbolism allows a politically and psychologically convenient distance to develop between the person and whatever is being discussed, even amongst those who are not signed-up constructionists, permitting an affected stance of 'caring', 'concern', even 'outrage' while not actually *doing* anything outside the symbolic realm. For example, various writers have noted the gulf between attitudes towards nature and 'real life' actions. As Arne Kalland remarks, in "southeast Asia the land is being stripped of its rainforests at an alarming rate, despite widespread belief that spirits and ancestors reside in the trees ... Nor has the Japanese sensitivity towards nature, expressed in poetry, gardens, and miniature trees, prevented environmental disasters occurring".[123] Such inconsistencies seem to occur almost universally.[124] And Anna Peterson notes that "political behaviour ... remains largely unaffected by expressed environmental values. In an October 2005 poll conducted by Duke University, seventy-nine percent of respondents favoured stronger environmental standards, but only twenty-two percent said environmental issues strongly affected their vote".[125] This suggests a growing gulf between symbolic and embodied aspects of life – the former being consciously foregrounded and the latter leading a mute, shadowy existence. This gulf is exploited commercially in 'greenwashing'; for example, a local firm has recently changed its name from '*** Waste Disposal' to '*** Environmental' – and complemented this change by respraying its trucks green.

Science and reality

In the physical and biological sciences, researchers are generally tethered to reality through the assessment of theory by experimental results, albeit indirectly and selectively. In the case of science fiction or novels, the relation to reality is in some respects absent; but because both writer and reader accept this lack of relation and frame it as 'fantasy', this does not constitute a delusion. On the other hand, in disciplines where researchers are judged only by their peers who work in the same area, and where few if any consequences follow loss of contact with reality, group delusions may develop. For example, as Medin and Atran observe, "At times within anthropology, the methodological point that anthropological observations are socially constructed has been elevated to a form of self-immolation that threatens to destroy the science part of anthropology as a social science and move it squarely into literature."[126] In contrast to engineering or medicine, unfounded beliefs and fanciful theories in cultural or social theory do not generally result in bridges collapsing or people dying, and in fact often pass unnoticed. To outsiders, however, the divergence from reality is often glaringly apparent; and this is the case in the area of sociology described as the 'sociology of scientific knowledge' (hereafter SSK) – an area which reached the peak of its popularity some time ago, but which remains highly influential in more subtle ways. As this debate has been well covered elsewhere,[127] I will comment on this area only as far as is necessary to indicate its relevance to the concerns of this book.

One of the problems in SSK writings is the confusion between *reality* and *the ways we represent* reality – or, in philosophical terms, between *ontology* and *epistemology*. If, as I argued above, our intoxication with the symbolic realm exceeds a certain level, then there is always a danger that we begin to perceive symbolism as the primary reality and materiality as a by-product. For example, Pinch and Collins reject "the intrinsic existence of accurate and fictitious accounts per se", and claim that "all of science is merely the "construction of fictions".[128] In the same vein, Latour and Woolgar claim that "scientific activity comprises the construction and sustenance of fictional accounts".[129] But in, say, medicine, the difference between an accurate and an erroneous account of appendicitis is that between life and death; so no physician would accept that these accounts are equally fictitious. Neither would a critical realist or a biological scientist, for whom scientific models are not 'fictions', but rather more or less accurate accounts of realities which are independent of their scientific or other description. As we have seen, science progresses through testing a theory against the realities it describes – a process which yokes theory to reality, allowing a theory to be developed or replaced by more accurate theories. Here, I am grossly oversimplifying and overgeneralising; but my account is, I believe, broadly accurate. Let us briefly examine a particular case in order to clarify these differences.

In the Nineteenth Century, cholera was common in European cities, killing many. The dominant theory of transmission was that it was due to 'effluvia' breathed out by the patient and inhaled by future victims. John Snow, however, pointed out that many of those who were close to the patient, and so presumably breathed in the 'effluvia', nevertheless did not succumb to the disease. He also noticed that those who *did* die of cholera often shared the same water supply. For example

> In Manchester, a sudden and violent outbreak of cholera occurred in Hope Street, Salford. The inhabitants used water from a particular pump-well. This well had been repaired, and a sewer which passes within 9 inches of the edge of it became accidentally stopped-up, and leaked into the well. The inhabitants of 30 houses used the water from this well; and among them there occurred 19 cases of diarrhoea, 26 cases of cholera, and 25 deaths. The inhabitants of 60 houses in the same immediate neighbourhood used other water; among these there occurred 11 cases of diarrhoea, but not a single case of cholera, nor one death.[130]

Snow therefore suggested that cholera was transmitted by contaminated water. To test his hypothesis, he examined the death rates in households supplied by two companies, the Lambeth Company, and the Southwark and Vauxhall Company. The latter sourced its water from a part of the Thames known to be contaminated by sewage; whereas the Lambeth Company drew water further upstream, from a part of the river believed to be free of contamination.

Examining the death rates for people using water from these companies, Snow found that among consumers of the Southwark and Vauxhall Company's water there were 315 cholera-related deaths per 10,000 people; whereas for the Lambeth Company, there were 37 per 10,000 people. Further epidemiological work also confirmed Snow's hypothesis; and more recent research has isolated a water-born bacterium, *vibrio cholerae*, which is responsible for the disease. Today, water companies are careful to separate the fresh water supply from sewers, and cholera is virtually unknown in the developed world.

There are, of course, hundreds of thousands of other examples which could be used to make the same point: that scientific accounts are not simply convenient 'fictions', but rather follow the empirical practice of relating theory to reality. The 'effluvia' and bacterial theories are not equally 'fictitious'; rather, one is indeed fictitious, and the other is not.

How could so many highly intelligent people become so woefully misinformed about the character of science and the nature of reality? This is a topic which deserves another book; but here I am concerned only to pick out the main features of the problem. The main culprit is perhaps the intellectually incestuous nature of academic life: if one only talks to others who hold the same viewpoints as oneself, and if one only reads journals which express one's own beliefs, then the result is the formation of a cult-like group whose views can become unrealistic. Such groups tend to become selective about which views they attend to, focusing, in the case of SSK, on science's limitations, uncertainties, gaps in the evidence, and so on, and ignoring science's overwhelming success. For example, SSK theorists tend to invoke Thomas Kuhn's book *The Structure of Scientific Revolutions* to support their view of science as changing unpredictably, even though Kuhn firmly believed that scientific paradigms were not arbitrary, due to social fashions, or discursively constructed, and in fact described himself as an 'unregenerate realist'.[131] Sometimes this sort of selective focus is 'built into' the belief system, which then becomes quite impervious to alternative perspectives. For example, I once offered a colleague of mine what I thought were some robust objections to his social constructionist interpretations, only to be told that my views were themselves "a wonderfully persuasive social construction"! Such belief systems are by their very nature as unchallengeable as those of religious fundamentalists; and Robert Shapiro's critique of creationism can be applied almost unchanged to constructionism, both sharing the belief in a (divine or social) 'creator'. For example, suppose that we wanted to demonstrate that Japan had triumphed over the United States in World War II:

> How would we go about it? First, we would have to discredit newspapers such as the *New York Times* which contain a detailed day-by-day account of the American victory. We might first collect typographical errors in *The Times* and instances when errata were published, retracting previous mistakes. After that, we would collect a list of unsound predictions: optimistic statements by economists, prizefighters, and election

campaign managers that were published in *The Times* and proved incorrect. We would put all these instances together and conclude that the *New York Times* was of no value whatsoever as a historical source.

We would then print an alternative newsletter that had the 'authentic' information, and give its publisher a high-sounding name such as the Japanese Victory Research Institute. In it, we would publish photographs of the raid on Pearl Harbour, transcripts of Japanese wartime broadcasts that claimed imminent victory, and current news concerning the spread of Japanese cars and Japanese restaurants in the United States. Finally, we might demand that this point of view be given equal time with the conventional one in public school history classes.[132]

Furthermore, if one believes that discourse creates reality, then clearly realities cannot predate the discourse which describes them. The consequences of such a belief are all too evident in Bruno Latour's discussion of the case of Ramses II of Egypt, whose mummified remains were transported to Paris by French medical scientists in 1976, 3000 years after he died. The spinal deformation they found clearly indicated that he had died of tuberculosis.

However, Koch's bacillus – which today is accepted as the cause of TB – was only discovered in 1882. "How", asks Latour, "could he have died of a bacillus discovered in 1882?"[133] He continues:

The attribution of tuberculosis and Koch's bacillus to Ramses II should strike us as an anachronism of the same calibre as if we had diagnosed his death as having been caused by a Marxist upheaval, or a machine gun, or a Wall Street crash. ... And yet, if we immediately detect the anachronism of bringing a machine gun, a Marxist guerrilla movement, or a Wall Street capitalist back to the Egypt of 1000 BC, we seem to swallow without so much as a gulp the extension of tuberculosis to the past.[134]

How does Latour resolve this conundrum? He argues that the Koch bacillus *can* be extended into the past "through the practice of science." To allow this to take place, the

mummy has to be brought into contact with a hospital, examined by white-coated specialists under floodlights, the lungs X-rayed ... and so on. [It is clear that] Ramses II's body can be endowed with a new feature: tuberculosis. But none of the elements necessary to prove it can themselves be expanded or transported back to three thousand years ago. In other words, Koch's bacillus may travel in time, not the hospital surgeons, not the X-ray machine, not the sterilisation outfit. When we impute retroactively a modern shaped event to the past we have to *sort out* the fact – Koch's bacillus's devastating effect on the lungs – with that of the material and practical setup necessary to render the fact visible. It is only if we believe that the facts escape their network of production that we are faced with the question of whether or not Ramses II died of tuberculosis.[135]

This is an example of what Roy Bhaskar has referred to as the 'epistemic fallacy', or the view that "statements about being can be reduced to or analysed in terms of statements about knowledge".[136] If the world is constituted through discourse, Latour asks, how can Koch's bacillus have predated the scientific discourses which describe it? If, however, we recognise that reality exists *independently* of discourse, then there is no problem: a reality can exist long before the associated discourse appears. A situation such as the case of Ramses II is only problematic if we get the discursive cart before the equine reality. The world is not 'constituted through language': it existed long before *homo sapiens* was around to describe it, write about it, or talk about it, and will continue to exist long after we and our languages have disappeared. There is therefore nothing problematic about suggesting that although the bacillus responsible for tuberculosis was only *discovered* in 1882, it has in fact *existed* for many thousands of years. Contrary to Latour's argument, the facts *do* "escape their network of production".

The confusions in Latour's work are equally evident in the book he wrote jointly with Steve Woolgar, *Laboratory Life: The Social Construction of Scientific Facts*. This is a study of research in the Salk laboratory – a world-leading centre of biological research. One might imagine that a reasonable grasp of biology would be a prerequisite for a sociological researcher in this area; but Latour and Woolgar tell us that "our observer ... has never seen a laboratory before and has no knowledge of the particular field within which laboratory members are working".[137] After a period of distinctly unproductive observation, these authors decided that examining the citations of papers produced by Salk researchers might offer some clue about the character of scientific research. Unfortunately, this tactic was no more successful:

Although citations revealed that items had varying impact, our observer felt that he had discovered little about why this was the case. Our reaction to this kind of problem is to engage with more sophisticated and complex mathematical analysis of citation histories, in the hope that some clearly identifiable pattern of citations will emerge. But our observer was unconvinced that this would alleviate his basic difficulties of understanding why items were cited in the first place. Instead, he reasoned that there must be something in the *content* of the papers which would explain how they were evaluated. Accordingly, our observer began to peruse some of the articles in order to ferret out possible reasons for their relative value. Alas, it was all Chinese to him! ... he felt entirely unable to grasp the 'meaning' of these papers".[138]

And there we have the crux of the problem: if researchers know nothing about the area they are researching, then their conclusions are likely to be devoid of insight, and about as useful as the efforts of practitioners of Pacific cargo cults to build 'radios' out of driftwood and coconuts. Whereas the Salk researchers use symbolism in a way that is securely if indirectly related to reality, Latour and Woolgar are floundering around in a symbolic morass of their own making.

All this may give the impression that I believe that science precisely and unproblematically models the real world; but I have already indicated that this is not so. Science interacts with the world only selectively, focusing on the lower levels of systemic hierarchies, and on what is static and predictable or can be made so through abstraction. Science is more comfortable and effective when it deals with elements and components rather than emergent relationships, and so tends to foster reductionist understandings, especially of human subjectivity and complex systems such as culture, ecosystems, and political and economic issues. Despite these important limitations, the most successful scientists can clearly distinguish between the real world and our symbolic understandings and descriptions of it. The Nobel Prize-winning physicist Richard Feynman, for example, tells how his father used to take him for walks in the woods, teaching him about the natural world.

The next day, Monday, we were playing in the fields and this boy said to me, "See that bird standing on the wheat there? What's the name of it?" I said, "I haven't got the slightest idea." He said, "It's a brown-throated thrush. Your father doesn't teach you much about science."

I smiled to myself, because my father had already taught me that that doesn't tell me anything about the bird. He taught me "See that bird. It's a brown-throated thrush, but in Germany it's called a halzenflugel, and in Chinese they call it a chung ling and even if you know all those names for it, you still know nothing about the bird. You only know something about people; what they call the bird."

"Now that thrush sings, and teaches its young to fly, and flies so many miles away during the summer across the country, and nobody knows how it finds its way," and so forth. There is a difference between the name of the thing and what goes on.[139]

The deceptive messenger

However, the view that we are 'constructed by language' may possess a certain ironic truth value in a world where more solid bases for identity have disintegrated. The anthropologist Veena Das has suggested that many of us are like Sandip, a character in a novel by Rabindranath. Sandip "exists only in language ... His words do not falsify an inner life or draw a veil over it – they are indeed functioning to hide that fact that there is no inner life to hide. ... Rabindranath ... compares him to the new moon ... simply an absence".[140] We are becoming like hollowed-out shells, our deep subjectivity and its embodied relation to the world atrophied, our situation camouflaged by an expanding virtual world. The soaps, movies, sporting heroes we identify with do not enhance our subjective life, but substitute for it, providing a ready-made realm of vicarious feelings, involvements, activities, and meanings behind which our embodied roots fade. These discordances between symbolism and embodiment – the incongruence which Rogers regarded as the basis of psychopathology – seem to be less apparent in societies that are in touch with reality in an intimate, material way. Hugh Brody reinforces Rogers' view that incongruence may cause psychological problems, pointing out that in industrial societies

the reliability of facts and the veracity of statements of personal history tend to be observed in the breach. This is not so among the Inuit or the Dunne-za nor, I suspect, in any other hunter-gatherer community.

Verbal untruth has the power to render us neurotic. ... A parent says she loves all her children to the same degree, even if this is far from the case. Or a husband insists he loves his wife, even when there is deep animosity between them. These are examples where dishonesty is somehow normal. Yet these lies are often detected as such at a preverbal level. Something deeper than words can understand that words are false. In this way, a tension or contradiction may arise between words and behaviour.[141]

Brody goes on to point out that what is said to a child, and its consistency or inconsistency with the child's embodied awareness, can have profound consequences for his or her identity and security:

> A child reads feelings from the subtlest of an adult's gestures. The child knows much of what is 'true'. … The disorders that psychology associates with the dissonance between what parents say to children and what children know to be reality – from deep insecurities to chronic anxiety to depression – are not to be found among the hunter-gatherers I have known. … The apparent sturdiness of the hunter-gather personality, the virtual universality of self-confidence and equanimity, the absence of anxiety disorders and most depressive illnesses – these may well be the benefits of using words to tell the truth.[142]

Our affluence has allowed us to erect a veil between dishonesty and its consequences. We can maintain the pretence that we are younger than we are, with help from skin lotions and cosmetic surgery; drugs can heighten our mood and dissolve our concerns for others; and we can convince ourselves of our environmental credentials with a monthly cheque to Greenpeace. Contrastingly, in those societies where the interweaving of human life and the physical environment anchors symbolism to the world in an urgent and immediate way, evasion and denial may immediately be life-threatening, making honesty a habit that is learned early in childhood. In how many 'Western' families could the following conversation, recorded by Edmund Carpenter in an igloo – have occurred?

> One day when Kowanerk and I were alone, she looked up from the boot she was mending to ask, without preamble, 'Do we smell?'
> 'Yes.'
> Does the odour offend you?'
> 'Yes.'
> She sewed in silence for a while, then said, 'You smell and it's offensive to us. We wondered if we smell and if it offended you.'[143]

Here is no relativistic comparison of alternative 'narratives' or 'constructions'. Words are used to refer to sensed realities in an immediate, undefensive, and honest way; and the realities that words are referring to are clearly understood as separate from the words themselves. In keeping with this distinction between language and reality, emotions may be felt but not talked about, just as the word may exist for a feeling which is seldom experienced. The story of our gradual loss of this direct connection between symbols and realities is a long and complex one that I will explore in the next chapter.

5
How the Mind Took Over the World

Evading reality

Reality can be inconvenient. It is hardly surprising, then, that we try to replace it with something more to our liking.

According to a story that was circulating on the Internet a few years ago, the Michaels family owned a small farm in Canada, just yards away from the North Dakota border. Their land had been the subject of a minor dispute between the United States and Canada for generations. Mrs. Michaels, who had just celebrated her ninetieth birthday, lived on the farm with her son and three grandchildren.

One day, her son came into her room holding a letter. "I just got some news, Mom," he said. "The government has come to an agreement with the people in Washington. They've decided that our land is really part of the United States. We have the right to approve or disapprove of the agreement. What do you think?"

"What do I think?" his mother said. "Jump at it! Call them right now and tell them we accept! I don't think I could stand another one of those Canadian winters!"

This sort of confusion between the nature of the world and the way we think and talk about it is not merely the stuff of jokes; it runs through the industrialised world like a fault line through an unstable mountainside. According to one self-help author:

> What we think about ourselves becomes the truth for us. I believe that everyone, myself included, is 100% responsible for everything in our lives, the best and the worst. Every thought we think is creating our future. Each one of us creates our experiences by our thoughts and feelings. ... No person, no place, and no thing has any power

over us, for 'we' are the only thinkers in our mind. We create our experiences, our reality, and everyone in it.[1]

This is an extreme example; but in more subtle forms our lives are riddled with such flights of fantasy. Today, it has become possible to live almost entirely in the symbolic realm and its materialisations, comfortably distanced from any interaction with features of the world that have escaped reconstruction. For example, Herbert Simon, the great information theorist, stated in his 'Travel Theorem' that "anything that can be learned by a normal American adult on a trip to a foreign country ... can be learned more quickly, cheaply, and easily by visiting the San Diego Public Library." Elaborating on this, he claimed that it "is well known that one can circumnavigate the globe, penetrating deserts and jungles along the way, without ever venturing outside one's own Western, industrialised, air-conditioned culture, or learning that there is anything different from it. I have had a thrilling view of Ulan Bator and the Gobi Desert from 30,000 feet, in the business-class comfort of a B747". Travelling to Europe with his wife in 1965, Simon visited many of the places depicted in Cezanne's paintings. However, they "learned nothing new; we had already seen the paintings."[2]

For those who have become accustomed to such symbolically groomed 'realities', returning to the real world can be a jarring experience, as Simon found when he tried to raise cattle. "Theories", he ruefully relates, "however plausible and 'obviously' valid, can be destroyed totally by the obvious facts of the real world". Although he was using "an unbeatable scheme for raising cattle profitably ... the cattle had a different scheme." This insight, however, apparently didn't generalise. Despite his complete lack of management experience, Simon later wrote "a textbook on municipal administration, to be used in a training course for city managers".[3]

Meanings change as symbolism works on real events and situations, illustrating what Paul Stoller refers to as "the anaesthetising influence of language".[4] There are gradients of meaning, of time, of distance. Temporally, today's grief becomes yesterday's tragedy, last year's 'events', and history's 'facts'; and we are encouraged to be sceptical about 'origin myths'. Pierre Nora, referring to "the conquest and eradication of memory by history", suggests that our "relation to the past [is] no longer a retrospective continuity but the illumination of discontinuity".[5] We examine the past as we examine a corpse: events such as the Holocaust or the Nakbah become notes in a history book, unconnected with lived experience. Headlines give more space to celebrities' affairs than to thousands starving in Africa. An event's immediate impact may be emotional; but the lasting traces take other forms. Raw emotion is hard to bear; so, we escape into symbolism, translating events, 'understanding' them, or externalising them through writing.

The result, as Michael Steinberg puts it, is that we "live within models of experience instead of experience itself".[6] Psychotherapists use this distancing as a way of alleviating psychological pain. Steve de Shazer's 'Write, Read, Burn' technique, for example, involves writing painful feelings down on paper; then reading them before burning the paper, so that the feelings are first externalised and translated into words, and then literally 'go up in smoke'.[7]

This retreat from the real world is a healthy and necessary ploy if it is used as a temporary respite so we can gather our strength before re-engaging with reality. However, if it becomes a culturally encouraged chronic escape, so that we build our lives in a realm of technologically assisted fantasy, then it becomes pathological and dangerous. As Steinberg also points out, it is "foolish to expect a discursive way out of a difficulty that was created in part by the assumption that the world could be embraced in discourse".[8]

How discourse and being diverged

During the Middle Ages, Europeans lived in a world that was largely mysterious and explicable only through the rather nebulous accounts provided by the churches. Lacking the technologically created haven from raw nature that we inhabit today, they experienced themselves as immersed more directly in the world, being thrown about and chewed up by events rather than analytically observing and controlling them. Religious texts allowed little control over nature, but perhaps offered the necessary illusion of understanding, since exposure to the raw unpredictability of the natural world may have been too overwhelming. As there was barely any integrated, coherent knowledge structure, contradictions could be allowed to exist as part of God's mystery – a notion we would struggle with today. The rigidity and oppressiveness of religious doctrine, however, was a brittle defence against a pervasive underlying uncertainty and powerlessness; and we can partly understand the replacement of religious orthodoxy by scientific method during the Enlightenment as the replacement of one means of reducing uncertainty by another, more powerful one. But whatever the differences between the religious dogma of the Middle Ages and the Enlightenment proto-science which partly replaced it, they were both bids to escape from the vagaries and ultimate death of the flesh into a more controllable symbolic realm. Descartes, for example, famously attempted to locate certainty in the midst of doubt; and his biographer describes him (in 1619) as "a young man who is ... unsure about the future, and perhaps even overwhelmed by the uncertainty and ambiguity with which he is trying to cope".[9] In most of us, such insecurities do not produce great philosophical systems; but we all cling to our fragile sources of certainty, whether these are science, or Jesus, or the *Daily Mail*. But while the certainties that science produces are felt reassuringly to be the 'building blocks' of the world, they ignore the emergent properties which become increasingly essential as one climbs the organisational ladder of life, leaving fundamental questions unanswered.

Language, too, was closely tied to the world; and before the sixteenth century, a word was experienced more as an intrinsic part of an entity than an element of a discourse which was applied *to* the entity. Knowledge, in Timothy Reiss's terms, involved accumulating "bits and pieces of meaning ... to approach understanding of the thing, of its place and that of the [person] in some divine plan ... The greater the accumulation of such meanings, the nearer the approach to a wisdom conceived as knowing participation in a totality".[10] The 'bits and pieces', however, did not need to be related within any coherent, consistent framework, as we noted above – a quality that, as we will see, is also characteristic of various indigenous epistemologies – so that the knowledge of the Middle Ages was more like a patchwork quilt than the relatively seamless tapestry of modern science. At the same time, the Aristotelian trust in sensory information as conveying something about the real character of an entity continued until Copernicus' 'rape upon the senses', as Galileo famously described it, announced the prioritisation of the conceptual realm which has accelerated ever since. Today, sense-related knowledge survives, for example, in our personal, often inarticulate, feeling-laden intuitions, and in the taken-for-granted skills involved in everyday activities such as walking, while conceptual knowledge is the publicly shared and greatly elaborated basis of technological society.

The Mediaeval world of multiple, unanchored bits of knowledge slowly coalesced during the Enlightenment to become a more constricted, abstract, and reproducible type of knowledge. Language, as a system intervening between the world and the individual mind, allows us to keep a dispassionate distance from the world, preventing us from being overwhelmed by its immediacy and diversity. A name operates a bit like a lasso: we have grasped a part of the wild world and set bounds on its variation while still not completely controlling it. But greater control comes when we connect the word with a system of other words, corralling the thing into a linguistic web. Consequently, we can inhabit it in some respects as if we were inhabiting the world itself, the distinction between them becoming clouded in our minds. Our increasing entrancement by language necessarily excludes – or 'occults', in Reiss's phraseology – whatever doesn't fit the new discourse; and the process of 'occultation' is *itself* occulted, so that whatever is outside the beam of light provided by our discourses fades from awareness. This produces a sort of epistemological unconscious, since not only do we lose sight of certain characteristics of the world and ourselves, but we also forget that we have forgotten, so that the residual, conscious knowledge comes to be mistaken for reality itself. As Reiss summarises this process, the "systematic discourse that names and enumerates becomes, replaces, the order of the world that it is taken as representing".[11]

This move towards a single, consensual, impersonal discourse did not take place overnight. For Kepler – often represented as the originator of modern astronomy – the mathematical understanding of planetary movement was only one of many possibilities within a 'ludic' whole. As Kepler put it, "I play ... with symbols";[12] although it is important to note that Kepler's symbolic 'playing' has an empirical dimension to it, and so is quite different to the symbolic play of a postmodernist. Similarly, others such as Newton, Boyle, Leibniz, and Locke dabbled in alchemy and other unscientific pursuits – but it is only with hindsight that we can separate the 'dabbling' from the 'serious' science. For example, Newton discusses the parallels between the growth of vegetation and the (alleged) 'growth' of metals within the earth's crust, claiming that "Natures obvious laws & processes in vegetation" apply equally to the 'growth' of metallic deposits in the earth, so that "metalls vegetate after the same laws".[13] With the benefit of a few centuries of hindsight, it is clear that the processes involved are quite different; but in the seventeenth century the outlines of the scientific knowledge were yet to crystallize out of the fog of conceivable patterns of relation.

There is a subtle anthropocentrism to this process in which concepts and the artifacts which embody them come to seem representative of the world as it is, rather than as the outcome of cognitive simplifications and selections. This trend towards a reliance on a web of concepts, abstractions, and language rather than on the changeable and diverse entities of the world itself has accelerated in the contemporary era, most extremely in postmodernism's claims to the priority of culture and language. But while postmodernism is not taken seriously outside academia, the industrial world nevertheless prioritises the 'web of concepts, abstractions, and language' in subtle but important ways. Knowledge is not just considered to be an understanding of the world 'out there'; it is also, more subtly, used as a prescription *for* the world, so that the world is reconstructed to conform to the way we think. The 'epistemic fallacy' – that is, the confusion of epistemology and ontology – has become sedimented into the world as the character of the world is made to conform to human knowledge, so that, in a sense, epistemology has become ontology. As Bhaskar remarks, "Hooke's Law ... is literally built into the construction of spring balances";[14] and Thomas Kuhn points out that the notion of energy as "an underlying imperishable metaphysical force, seems prior to research and *almost unrelated to it*. Put bluntly, [physicists] seem to have held an idea capable of becoming conservation of energy for some time before they found evidence of it".[15] Here, we can begin to detect the influence of the nascent cognitive/industrial system as it selects and moulds knowledge to suit its own internal dynamics. As Jesper Hoffmeyer remarks, Galileo's credo that the 'great Book of Nature is written in the language of mathematics' may be "an expression of human reason's wish that the world should always resemble reason itself".[16]

A similar resonance between latent ideology and the emerging mechanical consensus can be detected in the invention of the mechanical clock, which seems to have occurred almost simultaneously with its use as a widely applied metaphor in the seventeenth and eighteenth centuries, the clock materialising certain analogies with the functioning of the centralised state and the living body. Mechanism was an idea through which cognition could begin to grasp causality in various spheres, as Otto Mayr notes:

> The entire system was organised around a central authority that was directly linked to the multitude of organs or elements carrying out the system's various functions. The relationship between central authority and operating organ was conceived in principle as a cause-and-effect relationship ... God, the sun, the king, the brain (or the heart) and similar agents seen as central authorities were likened to the escapement.[17]

Selected from the myriad of mutual interactions in physical reality, this metaphor of central control, causally related to whatever is being controlled, became a fundamental one. The notion of causality, however, while it was understandably seized on due to its cognitive accessibility, was by the same token not easily applied to a natural world in which the origins of causal power seemed more diffuse. Only in recent decades have ecologists begun to come to terms with the dissonance between mechanical models of causality and the real world, catalysing the appearance of quite different models.[18] But the industrial world has in some respects outflanked this divergence between knowledge and reality by reconstructing the world to fit our understanding of it, so that aspects of the world that contradict the elemental, bottom-up explanations provided by science, including ecosystems, have faded under the onslaught of industrialism. Conversely, forms which are rare in nature, such as the perfect circle, have become commonplace as they are materially instantiated as vehicle wheels, gears, bearings, and so on.

This re-moulding of the world to fit cognitive principles, initially reflecting the search for stability in knowledge and prediction, therefore seems to have been more a product of a just-beginning systemic coalescence rather than of any grand human intention; but it became a cornerstone of philosophy and of the emerging industrial system. As so often seems to happen, intellect recognises a trend and recasts it as a fundamental principle, as in Fichte's claim that the physical world exists only inasmuch as "it arises through knowledge".[19] And Kant claimed that while "it has been assumed that all our knowledge must conform to objects", the apparent inadequacies of this knowledge suggest instead "that objects must conform to our knowledge".[20] We should note in passing that the exaggerated contrast between the two extremes posed in Kant's statement itself reflects an unconsciously assumed dualistic scheme in which only conceptually clear polar opposites are recognisable; and that this imposition of a conceptually-derived dualism suggests that the mind has *already* taken over the world. Here, the possibility that knowledge is partly the result of a more dialectical process of interaction among thought, perception, and reality is by-passed. For Kant, and for a plethora of post-Kantians, there is only one possibility, as Lee Braver relates:

> Gone is the inquiring subject (in both senses of the word) who must profess fealty to sovereign nature; gone is the soft clay of the Platonic-Aristotelian soul or the blank paper of the Lockean mind passively taking on reality's imprint. In its place is the Baconian scientist who boldly questions, even tortures, subservient nature ... and the Cartesian ideal of humans as the 'lords and masters of nature' made ontological. We do not find the order of nature; we make it.[21]

But if Kant overestimated the extent to which mind determines experience, his views nevetheless foreshadowed the more indirect determination of experience through the world's technological transformation. Here we have a pattern for future attitudes to nature as 'socially constructed', in which there is little recognition of any intrinsic natural structure or value, but only of what the mind – and by extension, technology – can make of it. A key aspect of the new mind-determined world was a focus on a few factors such as quantity or economic value, sweeping aside the previous emphasis on a divine but largely unknowable order that permeated the entire world. This ability to seize on one or a few factors out of the diversity of natural phenomena was a key *cognitive* capacity which was vital to the preoccupation with economic wealth and the tacit backgrounding of non-economic considerations such as psychological well-being, spiritual resonance, and ecological health. But as Whitehead pointed out, we tend to forget that our idea of a thing selects only a few of its properties, ignoring others, so that properties and qualities fade from the world as we rebuild it.[22]

As cognition increases its grip, so the world becomes reordered around a few primal industrialist metaphors such as 'monetary value', 'energy', and 'mechanism', and so becomes less fluid and unpredictable – at least within the area we have cleared. Knowledge seizes on what is static – or what can be made to seem static – and so has difficulty representing the movement, change, dynamism of the world. Plato was a precursor of many when he considered only the fixed and unchanging to be real; and as Mary Midgley notes, atomism arose in Greece "out of a determined – but ultimately doomed – attempt to find something in the world which was truly fixed and immutable".[23] This reductionist focus on tiny, unchanging particles left open the question of how the particles interacted; but so long as we are dealing with relatively amorphous materials such as steel, concrete, or gases, which possess minimal self-organisation, this is unproblematic. In nature, however, emergent properties are the rule rather than the exception; so representing nature as amorphous 'stuff' to be manipulated is only possible by doing (conceptual or physical) violence to the structures that are ignored. One way of evading this problem is to represent change by focusing on the beginning and end *states* rather than by looking at the *processes* involved: for example, we see a process of movement as a passage from one place to another, rather as we cross a stream by jumping from one stepping-stone to the next. Such conceptual stepping-stones include numbers, names, categories, and monetary value, all of which have the power to tether the melee of life and so reduce cognitively unmanageable movement. It is no coincidence, therefore, that James Buchan refers to money as 'frozen desire';[24] and Albert Borgmann refers to a "semantic ice age" as having "enveloped the globe":

> The Greek legend of Pygmalion and Mozart's *Don Giovanni* have scenes of life infusing stone – Venus animating Pygmalion's lovely statue to gladden his heart and God enlivening the grim statue of the commendatore to bring Don Giovanni to judgement. These stories have inspired Romantic painters and contemporary set designers to give us pictures of eloquence flowing into lifeless matter. The reverse has happened in the modern period. Eloquence and meaning began to drain from reality.[25]

We corrall diverse entities within fixed categories such as species, abstract 'forms', and 'ideal types', rather than dealing directly with the indefinitely large variety of real, fluid, self-organising entities. This is what Theodor Adorno referred to as 'identity thinking';[26] that is, the error of mistaking the thing itself – in all its variety, depth, and unknown-ness – with the *concept* we use to refer to it. Thus the bluebell becomes a saleable commodity within the economic system rather than a growing wildflower in the woods – a huge but mostly invisible transformation that is concealed by the constant name 'bluebell' and the perceptual sameness of the flower itself. Similarly, by referring to 'the Amazon rainforest', we divert attention from the shrinkage and degradation of Amazonia, since the name remains constant. Equally, if we *conceptually* understand the world as a cornucopia of 'natural resources', then its *physical* dismemberment will be less noticeable. The 'eloquence' that Borgmann sees as 'draining' from the world proceeds under cover of these conceptual constancies.

We also employ our portrayals of constancy and certainty in order to physically *build* a world that is in some respects more constant than the original, wild world. The conceptual world is a curiously static world, one more like the moon than the earth in many respects; and it is ironic that we search for life on other planets while at the same time destroying the life on our own. As we trim, lop, and channel life on Earth to fit simplified industrial schemes, 'life' as thus redefined begins to fit the way we think. The cars, houses, and washing machines that populate our urban environment, unlike trees, foxes, and poppies, remain the same size and shape; and there is nothing as variable as a sunset or a rainbow. Constancy is the background out of which the controlled variations of the industrial world emerge, in contrast to the continuous fluidity of the wild world. Tim Ingold notes one of the costs of this simplification in his discussion of cartography, pointing to the "discrepancy between truth and accuracy ... the more [we aim] to furnish a precise and comprehensive representation of reality, the less true to life this representation appears." He continues:

> In the cartographic world ... all is still and silent. There is neither sunlight nor moonlight; there are no variations of light or shade; no clouds, no shadows or reflections. The wind does not blow, neither disturbing the trees nor whipping the water into waves. No birds fly in the sky, or sing in the woods; forests and pastures are devoid of animal life; houses and streets are empty of people and traffic. To dismiss all this ... is perverse, to say the least. For it is no less than the stuff of life itself.[27]

As we try to pin down reality, to make it conceptually clear and cognitively accessible, a fugitive part of it slips away unseen, to hide in the shadows beyond the reach of thought.

But the conceptually elusive qualities we exclude from the world do not simply evaporate; and while we impose constancies we also export change and uncertainty to that occluded realm beyond the boundaries of what is scientifically understood. Outside the rather narrow range of our conscious awareness, this effort to achieve stability may actually be *de-stabilising* the world. For example, we use air-conditioning and heating to minimise changes in ambient temperature; but the indirect and unintended consequence of this use of energy is *more variable* climatic conditions. In general, holding one level of a system constant tends to cause change at other levels – an indirect effect which is extremely difficult to model. While the cognitive-industrial system can stabilise various conditions within a certain sphere of awareness and control, this stabilisation will have knock-on effects within a larger realm that extends beyond our awareness; and unseen interactions will eventually upset our carefully tended stability. The defensive function of language introduces similar instabilities in the social sphere, as words like 'community', 'love', 'progress', or 'empowerment' can only temporarily cover up and substitute for the fading of these qualities from the embodied realities outside language.

Substituting a symbolic world for the real one

Precision of understanding and prediction, then, is achieved at the price of driving out what is *not* precise, understandable, or predictable – which, as Ingold suggests, is the 'stuff of life itself'. Nevertheless, we tend to assume that the abstraction accurately represents the reality, so that, like Albert Speer, a Minister in the Third Reich who denied knowledge of the Holocaust, we feel that what we don't know doesn't exist.[28] This applies generally to the relation between the symbolic and real worlds. For example, we learn that language can express some of what we feel; but we learn little about what language *cannot* express. "Say what you mean!", we are told, the implication being that what we experience is always expressible in language.

This masking of anything that cannot be linguistically coded reflects an explicit identification with the symbolic realm, and an equally explicit distancing from embodied feeling. A mathematical formula is precise but generally unfelt, while the feeling we get when we survey a landscape is 'subjective' but conceptually elusive. So the formula is immortalised in books and buildings, while the memory of the landscape fades, and the prioritisation of symbols over material reality becomes a structural part of the modern world. Today, we take this priority for granted; but in sixteenth century Europe, it was a novelty that needed arguing. In Galileo's *Dialogues*, for example, he has Salviati say:

> Just as the computer who wants his calculations to deal with sugar, silk, and wool must discount the boxes, bales, and other packings, so the mathematical scientist, when he wants to recognise in the concrete the effects proved in the abstract, must deduct the material hindrances.[29]

The mathematician, supposedly, reveals the real character of things, the framework that is covered over and camouflaged by all the inconsistent, superficial, confusing variations of appearance. The 'reality' is the underlying form, we learn; and to reach this we have to discard 'the hindrances of matter' – and the senses which most directly register them. To take another example, what have a reservoir full of water, a buried coal seam, and a day's worth of desert sunlight in common? Not much, we might think; and yet each of these can be understood as offering us a certain quantity of energy, measurable, since Joule,[30] in the same units.

As Reiss points out, however, others were less single-minded than Galileo. Bruno and Campanella, for example, both saw knowledge as uniting the mind with "an all-embracing totality"[31] rather than operating in selective, reductionist terms. "It is first necessary to look at the life of the whole and then at that of the parts", suggests Campanella.[32] And critiquing the movement towards idealism which rejects direct experience of nature, Campanella has the heroes of his book, the Solarians, argue that the book-bound philosopher "has contemplated nothing but the words of books and has given his mind with useless result to the consideration of the dead signs of things".[33] As Reiss puts it, the goal of Campanella's Solarians is "a knowledge of things in themselves, without the mediation of signs, whether these be monetary, linguistic, or any other".[34]

Katherine Hayles refers to the practical consequences of the prioritisation of thought over reality, which Reiss and Campanella both criticise, as the 'Platonic Backhand':

> The 'Platonic Backhand' works by inferring from the world's noisy multiplicity a simplified abstraction. So far so good: this is what theorising should do. The problem comes when the move circles around to constitute the abstraction as the originary form from which the world's multiplicity derives. Then complexity appears as a 'fuzzing up' of an essential reality rather than a manifestation of the world's holistic nature.[35]

The 'Platonic Backhand', then, takes what is a *descriptive* abstraction and uses it to *prescribe* a particular situation – and ultimately as a template for a reconstructed material reality. In Albert Borgmann's terms, information moves from being information *about* reality to information *for* reality, and finally is taken *as* reality itself.[36] When this happens, we not only mistake a constructed order for the external order of the world; more profoundly, we see any divergence of the world from the abstracted order as *dis*order, and so perceive the Earth as an amorphous pile of materials, awaiting the ordering touch of industrialist processes. According to Jean-Pierre Dupuy, this subtle but radical change is a general property of scientific models:

> Science is the sole human activity in which the word 'model' has a sense opposite to the one given it in everyday speech. Normally we speak of a model as someone or something one imitates, or that deserves to be imitated. But a scientific model is an imitation from the start ... [A model] enjoys a transcendent position, not unlike that of a Platonic Idea of which reality is only a pale imitation. But the scientific model is man-made. It is at this juncture that the hierarchical relation between the imitation and the imitated come to be inverted. Although the scientific model is a human imitation of nature, the scientist is inclined to regard it as a 'model' in the ordinary sense,

[*for*] nature. Thus nature is taken to imitate the very model by which man tries to imitate it.[37]

Applying such models to the world, we risk the world's demonstration of their inaccuracy, as Marshall Sahlins notes. "Having its own properties, the world may ... prove intractable. It can well defy the concepts that are indexed to it. Man's symbolic hubris becomes a great gamble played with empirical reality."[38] What is different about the industrial world, however, is that we have used technology to extensively *rebuild* the world according to our concepts and categories, thus postponing everyday, trivial corrections and risking a much larger, eventual correction. Whereas a mediaeval peasant collected fuel to warm himself on a cold night, we build networks of power stations; and as we noted above, this exchanges a short-term individual problem for an intractable, biosphere-threatening one.

This is not to suggest that the 'realities' 'discovered' cannot be found in the not-directly-knowable structure of the world itself. Scientific concepts are not simply constructed through social processes, as we saw earlier: rather, they are arrived at by lengthy experimentation in different places and by different scientists, and are confirmed by a process that we might refer to as 'triangulation', involving a number of different methods. For example, that the notion of 'atoms' resonates with certain physical realities has been confirmed in many different ways, such as demonstrations of the relative proportions of different elements in forming compounds, the electronic behaviour of elements with different positions in the Periodic Table, or the fission products formed by disintegrating atoms. While scientific knowledge does not completely represent the material world, it nevertheless has a foot on reality itself; so in contrast to the claims of Kant and the entire tradition of idealist philosophy, the 'things-in-themselves' are not totally unknowable.

The problem, however, is the *partiality* of the knowledge generated by science. There are powerful psychological defences against recognising the limits of knowledge and the ultimate precariousness of our situation. One of these defences is to shelter within the illusion of certainty that science sometimes generates, in effect mistaking the comforting certainty and predictability of a model for the rather less certain and predictable character of the world itself. For example, in Dupuy's words, scientific models

have a life of their own, an autonomous dynamic independent of phenomenal reality. In constructing models, scientists project their mind [*sic*] onto the world of things ... because a model is so much purer, so much more readily mastered than the world of phenomena, there is a risk that it may become the exclusive object of the scientists' attention. Theories, indeed whole disciplines, can be organised around the study of a model's properties.[39]

True, this is an overstatement. Generally, the physical sciences are so intimately bound up with material realities that errors are quickly noticed, and the science modified to fit. And to the extent that the industrialised world is rebuilt in accordance with the models, it is hardly surprising that the 'reality' so constructed is consistent with the models it is based on. But both these processes – of developing models of the world and rebuilding the world – omit higher-order, emergent characteristics of the original realities, such as those involving ecological organisation. Consequently, the Platonic backhand relies on a sort of amnesia. Once abstractions are taken as primary, then real entities and situations err to the extent that they depart from these abstractions; so the colourful diversity of the emergent world is taken to be 'noise' or imperfection. Locke fully understood the inadequacy of such models:

> There is not so contemptible a plant or animal that does not confound the most enlarged understanding. ... When we come to examine the stones we tread on, or the iron we handle daily, we presently find we know not their make, and can give no reasons of the different qualities we find in them. ... Therefore we in vain pretend to range things into sorts, and dispose them into certain classes under names, by their real essences, that are so far from our discovery or comprehension.[40]

To exemplify this insight into the social realm, the statistical concept of 'normality' is used *descriptively* to reflect the degree of conformity to a norm; but in common parlance it also has *prescriptive* overtones which suggest that we *should* conform to the norm, as when we anxiously ask 'am I normal?' This concept of normality imparts a 'gravitational' pull towards the average, basic, 'economy model' individual with a minimum of idiosyncrasies, discouraging diversity. It also indirectly dampens the *complementarity* that would draw on a wider range of personality characteristics, attitudes, and abilities, and so characterises a weak and inward-looking social structure that depends for its survival on conformity rather than on the imagination that could be embraced within a more vibrant society.

The concept of 'intelligence' also provides an example of the way we turn descriptions into prescriptions. By studying a range of human behaviours, we learn to summarise these behaviours by describing them as more or less 'intelligent'. We then come to see 'intelligence' not as a simplified description of a multitude of human abilities, but as a *causative* factor and an *explanation* for individual differences in behaviour. This in turn can have social consequences if intelligence tests are used to *select* individuals with particular characteristics.

'Schizophrenia' is an even more dubious concept when understood as a causal entity. As Mary Boyle has effectively demonstrated,[41] the diverse symptoms of 'schizophrenia' do not amount to a coherent clinical entity, since they do not reliably occur together. Talking about 'schizophrenia' is therefore rather like referring to 'car breakdown syndrome', the symptoms of which might include steam coming from the bonnet, strange noises, oil leaking onto the road, and immobility.

Psychotherapists sometimes make a related mistake when they selectively interpret clients' behaviour in terms of their favoured theoretical perspective. Psychoanalysts discover repressed desires, family therapists recognise systemic influences, and cognitive behavioural therapists encounter unrealistic thoughts – a phenomenon that Bill O'Hanlon has labelled 'theory countertransference'.[42] This process of reinterpreting behaviour on the basis of a tacit symbolic taxonomy can seem seductively 'natural' when the symbolic framework is assumed to take precedence over the life-situations that it is employed to interpret. As Ignacio Martín-Baró reports,

> The first time I came into contact with groups of campesinos displaced by the war, I felt that much of their behaviour showed aspects of paranoid delirium. They were constantly alert and hyper-vigilant, and they mistrusted anyone they didn't know … yet, when I learned about what had happened to them and the real dangers still preying on them, as well as their defencelessness and impotence against any type of attack, I quickly began to understand that their hyper-mistrust and vigilance were not signs of a persecution delirium … but rather the most realistic response to their life situation.[43]

In such cases, the Platonic Backhand actively conceals the real, material causes of 'psychopathology', keeping diagnosis within a safely symbolic realm that distances the profession of psychiatry from the embodied realities it claims to explain. Robert Desjarlais and his colleagues relate that at a 1993 conference on 'Indigenous Peoples and Health', the official health agencies wanted to discuss psychiatric disorders and risk factors, while the representatives of indigenous groups spoke of their political oppression and the destruction of their natural environment. As these authors report, the "mental health problems of Latin American Indians today ... are inseparable from their continuing subjugation, forced acculturation, deteriorating local ecosystems, and invasions of their land by loggers, miners, and herders".[44] What we see here is the incompatibility between a particular symbolic system and its fully-socialised agents, on the one hand, and the embodied, ecologically-embedded experience of those who are still outside of this system, on the other.

On the other hand, conceptual systems and bureaucratic infrastructures tend to coalesce into a seamless whole, leading to an unrecognised 'naturalisation' of dominant modes of thought. There are plenty of historical precedents for this; and I have already mentioned the influence of the metaphor of mechanism. Furthermore, the Christian story of the creation is all too compatible with the industrialist separation of intelligence from matter, implying that the world is a sort of divine commodity; and as Prigogine and Stengers put it, "At the origin of modern science, a 'resonance' appears to have been set up between theological discourse and theoretical and experimental activity".[45] Temperton and Hobbs point to another type of confluence when they argue that "scientists from socialist countries have historically placed more emphasis on mutualism and symbiosis than their counterparts in capitalist countries;"[46] while Colinvaux dismisses the notion of ecosystemic behaviour as better explained by "private enterprise."[47] Likewise, Philip Pauly notes that the view that conflict between good and evil is embodied in a brain with two symmetrical halves has been a persistent theme in theology; and he also remarks that

> In Germany, united for the first time under the Prussian bureaucracy, most scientists described the brain as a set of functionally distinct departments; English medical writers, confronted with the psychiatric consequences of a class-based society, worried how the rational cortex could control lower, more primitive elements of the central nervous system. It was only in France, especially in the uncertain early years of the Third Republic, that anti-Catholic liberal scientists were determined to show that civilisation and rationality resided necessarily on the Left, while decadence and mysticism were on the right.[48]

In the same vein, it is notable that the competitive approaches of Newton and Liebniz seem to resonate with the political realities of their respective nations: in other words, Newton's physics of universal laws, constituted by one God and governing the behaviour of interacting atoms, resonated with the centralised locus of power in England; while Leibniz's 'monads', each an independent microcosm of the entire universe, reflected the independent states of Germany, each ruled by a prince.[49] Today, often-assumed notions such as 'self discipline' and the acceptance that the will should govern the emotions sit comfortably with a governmental system in which rationality and, especially, 'economic necessity' are assumed to take priority over less measurable felt values and priorities. Likewise, notions of 'skill' have been redefined to include behaviours that conform to corporate expectations such as "the ability to develop satisfactory personal relationships with others", and "sufficient political and economic literacy to understand the social environment and participate in it".[50] As Jonathan Payne remarks, such 'skills' are "specified at such a low level that they could only be understood as an attempt to create a cheap pool of malleable, submissive, semi-skilled labour".[51] Such convergences between political 'realities' and the ways we think and behave are unsurprising given the pressures towards consistency between different domains of the industrialist symbolic gestalt, and especially because the idealist notion that thought constructs reality is ironically materialised in the operation of the technological world. Using metaphors that themselves illustrate how mechanism has infiltrated our thought, we might say that the 'gravitational' or 'centripetal' tendency of the industrialist symbolic system is a powerful force for consistency between domains.

An example of this unsuspected, and sometimes quite indirect, convergence is touched on by Edward Sampson, who suggests that what he refers to as 'the cognitive perspective' – the priority that people give to their own thoughts about a situation rather than the objective properties of the situation – has widely infected contemporary society. In Sampson's words,

> The knower's psychological states, the ideas in his or her head, are held to be more important, more knowable, and more certain than any underlying material interests, social practices, or objective properties [of the world].[52]

This prioritisation of one's belief system, says Sampson, implies a picture of reality that gains its coherence and its order by virtue of the orderliness and universality of the building blocks of the individual mind. It is the order of human thinking and reasoning that grants an order and meaning to the world of reality.

Because of the disparity between reality and our understanding of reality, Sampson argues, the latter constitutes an ideology, in the sense of "a systematically distorted or false picture of reality".[53] He goes on to point out that although this 'false picture' appears to us to be reborn spontaneously within each individual mind, it gains much of its seductive power by being socially shared. "Human thinking is not simply something inside the head of an individual", as he puts it, "the 'I' remains forever firmly rooted in the 'we', to the social process"[54] – although this is denied (as well as tacitly affirmed) by the need to isolate the person from social context in the experimental laboratory and also by the need to use nonsense syllables to separate the person from external sources of meaning. That is, while cognitive psychology explicitly portrays the 'individual' as an autonomous thinker, it also – through its need to methodologically isolate the person – admits that we are contextually influenced. And having isolated 'subjects' from their social and ecological contexts, it then smuggles in a replacement context – that of industrialism. Within this new context, it is necessary that the person *appear* as the 'autonomous individual' while being silently incorporated into a consciously unacknowledged system, so that the individual-in-the-laboratory becomes a template for the new role of the person within the industrial machinery. As an illustration of this new role, Jonathan Crary describes the reaction time experiments carried out by J. McKeen Cattell at the end of the nineteenth century as "a sign of a fundamental repositioning of a perceiver in a relation of subjection to an objective world newly decomposed into autonomous and abstract stimuli", as the individual adapts to "machine speeds and rhythms that differed dramatically from those of the body".[55] Thus cognitive psychology serves two functions essential to the operation of industrialism – our isolation from the contexts we previously inhabited as embodied beings, and our repositioning as cognitive agents within the industrialist whole. Furthermore, this double process of repositioning has to be made to appear natural and unremarkable, so that we remain as oblivious to our industrialist metamorphosis as the bluebells in the market.

Denying nature's contribution

Nature separate from society, so some social theorists tell us, has no meaning.[56] Such statements enclose meaning within a realm that is specifically *human* and *symbolic*, denying nature any autonomous meaning, and ignoring the prior existence of nature for several billion years before the arrival of *homo erectus*. Social life and language, however, being historically and ontologically secondary to the natural order, have to be seen in the context of our birth as a particular sort of mammal, our utter dependence on the natural world for clean air, water, and food, and the mortality we share with all other creatures. True, technology can allow us a degree of distance from the demands of the natural order; and within the constraints of this order we have a certain amount of freedom to choose our lifestyles and construct our meanings. But to argue that nature's meanings are simply those we impose on or perceive in nature is an act of anthropocentric hubris that attempts to reverse inescapable ontological priorities.

Take, for example, the tree-sit at the proposed site of Manchester Airport's new runway in 1997. Although the tree-sitters were trying to defend the natural state of the woodlands, it has been claimed that

> the woodland was a social nature *from the very start*. Surveyed, mapped and culturally ordered by the environmentalists, the press, and the airport authority, the woodland was continuously spoken for, represented and scripted from the very moment it entered the public domain as an 'environmental issue'.[57]

Such statements exemplify a type of academic writing that has become fashionable over the past two decades, and that "takes it as axiomatic that 'nature' is indeed social through and through".[58] But the effects of 'surveying, mapping, and culturally ordering' in themselves have little effect on the woodland ecosystem. The changes they induce are entirely within the *symbolic* realm – a realm that remains separate from material reality until we take it as providing a prescriptive model to be realised through technological action.

A second example of this ontological inversion is Daniel Gade's account of *Nature and Culture in the Andes*. Gade prioritises cognition over reality when he argues that "even wilderness has become a human creation through our management of it and ultimately through the conscious decisions to preserve it as such".[59] Now 'managing' wilderness may change or damage it; but only in the most extreme cases could we describe it as 'a human creation' – in which case, it would no longer be wilderness. There is a further, underlying problem, too, which we noted earlier – that of mistaking the 'actual' for the 'real'. In other words, an adequate theory cannot be based solely on nature's current form. It should *include* this actuality, but also transcend it, seeing a particular landscape as a special case of a nature that may also take other forms.

Equally problematically, another writer suggests that the "Amazon jungle, as a matter of brute fact, was not just construed but actually created by humans".[60] To support such claims, Charles Mann refers to the broken pottery, half-buried in the earth, that indicates the presence of human inhabitants in previous eras. But this example would convince few people that Amazonia is "fashioned as purposefully as Disneyland". It is somewhat more convincing as an example of the recalcitrance of nature in the face of human attempts to transform it, since the pottery – and the pottery-makers – have long since been re-assimilated by the jungle. The presence of the human within nature does not make nature 'humanly constructed'; in fact, as Peter Hay remarks, it would be "just as logical to argue the opposite – that because trees grow in London parks and geraniums in its window boxes, London has ceased to be part of the realm of culture, and has become nature."[61]

All three of these examples tacitly assume that nature is malleable, passive matter, easily changeable by human action. A somewhat similar view is expressed by Stephen Budiansky, who points to the arrival of exotics such as "thistles, plantains, chickweed, dandelions" and a host of others as indicating that "human influence is woven through even ... the most pristine landscapes".[62] But ecosystems are not passive objects that are knocked sideways by a new arrival the way a pin is knocked over by a bowling ball; they are resilient systems that often successfully *incorporate* new members, maintaining ecosystem integrity even while the constituent species change.[63] To argue that the arrival of exotics makes landscapes 'fakes' is like arguing that introducing a new recipe into one's diet changes one's identity. There is an important difference between changes that are incorporated within ecosystemic functioning, on the one hand, and those such as urbanisation which destroy ecosystemic functioning. 'Romer's Rule'[64] can usefully be applied here: when considering changes, one has to look for the deeper constancies these changes are protecting.

The examples above take a current trend – the industrialisation of nature – and leap ahead to its conclusion of a fully 'humanised' world. In doing so, they share a common overestimation of the power of cognition to redefine the world and to 'remake' it as a 'cultural artifact'. True, the conceptual reordering of nature is a precursor to its material transformation; but the industrialisation of the world is not yet a fait accompli; and statements that because wilderness areas have been changed by human action they therefore become 'social' or 'artifacts' contains the same unconscious hubris as our claims to 'produce' oil, 'grow' cabbages, or 'have' babies, overestimating the power of conscious intention compared to the systemic powers of nature. Such 'backgrounding' of nature is formally built into economists' assumption that the 'market' is a closed system, separate and distinct from nature, and that the natural environment is both a limitless source of 'free' raw materials and a bottomless sink into which we can endlessly pour our waste.

The fading of the natural order and its replacement by a symbolically based system has not occurred only in a few, technologically or economically oriented areas of industrial life. Rather, it reappears – as we would expect given the widespread colonisation of contemporary life by symbolic 'rationality' – in a wide range of domains, including those we would least associate with such 'rationalities'. This historical change involved a certain reordering of attentional priorities, as perception came to be regarded less as a registering of external realities and more of a selective attention to those aspects of the world that are chosen to fit the developing industrial *gestalt*. As Jonathan Crary puts it, the

> idea of subjective vision – the notion that our perceptual and sensory experience depends less on the nature of an external stimulus than on the composition and functioning of our sensory apparatus – was one of the conditions for the historical emergence of notions of autonomous vision, that is, for a severing (or liberation) of perceptual experience from a necessary relation to an exterior world.[65]

During the nineteenth century, for example, the development of the 'realist' movement in European painting (which should be clearly distinguished from 'realist' social science) moved painting away from representing the world as it is toward representing the world as it appears to the viewer. Louis Sass notes that

> in the work of a romantic artist like William Turner, the external world increasingly serves to express the mood or sensibility of the artist ... in impressionism or cubism it is reduced to subjective sensations or formal structures of perception ... It is as if, in modernism, each artwork were aspiring to constitute an entire new universe on the basis of some unique way of seeing or representing.[66]

Such art, Sass argues, functions "as a haven of subjectivity, freedom, or order amid the darkness and chaos of a surrounding material world ... [or] to declare that there *is* no reality apart from what exists in our minds."[67] In keeping with this idea, when a critic complained to Picasso that he couldn't paint a tree, the painter agreed, but pointed out that he could "paint the feeling we have when we look at a tree".[68] Picasso expressed a perhaps more direct connection with reality in *Guernica*; but in keeping with the ideological inappropriateness of such direct expressions of reality, the reproduction of *Guernica* that hangs across the entrance to the Security Council was tastefully covered up when Colin Powell gave his speech to the UN in 2003 advocating war with Iraq.[69]

Art, of course, may not only emphasise our subjective reactions, but can in some instances entirely jettison its links with the real world and express a realm of autonomous subjectivity. Goux reports that Kandinsky felt that painting was unacceptably dependent on natural scenes; whereas music "has been the art which has devoted itself not to the reproduction of natural phenomena, but to the expression of the artist's soul and to the creation of an autonomous life of musical sound".[70] So Kandinsky tried to develop a form of painting that owed nothing to nature, suggesting that "we must begin at once to break the bonds that bind us to nature and to devote ourselves purely to combinations of pure colour and independent form".[71] As Mark Taylor puts it, "No longer determined by anything other than itself, autonomy becomes actual in the splendid isolation of self-related works of art." Furthermore, since, in Kandinsky's view, painting is 'deceitful' if it tries to portray anything in the external, three-dimensional world, it is "insistently two-dimensional. To achieve flatness, every trace of illusion must be overcome".[72]

Such attempts to isolate human meaning from the world are closely allied with other denials of natural structure and process. For example, Stephen Budiansky suggests that because the distribution of flora does not fall into recognisable clusters, then the notion that nature is other than randomly organised is an illusion. Drawing on Robert Whittaker's work, which found that populations of species could not be grouped into clusters whose constituent populations reliably co-varied, Budiansky concludes that "species evolve so as to avoid competition, dispersing themselves across environmental gradients"[73] – in other words, plants behave like self-interested individuals. But the fact that species do not obligingly fit into cognitively pre-ordained clusters does not indicate that they are *randomly* ordered; and the influences that determinine distribution are far more complex than Budiansky acknowledges. Nature does not humour our conceptual categories by obligingly arranging itself like cans of vegetables on a supermarket shelf; and as we lose sight of the way our cognitive categories diverge from nature, so nature apparently becomes more amorphous.

Budiansky is right to suggest that part of the problem is "the human instinct to seek out patterns and form generalisations";[74] but the reaction against this, towards denial of all patterns, leading to a mistaken belief in universal randomness, is just as misplaced. The problem is not that *all* patterns are humanly invented, but that *some* are; and the task is to recognise *which* ones. Books such as Budiansky's serve an ideological purpose – to suggest that humans are the *only* source of order in the world, and that any other supposed order is simply a projection of our need to perceive order and meaning. In much the same way, Bruce Hull and David Robertson claim that ecosystems are "transitory assemblages of biotic and abiotic elements that exist (or could exist) contingent upon accidents of environmental history, evolutionary chance, human management, and the theoretical perspective one applies to define the boundaries".[75] Nature, supposedly, is a jumble of miscellaneous species – at least until the infusion of order and meaning from 'human management' and a 'theoretical perspective'.

This denial is often a two-stage process. Firstly, we 'demonstrate' that nature is structureless, merely a "a kind of substrate that exists prior to our action";[76] then – secondly – this justifies us in rearranging this substrate according to our human meanings. But the claim that nature is without structure depends on methodology and preconceptions that make it impossible for its structure to emerge. For example, some theorists have dismissed evidence of intelligent behaviour in non-human animals on the grounds that it is *in principle* unlikely that nonhuman creatures could recognise meanings comparable to our own.[77] Tim Ingold, too, has pointed out how equivalent behaviours in humans and in animals are understood, respectively, in terms of 'cognitive strategies' and 'instinct'.[78] As Donald Worster documents, the tendency among contemporary ecologists to see nature as the sum of random behaviours among individuals can be seen as "a triumph of reductive population dynamics over holistic consciousness, or as a triumph of social Darwinist or entrepreneurial ideology over a commitment to environmental preservation".[79] As Albert Borgmann puts it, "the tendency of mainstream thought is to reduce the component of givenness and sheer presence to randomness and meaninglessness"[80] – a tendency which is, of course, reproduced in material practice in the reduction of the world to 'raw material'. Like the conquistadors of the New World, many find the notion that the world already has a perfectly good structure an unwelcome constraint on their ambitions to restructure it. The implications of the findings – referred to in Chapter 2 – of the extensive non-linearity and complexity of natural systems have not been widely recognised, perhaps because they point against the anthropocentric grain; for if our own cognitive powers are tiny when scaled against such enormous complexity, does this not suggest our need for humility and caution in our dealings with nature – two qualities that are remarkably absent from the current industrialist spirit? While in the fourteenth century Nicole Oresme was prepared "to accept the complexity of the natural order over the simplicity of the definitional, and to assume that the actual was ordered, even if that order transcended human understanding,"[81] industrialist hubris has long since forgotten that truth.

How quantification subverted reality

At the beginning of the fourteenth century, the widespread acceptance of the Aristotelian framework slowed the changes necessary for the beginnings of scientific thinking. Since the symbolic order had not yet come to dominate sensory awareness, the mediaeval cosmos was populated by an enormous diversity of different entities, each with its own individual peculiarities and characteristics; and the conceptual frameworks linking them were few and incoherent. Although bestiaries, for example, tended to group creatures according to their family, that was as far as any parallel with contemporary zoology went; for they were replete with fanciful and religious speculations about each creature, and owed as much to imagination and to social ideology as to empirical observation. Different species were viewed as essentially incommensurable, and so could only be represented conceptually in separate categories. Quality and quantity, too, were viewed as quite different, quantities being divisible and so expressible in terms of mathematically accessible continua, while qualities were irreducible to conceptually manipulable components and so could not be understood as present in measurably varying degrees.[82] This meant that properties such as heat or colour were effectively placed beyond the scope of the nascent mathematical sciences.

But even while the reluctance to question Aristotelian understandings damped down any movement towards change, these same understandings also provided the ground out of which would appear the first shoots of what we would today recognise as scientific ideas. For example, Aristotle's notion of justice as related to proportion provided a linkage of two essential social themes, and established the relevance of mathematics to cultural affairs. Nascent ideas such as these were incubated within scholastic circles, safely protected from the unmanageable diversities of the natural world; and it became accepted in scholastic circles that the apparently chaotic variation and inter-relatedness of nature doomed any attempt to root a coherent and reliable account of natural processes in nature 'in the raw'. As John Murdoch relates, the only reliable grounding for science lay "not in contingent *things*, but in *propositions* as the only possible bearers of the requisite universality and necessity".[83] Consequently, the concern of thirteenth- and fourteenth-century natural philosophers was hardly ever with natural entities themselves, but rather with the development of concepts and methods for addressing issues arising out of philosophical debate. The conceptual realm therefore began to float free of control and validation by the real world; and initially neither world impinged much on the other. During this period, there was, as Joel Kaye remarks,

> a concerted effort to cleanse philosophical discourse from the taint of its contact with contingent experience. Insights drawn from the experience of nature were quickly denatured – translated into propositional and logical terms deemed to be the proper subjects of scholastic debate [and] many of the most important works in natural philosophy [in the fourteenth century] contain not a single reference to personal observations of nature.[84]

The emergence of a partially autonomous symbolic realm is also suggested by changing attitudes to the human body, which for the first time began to appear less as an indissoluble facet of the self and more as a separate object to be explored from a detached perspective. The first official dissections took place in Italian universities at the beginning of the fourteenth century,[85] suggesting the emergence of an 'intellectual' part of the self which could look down on and examine physical embodiment. As Le Breton relates,

> A miniature from the treatise of Guy de Chauliac (1363) admirably expresses this symbolic topography, completely articulated around the relation to the body. ... Standing slightly away from the table where the body lies, the *magister*, with Galen's book in hand, merely reads aloud from the hallowed text. In his other hand, from a distance, he points out the organs he mentions[86]

What we are witnessing here is an act of translation, a demonstration that the two realms – physical and symbolic – which emerge from a previously integrated whole can be separated, and selectively reconnected only through appropriate texts. And there is no question as to which part we identify with. On the one hand, the body is governed by taboos and prohibitions, and the act of dissection itself is relegated to those with minimal education and status. The *magister*, meanwhile, stands aloof from physical contact with the flesh, relating to it only through the text. In this nascent symbolic realm, as Robert Romanyshyn tells us, we can dispense with the body, ridding ourselves of "all those extraneous enticing odours and sounds, textures and tastes, temperatures and rhythms which compose the world". Tasting our new symbolic freedom from materiality, the body "whose eyes would find fulfillment of their vision in a movement toward the world ... would be an obstacle". Consequently, the body "must shrink from the bottom up towards the head",[87] re-emerging within the symbolic sphere in a virtual form unencumbered by clumsy materiality. From this period on, Le Breton tells us, "the retreat and then the abandoning of the theological view of nature led to considering the surrounding world as ... an ontologically empty form that can only now be shaped by the hand of man"[88] – 'man', that is, redefined as a *symbolic* being.

Such techniques of distancing and translation allowed the science of nature, later to be imposed on the natural order with such catastrophic effect, to be sheltered from conceptually inconvenient physical realities as it gestated within the minds of philosophers and proto-scientists. The debates between natural philosophers therefore involved the drive towards consistency *between* propositions rather than the accuracy of propositions in representing the empirical world; and the emerging sciences generated a form of knowledge that rather than articulating our sensory connections to the world, tried to bypass them. This emphasis on internal consistency, as we will see later, remains a cornerstone of industrialist thought across a range of disciplines, later colonising more contemporary economic philosophy through the influence of writers such as Adam Smith. For example, as Richard Olson remarks, the "great significance of Smith's doctrine is that since it measures the value of philosophical systems solely in relation to their satisfaction of the human craving for order, it sets up a human rather than an absolute or natural standard".[89] Reinforcing the bifurcation of symbolism from nature that we noted above, Murdoch records the beginnings of this fateful preference for coherence over accuracy, remarking that while

> empiricist methodology was dominant in the fourteenth century ... this did not mean that natural philosophy then proceeded by a dramatic increase in attention being paid to experience and observation (let alone anything like experiment) or was suddenly overwrought

with concern about testing or matching its results with nature; in a very important way *natural philosophy was not about nature*.[90]

If thirteenth- and fourteenth-century scholars turned away from nature, there was nevertheless one important way that the emerging philosophical and scientific ideas were influenced by the outside world – namely, by trading relationships and the idea of money, which drove thinking towards the idea of a common metric through which everything could be valued. Aristotle had argued that "the measure is always a certain minimum unit of the things measured, and therefore is always of the same species as the thing measured ... [so that] all distance is measured by the foot or some other unit of length".[91] Elsewhere, however, he had suggested that although money does not share any of the properties of the things we barter and trade, it can act as a medium of exchange, facilitating everyday commerce and enabling people to procure the necessities of life. He also noted a crucial problem that resulted from money's disembedding from the material aspects of exchange: while money may enhance the harmony and smooth functioning of social life and trading, as explained in the *Ethics*, it also includes a dynamic that can malignantly colonise and redefine life, so that as Kaye notes, "in the *Politics* money is treated as a boundless and unnatural element, destructive of social organisation".[92] Aristotle attempts to resolve this ambivalence by distinguishing between money's 'natural' role in procuring the necessities of life and its 'unnatural' power to generate more money; and it is this latter use, he argues, that becomes an end in itself into which all human faculties are enlisted.[93]

Fifteen centuries later, Aristotle's commentator Albertus Magnus attempted to reconcile these inconsistent views, arguing that we can compare two things either through some common property or by 'accident'. For example, Albert suggests that things can be compared through their 'accidental' quality of *usefulness* – which today an economist would refer to as *utility* – although money "shares neither material nor form with the diverse things it measures".[94] Although the notion that some unrelated property such as monetary value could be used to confer a stable notion of utility on otherwise unrelated entities sat uncomfortably with the acceptance by other thirteenth century philosophers that comparability is only possible among things that have some quality in common, the lived practicalities of trade pushed ahead of philosophy in this regard; and the use of money as an intermediary even between quite different entities drove thinkers such as Jacob of Naples and John of Mirecourt to follow Albert's lead, arguing that human *need* or 'indigentia' could act as the go-between in assessing value.[95] This fateful if apparently insignificant step transmuted a calculus of *natural* relations into a rational *symbolic* system within which the qualities possessed by things became merely incidental, of interest only to the extent that they had roles to play within the expanding universe of wealth, commodities, and commerce. Thus interest in the things themselves paled as a relatively self-sufficient world of abstractions grew in influence and technological power.

However, this world of originally *human* abstractions quickly assumed an air of 'objective' reality as its human origins were gradually lost sight of. Whereas for Aristotle the judge was a necessary arbiter to ensure fairness in exchanges, for later interpreters such as Odonis, fairness became a property of the process of bargaining and rational estimation itself,[96] foreshadowing the contemporary faith in the benign properties of the 'market'. As early as the fourteenth century "the subjective orderer as a guarantor of equality in the marketplace was replaced by the concept of a dynamic, self-equalising system constructed around the 'instrumentum equivalens' of money".[97] As Kaye continues, "it is clear that the direction in fourteenth century theory was toward the elimination of the judge in the normal course of exchange and his replacement by the mechanism of market agreement".[98] Just as natural philosophy was only tenuously related to the natural world, so the 'market', which was ostensibly about human transactions, was also becoming an autonomous system, loosening its anchorage to human needs and human control.

Thus liberated into an increasingly self-sufficient conceptual realm, some philosophers – most notably the 'Oxford Calculators' – ran riot in their zeal to quantify almost any quality within reach, including not only those we would recognise today, such as heat, weight, or area, but also more theologically-oriented notions such as love, grace, or charity.[99] As the new abstract knowledge built up momentum, burgeoning commerce catalysed the rapid development of notions of quantity and monetary value. This created a good deal of 'cognitive dissonance', given the long-established embeddedness of everyday life in the natural world. As Kaye describes it,

> The distinction between natural order and market order created great tension within an intellectual culture whose habit was to unify and synthesise. The tension grew as the power and position of the market in society grew, until, by the late thirteenth century ... it was the conception of the *natural* order that began to give way. ... Where previously, for example, forests had been primarily valued by lords

for the status and pleasure they afforded, they came in the thirteenth century to be seen as income opportunities, as resources to be rationally exploited toward the end of profit.[100]

These traditional misgivings notwithstanding, however, "money was becoming the measure of all things" by the beginning of the fourteenth century.[101] Even the religious sphere was not immune to this mania for monetarisation: the cost of indulgences paid to Pope Clement V, for example, was fixed at one penny of Tours for each year of pardon conferred.[102] Thus the concept of monetary value, spreading into almost every area of life, and emerging from the system of trade between people rather than growing out of any *ecological* relation, marks the growing divergence between the two great systems that today order human life – a divergence that was, unsurprisingly, internalised within individuals. As Kaye remarks, money "was seen, often by the same person, as both a remarkably successful instrument of economic order, balance, and gradation, and at the same time as the great corrosive solvent, the overturner, the perverter of balance and order".[103] But – to echo one of Kaye's themes – the influence of money was seldom explicitly acknowledged, so that natural philosophy superficially appeared to be independent of mere worldly goings-on. Contrary to this belief, money in fact seems to have served – as it does today – as both a measure of value within a common metric and as directly exchangeable for material goods, in addition to its more covert function in colonising non-economic realms. While a quality such as aesthetic value existed more or less unequivocally in the symbolic realm and had no direct commercial implications, money was the consummate bridge between the developing symbolic system and the material world, opening the latter to invasion by an entirely alien system of valuation, exchange, and conceptualisation.

Not surprisingly, this led to much soul-searching about the conflict between the natural and Divine orders, on the one hand, and the growing economic realm, on the other; for how, in Kaye's words,

> could something so central to the good of the community be unnatural? Godfrey of Fontaines was one of a number of thinkers of his time who resolved this tension by recognising market exchange as a system governed by its own proper logic and possessing its own proper equilibrium – but a logic and equilibrium that were in themselves rational and natural. In doing so they expanded and reshaped the image of the natural order to comprehend the logic of the marketplace.[104]

Just how far this reshaping of the natural order would go was, of course, completely unsuspected.

Economic delusions

As we have seen, natural philosophy and proto-economic concepts such as 'usefulness' and 'need' relied on their detachment from the natural world in order to maintain their orderliness and coherence; and this detachment has been continued and expanded into the present era, based on our prioritisation of the conceptual realm over lived and sensed experience. Several centuries after thirteenth- and fourteenth-century philosophers had laid the cornerstones for an autonomous conceptual system, Descartes formalised the abandonment of sensation as a route to knowledge, laying the basis for a quite different order. In doing so, he sowed the seeds of a certain type of delusion which today is so much a part of our everyday lives as to pass almost unnoticed. As Reiss puts it, the effect "is to impose an intellectual structure upon the perceived world"; and this imposition solidifies into an assumed reality which will eventually take precedence over the natural order itself. Thus, as Reiss adds, one "would be tempted to argue that in later Cartesianism, if not in Descartes himself, a 'fictional' system becomes the real".[105] The sort of exact, precisely determined world envisaged by Descartes, and later by Laplace, is realistic only in a realm that has been purified not only of sensory connections, but also of feedback, randomness, non-linearity, and emergent properties; and despite industrial society's scientific anchorages to the real world, its coherence depends on its location within a safe haven of idealist regularity and predictability, distanced from the turbulence of the real, messy, emotional, unpredictable realm of nature. While science hand-picks those qualities of the natural world and human experience that do not upset this idealist pattern, the normal chaos of lived experience continues relatively untouched by scientific understandings. To the extent that scientific acceptability takes over from sensory experience as the defining criterion of reality, we inhabit two separate and only tenuously related 'worlds'; so that as Robert Romanyshyn puts it, "what we experience is not real and what is real is not what we experience.[106] In an age in which unbridled technological exuberance is closely followed by its shadow in the form of ungrounded religious fundamentalism, one wonders whether the human propensities underlying these extremes might be more constructively integrated within a single, more complete understanding.

But if the physical sciences relate to reality only in selective, albeit powerful, respects, economics has stretched the thread that binds it to reality beyond breaking point. Pioneering economists such as Walras, Jevons, and Pareto demonstrated a noticeable degree of 'physics envy' in their exceptionally crude attempts to emulate the emerging successes of the physical sciences, often translating physical and chemical concepts into economic terms with little concern that human social life might differ radically from the 'life' of atoms and molecules. These economists skated lightly over the differences between organic and inorganic phenomena, denying all those crucial emergent properties that distinguish a person from a corpse or a machine.[107] The reduction, then, is from the realm of life to the realm of mechanism – and, by implication, the tailoring of life to fit the symbolically expressible predictability of mechanical logic. As Robert Nadeau notes, the "strategy used by the economists was as simple as it was absurd – they took the equations from mid-nineteenth century physical theory and changed the names of the variables".[108] Robert Mirowski has documented the substitutions made in the highly respected work of Irving Fisher, for example: the individual replaced the particle, marginal utility or disutility replaced force, and total expenditure replaced kinetic energy.[109] Nineteenth century economics was riddled with such literal translations of physics into economics. As Jevons explained:

> Life seems to be nothing but a special form of energy which is manifested in heat and electricity and mechanical force. The time will come, it almost seems, when the tender mechanism of the brain will be traced out, and every thought reduced to the expenditure of a determinate weight of nitrogen and phosphorous [sic]. ... Must not the same inexorable reign of law which is apparent in the motions of brute matter be extended to the human heart?[110]

Within this enormously reductive scheme, the individual has been plucked of most recognisably human qualities. The broader cultural trend this reduction exemplifies is illustrated in its purest form by Condorcet's attempt to mathematise human behaviour, supposedly giving the human sciences the same levels of certainty as the physical sciences. Within Condorcet's calculus of voting behaviour, *homo suffragans* was an idealised unit, a "social atom that has been divested of all human qualities other than the human faculty of voting".[111] This drastic reduction of human being to fit monetary 'laws', themselves imported from physics, rapidly infected economic theory: Pareto, for example, criticised the "incoherent ramblings on solidarity [which] show that men have not freed themselves from those daydreams which have been gotten rid of in the physical sciences, but which still burden the social sciences".[112] Such views are still widespread. As a leading current textbook puts it, "economists ... begin with the assumption that people are self-interested utility maximisers – they are selfish and rational".[113]

In this vein, much economic thought since the nineteenth century promiscuously confuses an ideal realm that is pruned of mathematically inconvenient qualities and a realm of 'real life' that is alluded to only sparingly and selectively. This is not simply because the physical metaphors that were dragged into theory were hopelessly misplaced in the realm of organic life, but because economics' shaky ability to predict market behaviour necessitated that its predictions be of a rather nonspecific character. As Nadeau notes, for example, neoclassical economists "were obliged to view market processes as existing in an immaterial domain separate and discrete from material reality to make the case that the sum of income and utility is conserved in equations borrowed from mid-nineteenth century physics".[114] Similarly, Goux remarks on the "scission between two worlds – the material world of commodities and the ideal world of value and money", which has its roots in Plato's separation "between the tangible world and the intelligible world".[115] According to Goux, it is in the process of exchange, and the valuing process that underlies it, "that we must seek the root and the forms of the opposition between matter and consciousness ... and thus of the philosophical opposition between materialism and idealism".[116] The current market system instantiates many of the same idealist assumptions quite explicitly; and those that Nadeau notes include the belief that market systems exist in a domain that is separate and distinct from the environment; that the natural laws of economics will ensure that closed market systems will perpetually grow and expand; that environmental problems result from market failures or incomplete markets; that the external resources of nature are largely inexhaustible; that the environment is a bottomless sink for waste materials and pollutants; and that the damaging effects of industrial activity must be treated as 'externalities' – that is, they lie outside the closed market system.[117] If we disparage indigenous beliefs in comparison to scientific ones – say, for example, the Rock Cree belief that an animal killed is immediately reincarnated, so that more intense hunting has no effect on populations[118] – then one has to question whether the foundational beliefs underpinning the global economic system are any more defensible.

The industrialist fantasy and the material world it has spawned can be understood as dialectically constituting a system in which each plays a causative role within the accelerating spiral of industrial growth; and Goux suggests that affluence and idealism may be related in this way. The world in which the affluent can afford to live is the world of abstractions – the "economic, political, legal, and intersubjective processes" – rather than the realm of material realities; and he notes that "idealism, which affirms the primacy of the idea, consciousness, mind and the subordination of matter, will become the philosophical ideology of the dominant classes".[119] In this ideal realm, "the concept retains only what is common to diverse representations, effacing the differences among singular images. The concept … like the essence of the thing, becomes a metaphysical, detached quality. … Both currency and the conceptual terms perform an operation of abstraction, of universalisation, of reduction to what is held in common".[120]

This reduction of diverse qualities operates both conceptually and materially. Just as the relations among abstract physical variables are the subject matter of physics, so the interaction of components, built up from abstractions and elements, is the basis of the symbolic realm's reconstitution of the material world. The manufacture of a gear wheel, for example, requires the reductive purification of iron ore to produce pure iron, followed by the addition of other purified elements such as nickel and chromium; and the wheel is then forged and machined to embody ideal shapes such as the circle and the ellipse. And the complete gearbox will take its place within the material realities of the vehicle and the transport system, themselves ordered by economics. There are thus two ostensibly 'opposite' but in practice complementary movements that are fundamental to industrialism: the movement to abstract particular qualities from the flux of life, and the movement to materialise these abstractions and to recombine them within a new system that differs fundamentally from the ecological realm.

To view the divergence of economic theory from lived realities simply as a case of over-exuberant metaphorising is to misunderstand its significance, however. Many disciplines, awed by the success of the 'hard' sciences and their use of mathematics, have been drawn in that direction by an almost 'gravitational' pull, other examples being Freud's use of lightly concealed biological theories and the French postmodernists' bizarre incorporation of mathematics, both of which we discussed earlier. But we also noted above that some of the most basic scientific concepts originated from the demands of trade, so that multi-level interchanges between economics and science defined the emergence of a new 'reality'. As Mirowski remarks:

> I began to suspect that the fundamental issue was not simply the wholesale piracy of some physics by a doughty band of economists, but rather something akin to what Borges called 'universal history'. Perhaps what I had been doing was excavating a primal metaphor of human thought, a vein winding through both physical theory and social theory, changing from gangue to fool's gold over time, with chutes passing back and forth between physics and economics. Although it was ultimately called 'energy' in physics and 'utility' in economics, it was fundamentally the same metaphor, performing many of the same explanatory functions in the respective contexts, evoking many of the same images and emotional responses.[121]

Perhaps what Mirowski is driving at is something akin to the self-organising dynamic of complex systems, or the 'autopoiesis' explored by Maturana and Varela,[122] whereby a system develops an integrated direction or 'telos' around which the functioning of its subsidiary parts comes to be organised. Contrary to the conventional scientific emphasis on 'bottom up' determination, this would suggest the heretical possibility of 'top down' influences within systems. In keeping with the notion of a 'primal metaphor', the pushing aside of ecological, embodied, and material realities occurred in a range of spheres in addition to economics. As Goux notes, shortly "after Saussure had declared that linguistic values ... had no foundation in nature, shortly after Wassily Kandinsky and Piet Mondrian had abandoned the search for direct empirical reference in order to espouse pure painting, the economic system dispensed with the gold standard".[123] It is as if 'Western' society as a whole had lurched decisively towards the ideal, abandoning its roots in the ecological matrix from which it had emerged – a lurch largely based on the intellectual intoxication with pattern and relation. Thus Berkeley, echoing the Galilean approach which I referred to above, had two centuries earlier presciently enquired "whether the discriminations being retained, although the bullion were gone, things might not nevertheless be rated, bought and sold, industry promoted, and a circulation of commerce maintained".[124] This emphasis on the reality of mathematically-expressible *relation* rather than on the material realities they originally arose from allowed theorists from various disciplines to cast off from the real world inhabited by other creatures and to set sail into the realm of ideality and its technological materialisations.

Nevertheless, even while industrial life is ever more firmly embedded within an idealist realm, our embodiment still calls us back to ecological reality. James Buchan notes that in "the great speculations of the seventeenth and eighteenth centuries, tracts of the natural world are condensed into arbitrary money-values and traded feverishly for a while. Yet always some residual sense of quality – the fragility of a tulip or the dampness of a Louisiana swamp – at last breaks through the mere quantity and the bubble collapses".[125] Recent history – and especially the financial crash of 2008 – suggests that as our deviations from reality become more pronounced, so do the corrections that follow. Further from the centres of economic power, the clash between economic assumptions and embodied realities are an unavoidable part of day-to-day reality. Senegalese fishermen whose livelihoods have been decimated through over-fishing by EU trawlers; Inuit mothers who dare not breast feed their infants because of contamination by industrial chemicals; and forest peoples whose habitats have been destroyed by the ranching or palm-oil industries – such people do not have the luxury of retreating into a manufactured environment which exports its waste and labour to those less fortunate. Thus Nadeau notes that although 'externalities' are invisible to economists who point to the growth trends in GNP, when we include factors such as resource depletion, soil erosion, and forest loss in calculations of wealth, the "savings in most of the developing world have been negative since the mid-1970s".[126] Economists, then, live in a different world to that of most working people, so that the split between ideality and materiality is not merely a philosophical abstraction, but has become materialised geopolitically in the different life-experiences of the affluent and those whose main problem is staying alive. The extremity of this difference may be judged from the fact that according to Kofi Annan, to provide everybody in the world with adequate clean water, adequate food, safe sewerage, basic health care, basic education, and reproductive health care for women would cost less than 4 per cent of the combined wealth of the 225 richest people in the world.[127] The magnitude of these differences in wealth can be understood as due to the divergence between the systems the rich and the poor inhabit: that is, those who can afford to insulate themselves from the crippled natural sphere inhabit an environment of manufactured luxuries, while those who live on or outside the periphery of industrialism are forced to pay the costs of these luxuries. As the plunder of nature continues and the natural world becomes increasingly obliterated, we can expect two main trends: firstly, the gulf between richest and poorest will further increase; and secondly, the size of the group who can insulate themselves from environmental degradation will shrink as the effects become more severe so that, like a drought-stricken plant, the human population will be healthy only at its extreme tip.

How thought simplifies the world

Reductive simplifications such as those which underpin economics have now become so much an integral part of the industrialised world that we no longer regard them as simplifications. The symbolic world of predictability and coherence introduced by fourteenth-century natural philosophers has now colonised industrial 'life', becoming a material world that provides the affluent with an apparently safe haven from the vagaries of nature. Technology gives us tools which act as prosthetic devices, allowing us to manipulate the world at a safe distance. The spade and the saw extend our hands, the bicycle and the car extend our legs, and the library extends our memories. The house, as Drew Leder suggests, enlarges the limits of our bodies: its "walls form a second protective skin, windows acting as artificial senses, entire rooms, like the bedroom or the kitchen, devoted to a single bodily function", producing a "corporeal effacement" due to the diffusion outwards of our bodily limits. "As I gaze through the windows they are in focal disappearance, the means from which I look upon the world."[128] And on a larger scale, industrial society as a whole forms a protective shell that reaches beyond our perceptual and cognitive powers, so that we are oblivious to the need for consistency with the greater world outside. To change the metaphor: in a rowing boat, we are acutely aware of weather conditions; but in the *Titanic*, we continue dancing in the ballroom even while the iceberg looms. As technology enables us to expand outwards into the world, the world we build becomes part of us, taken for granted as we use, manipulate, and think with it. As we expand *conceptually* into the built environment, so we lose our *felt* awareness of the natural order, splitting thought from feeling.

As a glance outside the window generally makes clear, this process involves a drastic simplification. There are the straight lines of planks and contrails; the rectangles of houses and rooms; the circles of wheels and tyres; the elegant ellipses of power lines draped over pylons; the monocultures of wheat and barley; and the consistent textures and colours of concrete and tarmac. Artificial boundaries such as fences clearly separate a housing estate from farmland, my house from yours. Trains run at the same time each morning; and baked beans always come in the same-sized can. The world becomes a reflection of the conceptually colonised mind, estranged from the residual diversities of the natural world – the irregular curve of a hillside or river; the complexities of the branches of a tree; the nuanced seasonal changes of a natural landscape; the wheeling, unpredictable flight of a magpie; the endless varieties of cloud formations. Such conceptually indigestible aspects of the world, as we saw above, are simply dismissed as disordered. The sophisticated biochemistry involved in the growth of a blade of grass tests the understanding even of scientific specialists; yet we still claim that human intelligence is the pinnacle of nature's powers. As Samuel Butler put it, "[n]othing, we say to ourselves, can have intelligence unless we understand all about it – as though intelligence in all except ourselves meant the power of being understood rather than of understanding".[129]

Increasingly, the overriding simplification is a reduction to the balance sheet. In James Buchan's words,

> The land itself begins to change. The sloping meadow and the mill are no longer primarily sources of produce – of corn, hay, rye, barley, meat, wool, plums, walnuts, and flour – but of money. Forests are no longer pre-eminently places for hunting ... but the site of a money crop called timber. The place is no longer regarded as unique, for how can it be so, if it submits to calculation and comparison in the market of wishes?[130]

Qualities such as beauty are brought 'down to earth', becoming first 'attractiveness', and then 'sales potential'. Buchan remarks that "the sensation of beauty cannot survive in the age of money: for any beauty must be exploited, reproduced a million times over by every medium open to commercial ingenuity, till one can only cover one's eyes and stop one's ears".[131] Like the bluebell, beauty has been commodified, extending us into a commercial rather than a natural world.

What began as a minor symbolic extension of the world, then, expands into and finally replaces the order of the world, redefining humanity and sweeping aside natural structure as it does so. To the extent that we are colonised by this system, our behaviour will be consistent with it, so that even our well-intentioned attempts to restore wild landscapes often unwittingly end up simplifying them. For example, Franz Vera has shown that the old European terms for 'forest' – Latin 'forestis', French 'forêts', German 'wald', Dutch 'woud' – actually referred to a mix of woods, bogs, and grassland rather than to the sort of closed-canopy woodland the term brings to mind today.[132] In addition, the natural dynamics of 'woodland' may involve an alternation between woodland and savanna.[133] Such uncertainties are conceptually indigestible; and applying our relatively simple conceptual categories to complex ecologies simplifies and destroys them.

Systemic understanding does not come easily to the human mind. Given our inability to focus on all variables at once, there is not surprisingly a good deal of instability in our understandings of nature, as intellectual fashions flit around between alternative viewpoints, each of which is incomplete in a different way. At the time I am writing this, the emphasis on acidification of two decades ago has given way to a preoccupation with climate change, backgrounding other factors such as loss of biodiversity; and the focus on one or another factor lets slip the notion that these are merely the most visible signs of a deeper upheaval.[134] Today, the notion that wilderness is necessarily unpeopled has been largely replaced by the equally problematic 'realisation' that wilderness 'has always been inhabited.' Similarly, the notion of ecosystems as tending towards a harmoniously ordered 'climax state' has fallen out of fashion, leading to the even more unlikely notion that the natural world is simply a chaotic melee of stochastic processes – perhaps conveniently reflecting industrialism's need for a constant supply of structureless 'raw materials'. Likewise, the romantic idea of the 'noble savage' has given way to the dubious gloss that all peoples are equally environmentally destructive – so that, as Alf Hornborg sardonically puts it, "there emerges the new but implicit message that we have always been capitalists".[135] There is a curious reversal here: while the early colonists, battling through relatively wild nature, fantasised about productive, domesticated farmlands, today we live amidst a tamed nature, and wilderness is now, supposedly, a 'romantic fantasy' which, we are told, may 'never have existed'. Our seduction by such binary oppositions reflects a sort of cognitive laziness which compounds our necessarily limited ability to represent complex natural forms; and the examples given above illustrate cognition's predilection for plausible, clearly defined, and deeply misleading portrayals of complex situations.

A compounding factor in this tendency to mistake elegant cognitive models for an original reality is our difficulty in recognising very long-term changes – a difficulty pointed to by Peter Kahn's concept of 'environmental generational amnesia',[136] which I referred to in Chapter 2. This is closely related to to what Vera describes as the "shifting baseline syndrome" – a concept originally formulated by Daniel Pauly in relation to fish populations:

> [T]his syndrome has arisen because each generation of fisheries scientists accepts as a baseline the stock size and species composition

that occurred at the beginning of their careers, and uses this to evaluate changes. When the next generation starts its career, the stocks have further declined, but it is the stocks at that time that serve as a new baseline. The result is a gradual accommodation of the creeping disappearance of resource species, and inappropriate reference points for evaluating economic losses, resulting from over fishing, or for identifying targets for rehabilitation measures.[137]

As applied to landscapes, the shifting baseline syndrome involves "drift away from true natural conditions";[138] and because this occurs gradually over many generations, it is unnoticed. Communities therefore lose their memories of what constitutes a natural environment, and restorationists are left without a clear sense of what it is they are trying to restore. Of course, the notion of 'true natural' conditions is somewhat problematic; but while we cannot accurately specify how a natural system will vary over time, to assume that its current state is necessarily within the range of natural variation is likely to be unrealistic in the case of industrially impacted ecosystems. So it is that our horizons draw in to recognise the reduced world we mostly inhabit; and current, partial actualities are accepted, simply, as 'reality'.

Lost in translation: The replacement of embodiment by 'intelligence'

The intelligence of our ancestors crystallised out of the much more widespread biological intelligence that is inherent in life. Sounds, leaf movements, pawprints, and gradients of temperature and humidity are recognised and intelligently employed by diverse creatures and plants in order to survive; and this 'background' intelligence includes our own embodied awareness. In Frank Wilson's words, the "brain does not live inside the head, even though it is its formal habitat. It reaches out to the body, and with the body it reaches out to the world,"[139] recognising, as Edwin Hutchins puts it, that "symbols are in the world first, and only later in the head".[140] This symbolic continuity allowed our early ancestors to survive, creating a form of subjectivity that extended into the world and formed the basis of the embodied empathy that allowed them to predict the behaviour of other creatures; and it is still the form of intelligence used by those few indigenous groups that have escaped assimilation into industrialism.

A crucial milepost in the tearing apart of embodied intelligence and conscious rationality was the emergence of farming which – especially in its industrialised forms – involves the imposition of templates that are isolated from the characteristics of any particular landscape. Hugh Brody, who has lived among the Inuit, writes that farmers

> carry with them systems of control as well as crucial seeds and live-stock [and make] use of analytical categories that are independent of any particular geography ... the achievement of abstraction and the project of control are related. In addition, the deduction of conclusions from analytical, abstract theory is dependent on a precise rational process. This process depends, of course, on reason itself; and a foundation of reason is the law of the excluded middle. This is the seemingly straightforward proposition that nothing can be both true and false, or both itself and not itself, at the same time ...
>
> Hunter-gatherers, on the other hand, rely on a relative absence of exact or abstract categories that transcend geography and specific facts. Their knowledge is compounded of many specifics. At the same time, they believe that the boundaries around these facts, around the real, are unstable; one kind of thing can become another[141]

Indigenous societies are not built on the subjugation of embodied intelligence by rational precision; and consequently the latter plays a humbler role. For example, Jonathan Long and his colleagues, discussing the approach to ecological restoration taken in the White Mountain Apache reservation, observe that "because nature directs the recovery, practitioners may not plan a full course of treatment until observing how a site responds." In the words of one tribal member, "You go to a place and do some work for it. You let it rest, and then you come back to it to see what it has done. Then it thanks you."[142]

Likewise, Scott Atran, studying the relative merits of different approaches to forest management in Guatemala, found that although one tribal group, the Itzaj, care for the forest more effectively than other groups in the area, they have little in the way of formal rules or conscious principles governing conservation, instead relying on an "emergent knowledge structure". As Atran explains, an

> emergent knowledge structure is not a set body of knowledge or tradition that is taught or learned as shared content. ... The general idea is that one's cultural upbringing primes one to pay attention to certain observable relationships ... [For Itzaj,] there is no 'principle of reciprocity' applied to forest entities, no 'rules for appropriate conduct'

in the forest, and no 'controlled experimental determinations' of the fitness of ecological relationships. Yet reciprocity is all pervasive and fitness enduring.[143]

Note: "one's cultural upbringing primes one to *pay attention to certain observable relationships*." It is not a matter of imposing a *better* model onto the world, but rather of not imposing *any* model onto the world. Like the young Tim Ingold, Itzaj are taught to find order *in the world*, not in conceptual models; so knowledge remains *part of* embodied materiality, disappearing into the ecology like a camouflaged animal. As John Livingston puts it:

> It is the wholeness of the wild animal that makes ethical constructs unnecessary – indeed, probably unthinkable. Why create an abstract set of rules and guidelines when you are already doing all the right social things, and always have? ... Rules and guidelines are for domesticates.[144]

The 'wholeness' to which Livingstone refers encompasses both organism and habitat. Consequently, one need not conclude that there is an 'innate cognitive module' in order to account for the substantial overlap between various folk-biological taxonomies;[145] rather, this overlap can be understood as deriving from a shared sensory openness to the properties that the *world itself* reliably embodies. In contrast, the attempt to *impose* consciously derived principles, or government policies, or educational initiatives, or legal rights and duties is often unsuccessful; and as John Rodman remarks, "it is worth asking whether the ceaseless struggle to extend morality and legality may now be more a part of our problem than its solution."[146] Having lost the embodied continuity with nature that would directly anchor our behaviour in an implied ecological morality, we try to compensate for that absence by inventing religions, codes of morals, and legal systems to tell us *indirectly* how we should behave; and in so doing, we further enmesh ourselves within disembodied symbolic realms.

Tribal subjectivities, then, tend to be more continuous with the systems they inhabit, in effect sacrificing technological power in the service of integration between the cultural, psychological, and ecological spheres. While such subjectivities are often disparaged as 'folk psychology', they frequently contain a good deal of evolved wisdom. As David Geary puts it, "much of human behaviour, like that of other species ... can be understood in terms of evolved brain and cognitive mechanisms that operate automatically and implicitly and enable fast and generally accurate decision making and behavioural responding to the environment".[147] Such adaptations usually work well within the same type of context that they evolved within, but fail in alien environments. When an Inuit woman says of her removal from her land to a settlement, "That was the way they made our minds weak",[148] she was not speaking 'metaphorically': the embeddedness of experience in landscape is part of *reality*, so that the landscape is also a *mindscape*.

Such views seem odd to us urban dwellers, because we have *already* lost the embeddedness of thought in the world, so that thought and reality now appear – indeed *are* – distinct realms. This makes our embodied reactions indecipherable by consciousness, furthering our forced dependence on conscious reasoning. In Jacob von Uexkull's words, for example, mechanistic biology reduces the felt meaning of a meadow to "a confusion of light waves and vibrations in the air, finely dispersed clouds of chemical substances and chain reactions which control the various objects on the meadow". But for von Uexkull, "in nature the meaning-factors are related contrapuntally to the meaning-utilisers in its life", so that the flower is bee-like and the bee flower-like.[149] As we rebuild the world, so we sweep aside such interconnections and complementarities which have evolved over many millions of years, producing a world whose ecosystemic nature has been degraded to the level where it becomes more consistent with the fragmented taxonomies of consciousness. As Juhani Pallasmaa puts it, as "we construct our self-made world, we construct projections and metaphors of our own mindscapes. ... A landscape wounded by acts of man, the fragmentation of the cityscape, as well as insensitive buildings, are external and materialised evidence of an alienation and shattering of the human inner space".[150]

"An intimacy that secures complete understanding"[151]

Despite the dominance of the view that intelligence is an *individual* quality, some researchers have begun to move towards a more ecologically valid approach. Jean Lave, for example, has challenged the study of cognition as an isolable process by extending research into environments such as supermarkets, concluding that "'cognition' is seamlessly distributed across persons, activity, and setting ... this in turn implies that thought ... is situated in socially and culturally structured time and space".[152] The supermarket, too, is "a physically, economically, politically, and socially organised 'space-in-time'" which is "the product of patterns of capital formation and political economy. It is outside of, yet encompasses the individual".[153] And if

> the context of activity ... is included in the analysis of activity, then questions about *its* context are also relevant. [Consequently,] it is difficult to understand the context of arithmetic practices in the supermarket without considering the constitutive order which stages both the experienced dilemmas of the shopper and the supermarket as an arena.[154]

Thus thought and the contexts within which it occurs are interrelated through their *systemic* connections within the uber-context that is the economic system – just as embodied meanings are 'related contrapuntally' to *their*, natural, contexts. Science, too, is part of this uber-context, since it is concerned only with "that portion of the complex everyday world that we think we *can* know".[155] There is therefore a circularity implicit in studying 'normal' styles of thought which occur in an industrialised world which is itself largely a materialisation of the way we think – a circularity that increases our entrapment within industrialism and conceals other possible realms and other ways of thinking. This becomes obvious when we try to apply our models to the natural world, since – to use just one characteristic by way of illustration – the sort of linear thinking that often works adequately within an industrialised context works less well in situations involving nature. For example, attempts to control pests may lead to greater problems in the future, as noted by Holling and Sanderson:

> Successful short-term management (e.g. suppression of spruce budworm populations in Eastern Canada) left the resource (the forest) and hence the economy, more vulnerable to a system-threatening breakdown (more intensive outbreaks of pests persisting over larger areas). This growing vulnerability led to ever-increasing intensity of management, leading to higher vulnerability, and so forth.[156]

Those who live intimately with the physical realities of the natural world cannot afford to rely on types of thought that are inconsistent with the character of the landscapes they inhabit. While commercial farmers and the inhabitants of the industrialised world are able to think in terms of abstractions and cognitive categories because they have some power to impose these on the earth through technological devices such as fences, ploughs, chainsaws, and herbicides, the hunter-gatherer's survival depends on intimately *knowing* the world rather than *changing* it to accord with cognitive preconceptions. In effect, the pressing realities of the world trump the need for cognitive clarity and rationality. In Brody's terms,

> Hunter-gatherer knowledge is dependent on the most intimate possible connection with the world and the creatures that live in it. ... A fluidity of boundaries, a porousness of divisions, can be seen as useful and normal.[157]

This suggests a different form of reasoning to that which exists in the industrialised world, one that is "inductive and intuitive ... Reasoning is subliminal, and therefore has the potential to be more sophisticated, more a matter of assigning weight to factors, than can be the case with linear logic."[158]

Tim Ingold has a similar understanding of what happens between the Cree hunter and the caribou, referring to

> the feeling of the caribou's proximity as another living, sentient being. At that crucial moment of eye-to-eye contact, the hunter *felt* the overwhelming presence of the animal; he felt as if his own being were somehow bound up or intermingled with that of the animal.[159]

We are not simply talking here about differences in cognitive preferences, but rather of a more radical contextualisation of cognition; for the word thought *already* implies a distance from the world, the body, and behaviour. On the one hand, we have the narrowed, selective focus of science, based on a constant state of alert, cognitive discipline. On the other, we have a sensorily-based openness to the world as it is: rather than withdrawing into a realm of manufactured thought, language, and sociality, the hunter-gatherer engages with the world with all the faculties that have evolved for exactly this task, including those we push aside in our drive toward single-minded rationality. Among such peoples, Tim Ingold suggests, knowledge of the world is gained not by developing abstract models of the world but "by moving about in it, exploring it, attending to it, ever alert to the signs by which it is revealed".[160]

For example Ingold, whose argument is in this respect similar to my own, criticises Richard Nelson's understanding of Koyukon ontology. According to Nelson,

> Reality is not the world as it is perceived directly by the senses; reality is the world as it is perceived by the *mind* through the medium of

the senses. Thus reality in nature is not just what we see, but what we have *learned* to see.[161]

According to this, learning becomes a veil of interpretation that we place between ourselves and the world, filtering and selecting aspects of it, rather than a means of embracing the world more fully and openly. But while this may accurately portray thinking as it typically occurs in an industrialist context, it is a scheme which "flies in the face of what the Koyukon themselves, by Nelson's own account, are trying to tell us".[162] Development is not, as Ingold points out, about "acquiring schemata for mentally *constructing* the environment", but about "acquiring the skills for direct perceptual *engagement*" with the world.[163] Here not merely ways of thinking, but the entire frame through which we understand them, vary across cultures and landscapes. In other words, the development of a culture within a particular landscape – unsurprisingly – seems to harmonise thought and embodied experience with that landscape. But no – that description is still coloured by an industrialist perspective. A better way of putting it might be to say that thought and embodied experience never diverged in the first place, and so need no 'harmonising'. In Ingold's terms,

> In the hunter-gatherer economy of knowledge ... it is as entire persons, not as disembodied minds, that human beings engage with one another and, moreover, with non-human beings as well. They do so as beings in a world, not as minds which, excluded from a given reality, find themselves ... having to make sense of it.[164]

As Jim Cheney observes, "the notion of a living world is not part of an Indian *world view*, it is an everyday observation".[165] The mind, in other words, is an extension of the world, connected to it by feeling and sensory experience as well as by thought. Ingold gives many examples of peoples who live within what he refers to as a "sentient ecology", involving "a knowledge not of a formal, authorised kind ... [but one] based in feeling, consisting in the skills, sensitivities, and orientations that have developed through long experience of conducting one's life in a particular environment. ... Another word for this kind of sensitivity and responsiveness is *intuition*."[166]

Of course, the ordering/controlling mind becomes essential if we intend to technologically enslave the world to human needs and desires. Whereas a natural environment complements mind's evolved capacities, so that the fundaments of knowledge are already *in the world*, a manufactured environment presupposes a form of knowledge alien to the landscape on which one's design is imposed – otherwise, the design wouldn't need to be imposed in the first place. But even as we attempt to impose control by the human mind, control slips from our grasp as the system we have created develops its own emergent properties, many of which elude our conscious awareness.

What consciousness omits

Although conscious, rational thought patterns are prioritised in industrial society, the lived realities of our lives are still rooted in back-grounded, often unconscious awarenesses. We assume our ability to walk, organise our speech according to grammatical rules, interpret sensory information, to digest food, and so on, in rather the same way as we assume nature's ability to provide clean water and air or a forest's ability to regenerate. Conversely, 'rational' decision-making generally takes place within an assumed environment that is pre-structured to be consistent with rationality: the service station, the mall, the transport system have all been developed to embody and to complement very specific forms of decision-making. These industrialised environments are ordered by previous policy decisions that generally leave us with a few well-defined choices – to buy this or that, to take the train or the car, to watch television or play a computer game. Just as the ecological background to industrial life is physically distanced from us, so the modes of thought and feeling that are consistent with that ecology are pushed aside as styles of thinking consistent with the metaphors of causality, binary choices, and efficiency come to dominate.

Conscious thought, then, is foregrounded and conscious partly because it is consistent with and reinforced by the equally foregrounded social and physical environment of the industrialised world. In much the same way that water, having over many years carved a gully in the rock, is then channelled by that same gully, instrumental thought has constructed an environment that then constrains thought to be consistent with it. Consistency with this constructed environment then becomes positively valued and associated with concepts such as rationality, while the growing inconsistency with the natural order is dealt with by viewing this order as random and structureless, and the thought that is consistent with it as 'primitive' or 'irrational'. Given this, it is hardly surprising that we seem to have the utmost difficulty in envisioning ways of thinking and behaving that step outside indus-trialist 'necessity' and recontextualise us within the natural order. Research itself is invariably anchored with this same system of thought and material reality, so that even the attempts of pioneering researchers such as Jean Lave and Edwin Hutchins to break out of the laboratory get no further than the supermarket and the warship, respectively. Just as the natural world remains largely unknown and unrecognised except as a hazy background for industrial activity, the modes of thought and feeling that are consistent with it are backgrounded and almost unthinkable.

The differences between natural and fabricated environments are partly associated with their relative complexity. As Lave's research indicates, decision-making in supermarkets lends itself to an arithmetic frame of reference in which one compares the relative value of different items. For example, the question of whether a 24 oz packet of noodles priced at $1.02 is better value than a 32 oz packet costing $1.12 is one from which all values except economic value have been removed from the consumer's consciousness. In the natural world, however, things are more complex. Do I stop here for the night, or move on to the other side of the valley? How do the water resources, light, shelter, safety, freedom from predators and insect pests, and food availability compare? Should I introduce myself to the local inhabitants and seek their permission to camp here? All these and other considerations imply a degree of complexity that requires something other than a few simple calculations, something that is probably better described by terms such as 'feeling' or 'intuition' rather than 'calculation'.

As industrialism has developed into a global *system*, the character of individual attention has tended to move in the other direction – shrinking towards greater specialisation and a narrower focus – so that we have lost sight of the functioning of the whole system. As Ernest Gellner puts it:

> [I]n a complex, large, atomised, and specialised society, single-shot activities can be 'rational' ... they are governed by a single aim or criterion ... a man making a purchase is simply interested in buying the best commodity at the least price. Not so in a multi-stranded social context: a man buying something from a village neighbour in a tribal community is dealing not only with a seller, but also with a kinsman, collaborator, ally or rival, potential supplier of a bride for his son, fellow juryman, ritual participant, fellow defender of the village, fellow council member. All these multiple relations will enter into the economic operation, and restrain either party from looking only to the gain and loss involved in that operation, taken in isolation. In such a many-stranded context, there can be no question of

'rational' economic conduct, governed by the single-minded pursuit of maximum gain. ...

When there is a multiplicity of incommensurate values, some imponderable, a man can only *feel*, and allow his feelings to be guided by the overall expectations or preconceptions of his culture. He cannot calculate.[167]

Unlike contemporaries such as Zygmunt Bauman, Gellner sees the transition to 'single stranded' thought as necessary if we are to 'progress' towards a technologically advanced society; but he also recognises what is excluded by our 'single-stranded' realities, arguing that the industrialised world is "notoriously a cold, morally indifferent world. Its icy indifference to values, its failure to console and reassure, its total inability either to validate norms and values or to offer any guarantee of their eventual success is ... a consequence of the overall basic and entrenched constitution of our thought".[168] There are other costs, too: systems which eschew diversity in order to focus on the single most efficient or profitable approach have a strong tendency towards what Holling refers to as 'brittleness';[169] that is, they tend to put all their functional eggs in one basket, ignoring the need for resilience and alternatives, and so risk oblivion should the single pathway they rely on collapse for any reason.

We should not be surprised, then, at the findings of Ap Dijksterhuis and his colleagues, showing that while conscious calculation gives accurate responses to simple problems, accuracy declines with complexity; while for unconscious estimations, accuracy is independent of complexity.[170] Although some later studies have challenged Dijksterhuis' methodology and conclusions, there is accumulating evidence that the unconscious plays an important but generally unacknowledged role in complex decision-making.[171] Indeed, in one not untypical review of the capabilities of the unconscious, the authors conclude that "our nonconscious information-processing system appears to be incomparably more able to process formally complex knowledge structures, faster, and 'smarter' overall than our ability to think and identify meanings of stimuli in a consciously controlled manner."[172] But the possibility that unconscious judgements may in some circumstances be more appropriate than conscious ones hasn't really been tested properly, in part because of our slavish preoccupation with conscious decision-making, and also because Dijksterhuis and most other researchers have relied on everyday scenarios that are firmly rooted in the industrial world. These include choosing between cars, roommates, or flats, each with specified characteristics – in other words, situations that are often *designed* to be dealt with consciously. As was the case with the work of Lave and Hutchins, then, the assumptions that underly Dijksterhuis' research, the 'realities' they engage with, and the modes of thinking they imply are all consistent with industrialism, and cannot escape its 'gravitational' pull.

Likewise, feeling and intuition, which are generally rejected in the drive for rationality and efficiency, need to be viewed not as a redundant residue of our prehistory, but rather as including a systemic awareness that is desperately lacking within industrial society. Bechara and Damasio have suggested that rapid and effective decision-making can be the result of 'somatic markers' which guide and modify the decision-making process in ways that are often not conscious. Brain structures such as the amygdala and the ventromedial prefrontal cortex, according to Damasio, play crucial roles in orienting us towards the emotional states associated with different decisions so that we are more likely to produce decisions that are satisfying to us; and neurophysiological damage in these areas causes us to make decisions that ultimately hurt us.[173] This implies that listening to our 'gut feelings' may result in decisions that are wiser, in that they accord with contexts we normally disregard.

While a whole, emotionally-aware human being may function optimally within a whole, emotionally sophisticated society, however, the *homo economicus* model may in some ways be better adapted to a society which operates in 'single stranded' ways, and in which success is measured in terms of simple outcomes such as maximising productivity, earning a high salary, or reducing expenditure – although there may be an unnoticed price to pay for such reductions, as Tim Kasser has noted.[174] This underscores the point that styles of experience and intelligence need to be evaluated together with the contexts within which they occur. In a society of extreme specialisation, the need for a wide spectrum of skills has apparently diminished as we focus on our individual areas of expertise. But such specialisation allows the systemic functioning and emergent properties of industrialism to go unnoticed, so that decision-making and choice are in effect diffused into administrative structures we may come to feel alienated from and oppressed by. While the lives of our ancestors were subject to great physical hardships and restrictions, our own world is restricted in other, mostly unintended, ways. For example, the choice of route between two places is now governed largely by the road network, and our resting places en route have been decided for us by the placing of service areas, so that travel is now simpler, faster, less hazardous, and largely devoid of meaning and satisfaction.

We try to correct for the unarticulated losses that accompany modern life in piecemeal, and often commercial, ways. Lacking beautiful sunrises, we buy SAD lamps. Feeling powerless, we are vulnerable to advertisements for powerful cars, and replace the absent meanings of industrialised landscapes through drugs, television, and films. We overeat to fill emotional emptinesses; and having lost touch with a healthy diet, we buy vitamin supplements. Because our lifestyles no longer involve the physical challenges we evolved to survive within a wild world, we go to the gym to exercise. Lacking any well-developed embodied ethical sense, we lean instead on what is legal or politically correct.

Dualisms and conceptual distortions

Dualistic thinking, often identified as one of the villains of 'Western' thought, is embedded within industrial culture and materialised in our infrastructure,[175] so that the way we think, and often our actions, have historically been constellated around certain dualistic oppositions such as primitive/modern, human/animal, civilised/wild, and reason/ emotion. The primitive/modern dualism, for example, has for the last several centuries permeated European attitudes towards indigenous peoples, often justifying a brutal colonialism; while the civilised/wild dualism has served much the same purpose in subordinating wild eco-systems and their constituent creatures to a rampant industrialism. The reason/emotion dualism ensures that reason becomes integrated within powerful social institutions such as science, technology, and the law; while emotion mostly inhabits the shadowy, often unarticulated realm of private subjectivity, where it is exploited by advertisers who appeal to our self-doubts, weaknesses, and insecurities.[176]

Such dualistic thinking is often less pronounced in non-industrial societies. For example, the Northern Ojibwa view various entities such as the sun and the wind, that we would refer to as 'inanimate', "as both grammatically and conceptually living things" – a view that, as Irving Hallowell points out, "sharply distinguishes the Ojibwa worldview from our own;"[177] and indigenous peoples generally see more continuity and less contrast between humans and nonhuman creatures compared to inhabitants of the industrialised world. These cultural differences are reinforced and partly materialised in the differences between a natural landscape and one that already embodies the dualistic thinking of our ancestors. In one part of the campus where I work, for example, everything is culturally (or socially, or linguistically) constructed, and biology is just another 'discourse'. Two hundred yards away, biology rules, with barely a mention of culture or social factors; so a form of culture-nature dualism is materialised within the geography of the campus and no doubt unconsciously internalised by students. The freeing of thought from material and ecological realities, seen in its purest form in those academic disciplines not directly connected with manipulating the world, often results in conceptual polarisations that are quite unrealistic. As we noted earlier, the quite reasonable recognition that perception introduces various cognitive and attitudinal biases is often extended by some academics into an implicit claim for the necessary priority of thought over perception, so that – as Derrida puts it – "blindness seems to illuminate the 'inward eyes'".[178] This exaggerates the gulf between perception and reality and resurrects the spectre of Kant's notion of the unreachable, unknowable 'things in themselves'. But recognising that biases, limitations, and distortions are unavoidable aspects of perception and cognition does not imply that we should abandon seeing and thinking; for if these biases and distortions were were fatal problems, the human race would have long ago become extinct. The major problem underlying the industrialist loss of contact with reality is not so much the intrinsic shortcoming of our sensory apparatus as the wilful exclusion of incoming sensory information, inviting an eventual and far greater correction than that which would result from any immediate sensory error.

In some disciplines, the reaction against dualism introduces its own distortions, as suspicion even of real differences leads them to 'transgress boundaries', develop 'fluid identities', and 'transcend' biological definitions. The term 'dualism' has become a trendy term of abuse in some departments, and when this is combined with an unbalanced prioritisation of language and culture, even naturally occurring and evolved differences such as male-female are in danger of being labelled 'dualistic'. Thus Donna Haraway writes approvingly that cyborgs dissolve the boundaries between sexual and genetically engineered differences;[179] and Judith Butler maintains that sexual difference itself is *discursively* formed rather than biologically given.[180] Turning to another disciplinary area, J. Baird Callicott argues that the distinction between wilderness and domesticated land is dualistic, urging us instead towards "the sustainable development alternative".[181] Such reactions against dualism may be as insidious as dualism's more direct effects, dissolving ecological structures and so unwittingly generating the sort of conceptually bulldozed landscape desired by industrialism.

The natural order is *based on* "dual and opposing tendencies"[182] which, while they may *appear* conflictual or opposed at one level of the organisational hierarchy, complement each other at higher levels. Obvious example are triceps and biceps, male and female, and predator and prey. Dismissing such differences as 'dualistic' is as unhelpful as the dualistic thinking it claims to oppose, and implies a destructive homogenisation of organisational levels.

A dualism is often a piece of 'frozen' reality, a distinction that in an ecosystemically healthy world would come and go fluidly, alternating with the reunion of the separated elements. Thus we may focus momentarily on the differences between men and women, and a moment later on the empathy and complementarity between them. This fluidity implicitly recognises the inability of consciousness to capture the whole of reality; so we flit between partial representations in order to construct a sense of the whole. But this dynamic process unravels when we focus on isolated, frozen fragments, which of course are unrealistic and therefore easy prey for deconstructionists. Just as the isolation of fragments of ecosystems destroys the natural order, so the isolation, freezing, and analysis of isolated elements of thought undermines the usefulness of thought. Rabbit and lynx are mutually dependent, inter-related elements of a system – as are man and woman; and the differences between them only make sense if we view them in their systemic context. Arguments that forget the systemic embeddedness of elements that we cognitively isolate unknowingly import the reductionist focus of consciousness into our representations of social or natural reality, forgetting the complementarities that are essential to life. Rather than trying to abolish the differences that *seem* destructive when viewed in isolation, but are actually the basis of higher-order organisation, we need to cultivate the relationality that makes difference a positive force.

Forgetting consciousness's reductionist focus causes havoc across all areas of life. Often, we transfer issues that need to be dealt with at higher hierarchical levels to lower ones – which is mystifying in a number of ways. To take an example from family therapy: if Sheila complains that Bruce doesn't help with the washing-up, it may be that the issue is not really about who does the washing-up at all, but instead concerns higher-order rules, such as who *decides* who does the washing-up – or even *who decides who decides* who does the washing-up.[183] Similarly, when Steven Vogel questions whether "the carbon dioxide we produce when we exhale is … more natural than the carbon dioxide we produce when we burn fossil fuels in internal combustion engines",[184] the misses the point that the meaning of an action depends on its higher-order context. Emitting carbon dioxide may be part of a natural, ecosystemic process, or it may be part of a process that destroys ecosystems; and the pretence that we can separate the act from the context is a conceptual distortion that further muddies the necessary distinction between healthy and unhealthy ways of living.

The denial of difference can also take the form of subsuming one pole of the dualism into the other. Thus some theorists argue that there is no tension between nature and culture, since nature is essentially cultural. Similarly, the dualism which separates feeling and thinking is 'eliminated' in rational choice theory by redefining feeling as rational (of which more later); and that which separates the wild and the civilised is demolished by measuring, observing, and understanding the wild so that it is subsumed within rationality – a tactic also implicit in Freud's famous statement that "where Id was, there will Ego be". This parallels the wildlife biologist who 'understands' the navigational skills of the crane, or the anthropologist who 'decodes' the significance of an indigenous ritual. Here, the dualistic rejection of something beyond rationality as 'alien', 'wild', or 'primitive' has been replaced by the thing's assimilation into rationality. This can be regarded as a second phase in the colonisation of the world: if the first phase involves the destruction of the alien and its forced enslavement, the second is more subtle – the expansion of the boundaries of the colonising, rationalising power to *include* the alien. For example, the indigenous war-dance is no longer regarded as a threat, but as a colourful tourist attraction. This act of reframing preserves the skeleton of 'otherness' while sucking the meaning out of it: the ritual still appears the same, as does the bluebell in the market. While the act may remain the same, the context has radically changed; and since the context is generally disregarded, we are seduced by the perceptual immediacy of the act or entity into assuming that the original is being preserved.

Denying the outside world

Descartes' rejection of sensory awareness can only be described as paranoid:

> I shall suppose ... that ... some malicious demon of the utmost power and cunning has employed all his energies in order to deceive me. I shall think that the sky, the air, the earth, colours, shapes, sounds and all external things are merely the delusions of dreams which

he has devised to ensnare my judgement. I shall consider myself as not having hands or eyes, or flesh, or blood or senses, but as falsely believing that I have all these things.[185]

This scepticism about the senses has cast a long shadow over philosophy, as expressed in the following joke:

> A philosopher and his wife go out for a country walk in the spring sunshine. "Oh look", says the wife, "the sheep have been sheared!" "Yes", replies the philosopher, "on this side".[186]

How different the world might now be had Descartes instead possessed a benign faith in the senses! According to his biographer, his scepticism also pervaded his personal life, and he is reported to have had "a penchant for misunderstanding those who disagreed with him, attributing motives to their alleged mistakes that were less than complimentary, and thus adopting the moralistic posture of someone who had been deeply wronged".[187] This 'radical doubt' has the effect of driving subjectivity into a heavily defended corner of the individual mind labelled 'rational thought' – a sort of intellectual fortress from which the individual peers out suspiciously onto an alien terrain. But Descartes' retreat into the mind is only the most well-known example of a trend that originated in early Greece and had been gaining momentum since the end of the thirteenth century, Montaigne being one of several who had anticipated all three of the arguments for scepticism that appeared in Descartes' first Meditation.[188] Whereas for Aristotle, as well as his followers during the Middle Ages, sensory information was an acceptable means of accessing reality, later philosophy has tended to lean towards Plato's view that if we investigate something through the senses, "the soul itself strays and is confused and dizzy, as if it were drunk".[189] This slide towards cognitive solipsism has a long history, then; but Descartes' was probably the most definitive lurch in this direction, replacing the world-orientation of Aristotle by the nominalism championed especially by William of Ockham, for whom all *relations* were products of the mind. As Deely[190] has argued, this foreshadowed the idealism of most later philosophy, in which the structure of the world is put there by the mind, rather than being *recognised* by the mind through the intelligent interpretation of sensory information. It is also an orientation that is implicit in the selectivity of much science, engaging with the world only through certain carefully-filtered measures and concepts, so that as Gaston Bachelard put it, "In the formation of a scientific mind, the first obstacle is primary experience".[191]

This fateful decision, Deely argues, opened up a 'chasm' between the world of thought and the world of things, leading to the supposed 'problem' of the external world – that is, the question of how we can know anything about a world whose character is forever beyond the reach of our minds. Consequently, "[p]hilosophy after Descartes became a dead-end path to solipsism".[192] Academic postmodernists, following this idealist tradition associated in various respects with Fichte, Hegel, and Kant, and over-reacting against the naïve realism of correspondence theories of nature, have over the past half century moved towards the Kantian position that any objective 'reality' was unknowable to the point of nonexistence, and that reality is – in varying degrees depending on the philosopher – created by language, thought, or social consensus. As Roy Bhaskar puts it, "from now on any structure ... had to be located in the human mind or the scientific community. Thus the world was literally turned inside out in an attempt to confine it within sentience".[193]

The 'road not taken' by this movement is one in which sensation connects us, through 'signs', to the real world, to Kant's supposedly unreachable 'things-in-themselves'. True, there are always limits to what we can know about the world; but that is hardly an adequate reason for deciding that the quest for knowledge about the world is a waste of time. We can only know the world in terms of concepts, so we are told; but the concepts themselves, contrary to most philosophy, are ultimately derived from our sensory awareness of *things in the world*. In his exposition of Alfred North Whitehead's critique of idealism, Thomas Hosinski explains that

> In our common experience we find something given to us at the outset of experience, and that datum has a vector character; that is, it is directional, pointing to something other than us. The datum has, in other words, an 'objective content'. But modern philosophy has so construed that act of experience that the objective content of the datum has been stripped away and the act of experience reduced to the private, subjective entertainment of 'universals' with no particular referent. Subjectivity then becomes a prison from which it is exceedingly difficult to make contact with the world.[194]

In the same vein, Deely notes that although both empiricists and rationalists assumed that "all awareness terminates in images produced by the mind", he adds that Descartes' contemporary Poinsot had observed in 1632 that "the view that external sense already depends upon an image ... is at odds with our direct experience of the difference between physically present objects with which we can interact causally as well as cognitively ... and our contrastive awareness of objects not physically present".[195] For example, while I can *imagine* a pint of beer in front of me on the desk, my experience of a *real* pint of beer on the desk is a quite different one. Furthermore, the "view that external sense depends upon an image puts things of the environment behind a seamless phenomenal veil, placing them forever beyond our cognitive reach":

> Only by misinterpreting the sense-data as 'ideas' (in the terms of Descartes and Locke), or as a 'phenomenal veil' (in the later terms of Kant) ... do we open the way to mainstream modern idealism according to which mental construction enters into the first moment of objectification and permeates it throughout.[196]

While the various versions of idealism are equally unsatisfactory as guides to a healthy relation to the world, there is a curious and ironic accuracy if we take them as describing industrial society's detachment from reality. While one might view this consistency between idealism and industrialism as reflecting Mirowski's "primal metaphor of human thought",[197] it is perhaps more accurately understood as expressing industrialism's systemic tendency to entrain previously independent areas as it moves towards complete Earth-domination.

A case study in symbolic dominance: The conquest of the Americas

However one represents the coalescing elements of the proto-industrialist European order, the world-changing potential of this order did not become apparent until after the end of the fifteenth century, when the full force of its virulence was felt by the native inhabitants of the Americas in the greatest genocide the world has ever witnessed. The drive to impose newly emerged European ideological templates on discrepant natures and civilisations signalled the sweeping conquest of an embodied world and its symbolic dimensions by forms of symbolism that had cast off from embodied life, and so were utterly alien to indigenous life and culture, as Tzvetan Todorov makes clear in his account of the conquest. Columbus, for example, carried with him a *preestablished* body of 'knowledge' about the lands and peoples he expected to find; and no amount of experience could change this knowledge. As Todorov puts it, "Columbus performs a 'finalist' strategy of interpretation, in the same manner in which the Church Fathers interpreted the Bible: the ultimate meaning is given from the start ... what is sought is the path linking the initial meaning ... with this ultimate meaning."[198] Thus a receptive openness to the world has been replaced by an expanding germ of 'faith' that will impose its order on reality regardless of the intrinsic character of that reality.

For example, when the local people explained to Columbus that the land we now know as Cuba was an island, this conflicted with his conviction that he had reached the mainland; so he dismisses his informants as "bestial men who believe that the whole world is an island and who do not know what the mainland is".[199] Furthermore, he insists on pain of mutilation that his men swear an oath affirming their belief that they have reached the mainland. Here we have an almost complete domination of sensorily accessed physical realities by the symbolic realm of beliefs and words – *almost* complete because by all accounts, Columbus was a skilled navigator who knew how to interpret meteorological indications.[200] Outside this realm of technical expertise, however, "the finalist strategy prevails in his system of interpretation: the latter no longer consists in seeking the truth but in finding confirmations of a truth known in advance".[201] As Bartholomé de Las Casas – the cleric who accompanied Columbus on his voyages – comments, "It is a wonder to see how, when a man greatly desires something and strongly attaches himself to it in his imagination, he has the impression at every moment that whatever he hears and sees argues in favour of that thing."[202] Columbus' journey to the New World was not a voyage of discovery, but of *validation*: as Todorov puts it, "he finds it where he 'knew' it would be".[203]

This prioritisation of the symbolic over the empirical contrasts with the attitudes of the native peoples who were destroyed by the conquistadors. "For the Aztecs", Todorov tells us, "signs automatically proceed from the world they designate, rather than being a weapon intended to manipulate the other."[204] The Spanish, however, had already been inducted into a realm that was primarily symbolic. Cortés, for example, frequently used violence less as a method of achieving his goals than as a means of cultivating certain impressions. He also employed the "language of pretense", exploiting the indigenous peoples' beliefs to achieve his own ends,[205] and using armaments such as a (non-functioning) catapult for psychological as much as military advantage. As Todorov notes:

> The very use Cortés makes of his weapons is of a symbolic rather than a practical nature. ... He conceals a mare at a certain point, then brings in his Indian guests and a stallion; the latter's noisy manifestations terrify these persons, who have never seen a horse. Selecting a moment of relative calm, Cortés has the nearby cannons fired ... Cortés' behaviour irresistibly suggests the almost contemporary teachings of Machiavelli ... in the world of Machiavelli and Cortés, discourse is determined not by the object it describes, nor by conformity to a tradition, but is constructed solely as a function of the goal it seeks to achieve.[206]

Within this symbolic scheme, the role of organic entities is fundamentally changed as they become inducted into the new order – in either of two ways, Todorov suggests. Firstly,

> Enslavement ... reduces the other to the status of an object, which is especially manifest in conduct that treats the Indians as less than men: their flesh is used to feed the surviving Indians or even the dogs; they are killed in order to be boiled down for grease, supposed to cure the wounds of the Spaniards: thereby they are identified with animals for the slaughterhouse; all their extremities are cut off, nose, hands, breasts, tongue, sexual organs, thereby transforming them into shapeless trunks, as one might trim a tree ...[207]

In the second means of induction, in 'colonisation', the Indian is inducted into the system as a *subject* rather than an object, so becoming an active agent who can himself or herself spread the colonising system. Here aspects of the world are exploited more efficiently, since it is not only their material parts that are used, but also some of their higher-order abilities. This is an important principle of colonisation that is widely employed today: individuals are generally not coerced into conformity with industrialism; rather, the higher-order functioning of the industrial system can absorb and employ our individual qualities and intelligences within the system, so that a pervasive illusion of 'freedom' exists at those levels we are most aware of. One of the basic principles of the proto-industrialist order is that existing material relationships and organisation must be commandeered by the new symbolic system; and Las Casas' role is clear in this respect: he "does not want to put an end to the annexation of the Indians, he merely wants this to be effected by priests rather than by soldiers."[208] The strategy employed is that *earthly* problems be translated into and solved through *symbolic* means, as an excerpt from Las Casas' account of a massacre makes clear:

> And just as the young man came down, a Spaniard who was there drew a cutlass or half sword and gives him a cut through the loins, so that his intestines fall out ... The Indian, moaning, takes his intestines in his hands and comes fleeing out of the house. He encounters the cleric [Las Casas] ... and the cleric tells him some things about the faith, as much as time and anguish permitted, explaining to him that if he wished to be baptised he would go to heaven to live with God. The poor creature, weeping and showing pain as if he were burning in flames, said yes, and with this the cleric baptised him. He then fell dead on the ground.[209]

Here we have a prototype for all those psychotherapy sessions where the solutions offered seem elegant and convincing in the terms of the theoretical model used, but largely irrelevant to the economic, political and material circumstances which lie at the root of the problem. The role of the symbolic, however, is not simply as a *replacement* for embodied forms of organisation; it is also used to *conceal* them. Thus in the Ordinances drawn up under Philip II concerning the Indies, Todorov tells us that "it is not conquests that are to be banished, but the word *conquest*; ... one is to act *under cover* of commerce, by *manifesting* love, and without *showing* greed".[210] Here, the process of colonisation is split into two complementary parts: the *mechanisms* underpinning colonisation are hidden beneath a symbolically-contrived *appearance* of humanity and compassion – just as a parallel complementarity exists today between the glossy human façade presented in advertising and the multiple exploitations inherent in capitalist activity, or between the inspirational speeches of presidents and their actions. Industrial humans, uniquely, inhabit both these realms, being split between embodiment and the industrialist symbolism which destroys it. As Todorov notes, "man has just as much need to communicate with the world as with men ... this victory from which we all derive, Europeans and Americans both, delivers a terrible blow to our capacity to feel in harmony with the world, to belong to a preestablished order". And as he adds, "During the centuries to follow, they would dream of the noble savage; but the savage was dead or assimilated, and this dream was doomed to remain a sterile one."[211] Our embodied yearning for a different world, currently unrealisable, remains; and the yearning itself is reinterpreted as unrealistic nostalgia or childish fantasy, sealing the repression of an important dimension of ourselves.

How the world invites us to extended subjectivity

'Primary' qualities such as motion, shape, and number, Locke tells us, are inherent in 'things', independently of how we perceive them. 'Secondary' qualities, on the other hand, are supposedly supplied by the observer; and these include colour, taste, smell, beauty, and sound. Locke, like his predecessors such as Galileo and his critics such as Berkeley, is teasing out the differences between what is *within* the person and what is in the *outside* world – the fundamental distinction upon which almost all philosophy, and our entire industrial way of life, is built. It is necessary, says Locke, "to make the difference between the qualities in bodies, and the ideas produced by them in the mind, to be distinctly conceived, without which it were impossible to discourse intelligibly of them".[212] If we accept the assumption that we are detached from the world, then qualities must necessarily be *either* in the world *or* in us.

But as we noted above, qualities such as colour can be the outcome of a process of recognition that involves *both*. As Deacon suggests, something in my sensory apparatus recognises something about the tree which I refer to as 'greenness':

> the pattern of electromagnetic waves reflecting off an object and entering the retina and the pattern of neural signals ramifying through circuits of the visual areas of the brain are both part of a causal chain on which the experiences of colour are based. The colour does not inhere in the object alone, nor is it merely a mental phantom. Something intrinsic to the object is re-presented in the pattern of light waves and again re-presented in the pattern of neural signals. But it is also re-presented in the experience of colour. There is no jump from the material stuff to the mental stuff in this process.[213]

Colour is thus not something we 'create' and then 'project' onto the world; but neither is it something that exists, fully formed in the world itself, awaiting 'discovery'. The notion of 'colour' does indeed reflect an aspect of the world 'out there' – but it is an aspect that is interpreted and categorised through our sensory faculties, and so is an 'emergent property' of an *integrated system* of 'self-and-world'. Conceptually, we try to divide this integrated system into an 'emitter' and a 'receptor'; but seeking the 'meaning' in one or other of these is rather like asking whether a car is powered by the engine or the fuel. The nervous system has evolved to operate as part of a larger system which includes *what* we perceive; and perception, intelligence, and the 'objective' world, which psychology likes to separate, are bound up together in this system:

> An expanse of smooth gravel is a sign that you are close to a river. Cottonwoods tell you where the river bank is. An assembly of twigs in a tree points to ospreys. The presence of ospreys shows that there are trout in the river. In the original economy of signs, one thing refers to another in a settled order.[214]

This is not an order that we impose *on* a world that is blank. Rather, it is an order that exists *in* the world, one that we *participate* in by virtue of having a sensory apparatus that has evolved to complement the order of the world. As Albert Borgmann suggests above, cottonwoods are not just an arbitrary sign of water, in the way that the symbol '3' arbitrarily indicates three of something; we recognise that cottonwoods and water occur together for reasons that have nothing to do with our imposed symbol-systems. While we may seldom be able to access reality directly, we can often perceive the paw-prints of reality, so that 'things in themselves', however indirectly, become accessible to our intelligence. Perhaps we should follow the example of Socrates who, "in order to cure Phaedrus of his infatuation with letters ... takes him out of the city into the world of natural things to remind him of the eloquence of the oak and the rock."[215] Thus as Tim Ingold argues, "perception is a mode of *engagement* with the world, not a mode of construction of it".[216] 'Secondary' qualities, then, are neither in the world nor in the person: they are emergent properties of our participation in the world.

If we view the human person not as a discrete bundle of character traits and abilities, but as potentially open to qualities and structures in the world, then as Roger Brooke puts it, we "discover that the self is not an entity but a capacity that emerges through the revelation of the world. The spirituality of the self, for example, is a capacity that emerges through the world's revelation as a temple; what is found is not one's spirituality but a temple".[217] When we're somewhere beautiful, the landscape arouses resonances that challenge the normal separation between the individual and the world. We feel our subjective awareness transcend its narcissistic concern with 'our own' feelings as it is drawn into the world, which we then experience as beautiful, frightening, or mysterious. As Arne Naess argued, when it is not fragmented by abstract thinking, experience is integrative, so that joy is "not *my* joy, but *something joyful* of which the I and something else are interdependent, non-isolatable fragments".[218] Similarly, we learn to be loving individuals by opening our eyes to somebody's loveliness – not by discovering such feelings hidden away in some crevice of our own personality structure.

Although such integrative experiences occur somewhat against the grain of industrial society, they are often precipitated by natural surroundings. Psychological studies of the motivations underlying ecological restoration, for example, suggest that behind the notion that a volunteer is 'doing something to' an area of land, a more complex, and more reciprocal, drama is being played out. Thus in Irene Miles' study of motivation among volunteer restorationists, the two clearest factors which emerged were: 'Meaningful Action' – indicated by such items as "feeling I am doing the right things", "a sense of accomplishment", "feeling I can play a role in nature"; and 'Personal Growth' – indicated by items such as "being a part of something profound"; 'changing my life"; "restoring or contributing to my spirituality". What this suggests is that volunteers are not only motivated by concern for the landscape, but also by the need to express something about themselves. As one volunteer insightfully remarked, "there is a sense of communion; it is fulfilling and self-transcending".[219] While the restoration of the landscape is important, so also is the restoration of personal meaning, wholeness, and a sense of being part of nature – precisely what has been lost through our dualistic separation from the world.

It misses the point, I think, to suggest that Miles' results imply that people volunteer for restoration programmes in order to fulfil covert egoic needs as well as to benefit the environment. What is implied is, I think, rather more radical than this, amounting to a rejection of the deeply embedded dualistic perspective which separates the individual from the world and individual needs from those of the ecosystem. What we are restoring is not just the wholeness of the landscape, nor our own wholeness, nor even both at once – but rather the emergent wholeness that results from our own reintegration *into* the landscape. In this sense, ecological restoration can be understood as a *symptom* – that is, a behaviour which expresses something unconsciously problematic about our current situation, and is simultaneously an attempted solution to it. Since the problem is embedded within the symbolic forms available to us, which *assume* our separateness from the world, solutions can *only* be expressed through embodied action. There are parallels here with the theory of alcoholism put forward by Gregory Bateson, suggesting that some people self-medicate themselves with alcohol in order to overcome the dualistic division between body and mind.[220] One wonders whether our cultural obsession with love relationships, being one of the few remaining opportunities for expressing such personal transcendence, is also motivated by a need to escape from our paralysing isolation within the egoic self.

Ethnographic reports such as those of Ingold and Atran give us a poignant glimpse of a form of selfhood which is open to the world and embedded within it – one which we have lost, but which is still latent within our physical being. This world was still alive among many sixteenth century Europeans, despite the encroaching symbolic transformation; and Michael Jackson describes the Europe of this era as "bristling with signs, blazons, omens, and figures".[221] A 'biosemiotic' understanding, which sees our awareness of signs as vital to this embeddedness, is a crucial part of this almost-lost aspect of selfhood. This biosemiotic approach is radically different from that of linguists such as Saussure, who argued that although signs may retain some slight relation to the physical world, they are pre-eminently part of a *linguistic* system that has only an arbitrary relation to reality. Jeffrey Wollock criticises this view, according to which

> the natural world provides no meaningful input either to the speaker, the hearer, or the language. We are stuck with a self-referential semantic and pragmatic system with no touchstone to any reality outside of it ... Nature, if considered at all, is considered only as an abstraction, a sociolinguistic construct.[222]

As Jim Cheney expresses the insight of the Zen Buddhist philosopher Dögen, "Whoever told people that 'Mind' means thoughts, opinions, ideas, and concepts? Mind means trees, fence posts, tiles, and grasses."[223] Likewise, Charles Sanders Pierce, the American philosopher who pioneered the theory of signs which is now known as 'semiotics', realised that signs are commonplace and essential aspects of nature. For the blackbird, a particular sound signifies the presence of a worm beneath the surface, just as yellow eyes signifies jaundice to a physician. Signs, therefore, involve three components; the *sign vehicle* (the sound or yellowness), which signifies *something* (the worm or jaundice) by means of an *interpretant* (the sign's significance to the blackbird or physician). It can immediately be seen that this system overturns the tacit anthropocentrism of most communication theory, since it applies equally to blackbirds and physicians. In Donald Favareau's terms,

> the world of sign relations per se did not start with the advent of homo sapiens – and … a sign relation is not something that was created *ex nihilo* by the minds of human beings – but rather … the minds of human beings are themselves the products of a *de novo* use of absolutely natural and biological sign relations.[224]

The evolutionary process depends on and perpetuates an alignment between organisms, including humans, and their environment; and it is clear that any departure from this alignment will adversely affect the survival of the species. As Francois Jacob puts it,

> No matter how an organism investigates its environment, the perception it gets must necessarily reflect so-called 'reality' and, more specifically, those aspects of reality which are directly related to its own behaviour. If the image that a bird gets of the insects it needs to feed its progeny does not reflect at least some aspects of reality, there are no more progeny. If the representation that a monkey builds of the branch it wants to leap to has nothing to do with reality, then there is no more monkey. And if this did not apply to ourselves, then we would not be here to discuss the point.[225]

Knowing relevant aspects of the 'things in themselves' is therefore necessary for survival; and a Kantian – if true to his beliefs! – would therefore be significantly less likely to survive than a biosemiotician or critical realist, at least outside academia. This need to view symbolisation as clearly reflecting reality was indicated by Gestalt psychologist Wolfgang Kohler in 1913 when he asked: "How is it that visual illusions are pervasive but do not prevent a successful interaction with the environment? … This fact might be easily explained if we regard as … the biologically primary reality, *not sensations but, for the most part, things.*"[226]

Too great an intoxication with the products of mind alone leads us into the predicament of social constructionists or idealists such as Hegel who, as Mary Midgley tells us, was unfortunate enough to prove "a priori that there could only be seven planets just before the discovery of the eighth".[227] Even quite elementary forms of intelligence depend on the interpretation of signs: Darwin, for example, showed how the earthworm was able to distinguish the shape of a leaf in order to be able to drag it into its burrow. Thus the interpretation of signs is essential for the alignment between organisms and the external world; and this is why the natural world is best described as a system of sensitively interacting entities rather than as a diverse collection of separate ones.

This responsiveness to the outside world crucially differentiates the brains of animals – including humans – from the 'brains' of computers. As Deacon explains, "though it is common to find researchers referring to simple brains, such as insect brains, as computers, the representational architecture of computers, on the one hand, and simple brains, on the other, are essentially inverted".[228] The computer, like the person in the Chinese Room, simply juggles symbols without understanding their significance, whereas the animal brain essentially *represents* some aspect of the world. This representational or meaning-producing function is vital to the survival of the animal: there is no point in identifying that cat-shaped shadow in a tree if I don't also understand its significance for my chances of surviving until tomorrow. This, as the reader will by now recognise, is one of the central arguments of this book: that in the industrialised world, a divorce – as Donald Favareau puts it – between "knowers and the world they know [is] the bewitchment of a symbolic overcoding system that itself no longer recognises its own grounding in the relations of the material world."[229]

Local and bureaucratic knowledges

The dissonance between approaches which impose meaning *on* the world and those which discover meaning *in* the world is one that has spread to almost every aspect of human activity. For example, Nick Totton has drawn attention to the similar changes occurring in two apparently quite disparate areas, both of which have traditionally supported a diversity of approaches, and both of which are now being subsumed into larger bureaucratic forms of organisation; firstly, potato farming in the Andes, and secondly, psychotherapy in the UK.[230]

In his study of Andean potato farming, Jan Douwe van der Ploeg examines tensions between the traditional 'local knowledge system' and the scientifically informed industrial farming that is rapidly displacing it. In the former, the ecologically and culturally embedded specifics of local knowledge take precedence over universal laws and invariant principles, so that culture and practice reflect locally developed adaptations to the landscape rather than being subservient to centrally organised systems of knowledge. Thus land, knowledge, and cultural customs exist in a fluid relationship that is responsive to changes in any one of these components, and "diversity both permeates and is created by the process" of peasant farming. A constant interplanting and exchange of potato cultivars between farmers keeps the genetic stock responsive to local conditions. Due to their fluidity, the concepts used seem vague and 'inaccurate' to a scientifically-trained outsider; but they are in fact used and communicated with precision, for "interpretation and communication can only be active processes: concepts must be weighed against each other every time a specific plot is being considered".[231]

In contrast, the system of industrialised farming does not involve the interaction of place-specific knowledge, cultural traditions, and a diversity of cultivars, but rather is based around the imposition of a top-down system of determination that is ultimately derived from economic considerations. Choice of cultivar is not made through a history of experimentation under a variety of local conditions; rather, genotypes are developed so as to give the maximum possible yield under *standardised* conditions. For example, a particular cultivar may require a certain nitrogen level, delivered according to a precise time-schedule, which is itself derived from the specific genotypical cycle; and this in turn dictates the water requirements.

Just as the growing conditions of scientifically developed potatoes need to reproduce those under which the cultivar was developed, so the behaviour of the farmers needs to strictly follow centrally planned schedules. The accumulated knowledge of the farmer in his interaction with his plots becomes irrelevant as he is transformed into the invisible agent of a larger scheme over which he has no control. As Van der Ploeg puts it, invisibility "seems to become especially reinforced when all your careful attention and love for the land are at once declared insignificant by the introduction of general schemes to be followed in production and by the introduction of 'miracle seeds'".[232] Just as the diversity of potato varieties and of the individual plots on which they are grown is discarded by the imposition of certain genotypes and conditions, so the individuality of farmers is subordinated to the agricultural processes they operate.

This conflict between forms of knowledge that are determined by abstract principles, economic requirements, and bureaucratic necessities, on the one hand, and those that are integrated within local cultural and ecological conditions, on the other is a pervasive feature of the interface between industrialism and the natural order, and can also be seen in areas such as education and health, where target culture, the increased power of managers over practitioners, and the deflection of aims away from human welfare towards profit generation are the symptoms of an assimilation to commercial interests. For example, many of the features that we noted above in relation to Andean farming – standardisation, top-down organisation, and the invisibility and powerlessness of the individual within a prescriptive system – also apply to the 'professionalisation' of psychotherapy in the UK and elsewhere, 'cleansing' it of unique or niche-specific approaches and leading to a "monoculture of the mind", to use Shiva's integrative metaphor.[233] Indeed, a former chairperson of the United Kingdom Council for Psychotherapy described the process of professionalisation as 'cutting back' the weeds, 'the sprawling plants that obscure each other's light and deprive each other of nutrients'.[234] The aim is the elimination of wildness, whether in the psycho-professions, agriculture, society, or forms of thought, exemplifying Mary Midgley's description of rationality as a sort of weedkiller of the mind.[235]

An overriding difference between local know-how and bureaucratic knowledge is that the latter implies the practitioner's psychological distance from whatever the knowledge is about. Rather than participating in the activity in a fully embodied and therefore personally integrative way, the feelings, sensory awarenesses, and embodied knowledge of the individual are sidelined and replaced by a following of instructions formulated elsewhere, resulting in the emotional detachment of the practitioner and the destruction of cultural meaning. Likewise, the spontaneous interplay between the subjective awarenesses of therapist and client, like that between peasant farmer and the plants he grows, are replaced by a technique defined by pre-established patterns. In the UK, for example, one therapeutic approach – cognitive-behavioural therapy – is favoured by the Health Professions Council, potentially reducing the variety of approaches available to the individual, much as the Andean farmer's choice of potato variety or agricultural practice is also centrally ordained. In effect, meaning and human well-being become irrelevant within a technical system that has only *economic* aims.

Often, the transition between these two is indicated by the prominence of theoretical analysis over practice, of mind over embodiment. When something becomes the object of conscious analysis, an important shift occurs into the symbolic domain and away from forms of organisation that emerge from embodiment and ecological organisation. For example, Alf Hornborg notes that when "a person begins to talk about his or her own 'culture', it is a sign that another life-world is being objectified and decontextualised".[236] Likewise, industrial society's increasing reliance on consciously formulated and often legally enforced principles governing many aspects of individual behaviour represents a 'hollowing out' of culture which leads to its detachment from embodied intelligence. The response which Hornborg suggests to this forced assimilation into an alien realm is to refuse the language of the assimilator – the language of principles, techniques, concepts, and other abstractions – and to obstinately remain within the realm of local, ecological connections.

Local knowledges are continuous with local forms of social organisation, as Van der Ploeg's study makes clear. This is not to say that local organisation is a guarantor of social and psychological health: one can have local as well as national tyrannies. But local organisation avoids one prevalent form of tyranny: the imposition of organisation from outside, suppressing individual and communal needs and preferences. This tyranny reproduces at a social level that of the symbolic over the embodied, of words and principles over the forms they once claimed to represent, so that the administrative, the symbolic, and the psychological aspects of alienated life coalesce within a single system. Václav Havel, in his influential analysis of the pre-1990 east European communist system, argued that

> Between the aims of the post-totalitarian [i.e. communist] system and the aims of life there is a yawning abyss: while life, in its essence, moves towards plurality, diversity, independent self-constitution and self-organisation, in short, towards the fulfilment of its own freedom, the post-totalitarian system demands conformity, uniformity, and discipline. While life ever strives to create new and 'improbable' structures, the post-totalitarian system contrives to force life into its most probable states.

As ideology loses touch with reality, argues Havel, it

> becomes reality itself, albeit a reality altogether self-contained … The significance of phenomena no longer derives from the phenomena themselves, but from their locus as concepts in the ideological context. Reality does not shape theory, but rather the reverse.[237]

But these characteristics did not only apply to the East European countries that Havel focuses on, since the communist system was "simply another form of the consumer and industrial society, with all its concomitant social, intellectual, and psychological consequences". Consequently, this system stood "as a kind of warning to the West, revealing its own latent tendencies".[238]

These examples encompass themes that will by now be familiar to the reader. They demonstrate the drive of industrialism to replace the existing untidy, mysterious, diverse, fluidly changing ecological world by one that is systematic, rationally ordered, internally consistent, and cognitively comprehensible. This objective begins as a form of symbolic organisation, but becomes a system of material control through its application as a technology. Just as the notion of wildness is replaced by the more cognitively accessible notion of wilderness, and then by the quantifiable notion of biodiversity[239] so, as Havel points out, social movements that originally sprang from human needs become something alien to these needs as they are assimilated into symbolic frameworks. Thus the internal consistency of the system becomes complete as rational ordering generates a complementary material world that is rationally ordered; and thought, behaviour, and 'reality' seem perfectly attuned.

Symbolic substitutions

While there is a manufactured consistency between material realities and the dominant symbolic order in already-colonised areas of the world, this is often not the case in areas where the process of industrialisation is still in its most physically destructive phase; and denying expression to those peoples whose cultures are continuous with local ecologies becomes essential to industrial 'progress'. For example, one of the Penan tribe, describing their attempts to inform the timber companies about the effects of logging on their way of life, said that "talking to representatives of timber companies or the government was like talking to a drawing: they neither hear nor respond." And when Mahathir Mohamed, Prime Minister of Malaysia, visited the Taman Negara National Park, instead of meeting the local Batek people, he met specially imported, ceremonially dressed people from Sarawak.[240] In such cases, a narrative that falsely claims authenticity is slipped into place to cover up the personal, cultural, and ecological price of industrialist colonisation.

Academia is often complicit in this isolation of industrial narratives from lived realities. In one paper, for example, we read that international conflicts can best be resolved by separating them from the underlying grievance and the ongoing injustices. Thus the emphasis moves from actual injustices to "shared perceptions of injustice", from histories of oppression to "a historical perspective that emphasises past episodes or periods of mistreatment and exploitation", from physical realities to "mental representations".[241] This strategy tends to cement existing injustices by putting them on one side while a strictly *symbolic* and decontextualised 'solution' is proposed which 'reframes' the injustices so that they *appear* less unjust. As Kamyar Arasteh argues in reply to this paper, "reducing conflicts to problems of perception is fraught with dangers that are ultimately directed toward the victims".[242] Michael Humphrey makes a similar point, showing that material and political disenfranchisement is often dealt with in post-conflict situations as a matter of *psychological* injury:

> Grievances are looked at as obstacles to an individual sense of well-being which is amenable to 'emotional management'. Thus psychosocial management promotes self-disciplining while using the 'therapeutic language of self-actualisation, participation, empowerment, and self-esteem' (Pupavac 2004: 156). In post-war Bosnia, for example, people were seen as having 'a subjective poverty problem' the solution to which was a psychosocial adjustment aimed at lowering people's material expectations.[243]

While it is possible to *symbolically* dis-embed the abstract elements of a conflict from material realities, these material realities are in fact embedded in people's lives and experience; so sidelining them becomes a form of repression. In cases of systematic violence, occupation, and ethnic cleansing, as Arasteh concludes, "no amount of cognitive restructuring can reduce the threat of victimisation, and to effect change it is necessary to address the underlying injustice and prevent perpetration of the crime".[244] The attempt to substitute cognitive for 'real world' action is a classic example of the Platonic Backhand; and as Edward Sampson puts it, "reified cognition and reified psychological processes take what is empirically observed, abstract it from the particular socio-historical conditions of its constitution, and grant it timeless, objective standing".[245]

The distancing of the symbolic from the material, however, is not simply a conscious strategy employed by politicians: it is also a basis of the way we think, learned during our school years. We quickly learn the value judgements – and the rates of pay – associated with manual and mental work. Once we begin to substitute the cognitive realm for the real world, we tend to engage in cognitive 'actions' as a substitute for real-world actions. The seeds of this substitution occur in early childhood: as Piagetian researchers have shown, children gradually substitute a specific type of symbolic thought for sensory-motor action.

Psychology has taken the lead here; and as Sampson remarks, "the tendency for psychologists to reduce conflicts to subjective misunderstandings, to misperceptions, and to psychological factors within individuals ... serves primarily ideological functions by eliminating from our analysis the contradictions ... among groups". Thus conflicts between, say, loggers and peoples such as the Penan are, in Sampson's terms, reduced to 'subjective misunderstandings', to be dealt with in the symbolic realm. This tactic serves the status quo, since existing "arrangements of power and domination ... are served when people accept a change in their subjective experience as a substitute for changes in their objective reality."[246] As Sampson summarises the problems with this approach,

> The cognitivist perspective offers a portrait of people who are free to engage in internal mental activity – to plan, decide, wish, think, organise, reconcile, and transform conflicts and contradictions within their heads – yet who remain relatively impotent or apparently unconcerned ... about producing actual changes in their objective social world. In substituting thought for action, mental transformations for real-world transformations, cognitivism veils the objective sources and biases of social life and relegates individual potency to the inner world of mental gymnastics.[247]

Thus freedom becomes understood as freedom to *redefine* what is happening rather than freedom to *influence* what is happening. As Sampson points out, this is a definition that is adopted by people who have abandoned any hope of changing material reality, and instead prefer to adjust themselves to a situation as comfortably as possible. There is therefore a certain hopelessness, a defeat, embedded in our acquiescence to a symbolically-defined 'reality', as 'freedom' becomes a purely symbolic freedom defined against the context of an assumed embodied unfreedom. The symbolic realm substitutes for and provides a cover for what is being drained from embodied life, so that embodied restrictions governing our behaviour are made to seem more palatable by our freedom to consume, to vote for one of several indistinguishable political parties, and to 'freely' utter opinions that change nothing.

It is not difficult to see how this substitution of 'virtual' satisfactions for real-world ones is used today – for example, by the media and the 'leisure' industries. Feeling powerless to make real-world changes, we watch Bond films, play at martial arts on computers, or drive cars that can do more than twice the legal speed limit; and virtual realities allow a painless escape from a degraded world into (another), quasi-material 'new world' we can design ourselves. As Katherine Hayles puts it,

> In a world despoiled by overdevelopment, overpopulation, and time-release environmental poisons, it is comforting to think that physical forms can recover their pristine purity by being reconstituted as informational patterns in a multidimensional computer space. ... The sense that the world is rapidly becoming uninhabitable by human beings is part of the impetus for the displacement of presence by pattern.[248]

Even the material aspects of life are orchestrated by symbolic needs that have little to do with real-world needs. Buying clothes has as much to with status and fashion as with warmth and protection. And when our powerlessness occasionally causes frustration to boil over, we are taught to look inside rather than at the external causes, perhaps taking an anger management course or learning to meditate, deflecting attention from from the causes of these problems and instead focusing on the need to adjust and adapt to unchangeable situations. As Sampson argues, "the psychological villain or misfit is the person who declines to conceal an objective contradiction with a subjective fig-leaf".[249]

Digital life

Analogy – a close relative of empathy – is a fundamental aspect of intelligent behaviour; and analog information often relates us to the world in a way that digital information doesn't. This is not to say that digital codes are uncommon in nature, only that they do not convey the subjective significance of natural phenomena as adequately as analog codes. Nor is it to claim that all analog information enlarges our sensory reach; for even analog instruments such as the mercury thermometer or the speedometer are instruments of conversion from sensation to intellect, recoding the visceral feelings of heat or speed as numbers in the mind. Nevertheless, analog information often retains a sensory dimension: the hour-hand of an old-fashioned watch moves in a way that mirrors the movement of the sun relative to the earth; and a weather vane enables us to 'sense' the wind direction. Analogue information can be a bridge to whatever the information is *about*, amplifying our senses, enlarging our Umwelt so that the world we inhabit seems a richer, wider, more sensuous place, and extending subjectivity beyond our physical boundaries. The signs of nature are generally based on physical relationships: the sound of a displaced rock may be a sign of danger, and bent grass indicates the direction of a creature's progress across the prairie in a sense that is not merely arbitrary. Animals, including humans, are predisposed to recognise the significance of particularly conjunctions. For example, Öhman and Mineka have shown that monkeys easily learn to fear snakes, but not flowers.[250] Similarly, particular percepts may be more easily recognised and coded than others; and cross-cultural research has found that names for 'focal' colours such as red, blue and green which are more 'perceptually salient' are more easily learned than those of other colours, suggesting that colour naming is not simply an arbitrary linguistic imposition. As Eleanor Rosch puts it,

> Because it was the same colors that were the most codable in a large sample of languages which possessed the full complement of basic color names ... and because even speakers of a language for which these colors were not more codable remembered those colors better than nonfocal colors, it would appear that the color space, far from being a domain well suited to the study of the effects of language on thought, is rather a prime example of the influence of underlying perceptual-cognitive factors on the formation and reference of linguistic categories.[251]

Our senses have evolved to recognise the world as diverse, often continuous, and containing multiple analogies. Recognising these analogies is part of our evolved repertoire of survival skills; and they are often more a matter of degree, nuance, and relative prominence than of digital 'yes-no', 'either-or' presence. A colour may be 'reddish-brown' rather than being clearly either red or brown; and a robin is more clearly a prototypical bird than a penguin.[252] Although we are generally socialised into thinking and speaking in terms of clearly defined categories, our experience often reflects the gradations of the natural world; and the view that language imposes structure on thought may, if pushed too far, represent a sort of ideologically-tinged wishful thinking rather than truthfully representing human experience. Similarly, the concept recognition tests often employed by psychologists ("Divide these items into living and non-living") reflect the same digital assumptions as the linguistic determinism that has become so influential over recent decades. It could be argued, too, that the urban world is more easily codable in digital terms than the natural world, and so is consistent with the digital emphasis of communication theory. We no longer ramble across subtly changing landscapes, but instead walk *either* on the pavement *or* the road, past traffic lights that are *either* red *or* green. I am *either* on private property *or* on common land, and looking *either* at a field of sugar beet *or* of cabbages. While such clear differences *do* exist in nature, they are rarer than in the contemporary world. Generally, we industrialised peoples are educated to think in digital terms; and through the 'Platonic Backhand', we reconstruct the world to accord with this digital preference.

As Rosch argues, "the prevailing 'digital' model of categories in terms of logical conjunctions of discrete criterial attributes is inadequate and misleading when applied to most natural categories;"[253] and she proposes an alternative analog model. The cultural narratives of non-industrial societies have a strong analog flavour to them, often involving layers of metaphor that are difficult to code in more digitally-oriented understandings; and as Hugh Brody notes, the "analogue nature of myth mirrors a sense that the world itself defies digital ways of speaking".[254] We can in a very general sense point to a difference between analog and digital understandings: the former involves a subjective continuity between the individual and the world, while the latter involves the coding of some aspect of the world into an autonomous abstract system.

In a sense, we can 'identify' with the weather vane as it is blown about by the wind – we feel ourselves being *extended* into the vane as if it becomes part of our body. Similarly, if

> I can see houses more than four kilometres away on the other side of the fjord, it seems as if part of my self reaches out over such a large area. And if ... a lightning bolt strikes the other side, 'I' will see it in an instant, even before 'I' hear the thunderclap. 'I' exists, so to speak, in places over there.[255]

The thunderclap, then, is a 'sign' of the lightning, just as a black cloud is a sign of an approaching storm; and by recognising such signs, we can integrate our own intelligence with that of the world, extending subjectivity beyond our physical boundaries. Thus analog information makes us part of a world we empathise with, that we become part of; and our intelligence expands as it suffuses the world.

The movement towards a technology based on digital rather than analogue information, however, is consistent with industrialism's drive to develop a world that has thrown off its natural origins. We are educated to think abstractly rather than concretely, discarding context and relevance in order to move within a world of words and numbers. While coding may originally have been viewed as representing a more basic material realm, it has become increasingly autonomous and hostile to natural form. For example, genomic understandings of inheritance are now less models of a natural process than the basis of an industry focused on the creation of lifeforms that nature has unaccountably failed to produce. And whereas a map makes us part of a world we can also see and hear, our satnav makes us reliant on the instructions the machine gives us: rather than feeling ourselves extended into the world, we are detached from it, and our subjectivity reaches no further than the screen in front of us. The effect is to distance us from the world and from our own bodies as we become reliant on instruments that replace rather than amplify our sensed understandings.

The consequences of this distancing may be indirect or overt, trivial or catastrophic. We may have a hard-to-articulate feeling of numbness in relation to the outside world, or an anxious sense of disorientation, like the girlfriend with whom I once went hiking in Utah who began to panic when she could no longer see any buildings. Perhaps, like the pilot of a fly-by-wire aircraft that suffers a computer malfunction, we may experience a terrifying and possibly fatal loss of control.[256] Furthermore, if we *collectively* lose our sense of embeddedness in the world, so that society abandons its natural bearings, the result may be an illusion of security similar to that which passengers in the *Titanic* may have felt before the intrusion of external reality in the form of the icy waters of the north Atlantic. While the digitisation of life, then, may have obvious technological advantages, there may be a more subtle and far-reaching price to pay in the longer term. If we are drawn into a world of calculation and abstraction that maps onto our experience only cognitively, a distance grows up between the world we *think* we live in and the world we *feel* we live in, as we become experientially disembodied. As one anorexic put it, "reading recipes made up for my lack of eating"[257] – which, of course, it didn't.

The effort to substitute coding for embodiment has nowhere been more intense than in evolutionary biology. Developments in information theory after the Second World War provided fertile ground for the view that the reproduction of life-forms depended on a digital 'code'. The DNA 'double helix' discovered in 1953 was thereafter enthusiastically proposed by many as the central vehicle of information transmission in embryogenesis, despite the fact that much of the genome has been found not to directly code for protein and to have functions that are more elusive than previously realised. As Lily Kay explains:

> Within these new sciences of communication, *information* was meta-phorised from being an entity that carried meaning to being purely

a stochastic process – a decision-making capacity based on binary choices – that produced meaningless signals, just syntax. … Technically speaking, information no longer informed; it was independent of subject-matter and had no semantic value, so that two messages – a random collection of letters and a Shakespearean sonnet – could have the same information content. Notions of 'alphabet', 'code', and 'language' – idioms central to the information discourse – were similarly metaphorised.[258]

In Richard Dawkins' view, rather than DNA being viewed as one player in the drama of evolution, it becomes the end-point for which the embodied animal and the process of evolution itself are mere vehicles[259] – in other words, another version of the view that the 'language', rather than expressing and participating in embodiment, instead controls it. Thus the DNA is the 'ideal type', the model, made flesh and thrown into the world in order perfect itself, rather along the model of the human soul. As Susan Oyama puts it, the gene

is hailed as Nature's Chosen (Selected!) Molecule, the agent into which the evolutionary Word is breathed, the worker of ubiquitous secular miracles of life … The dominance of the language of *language* in genetics, in fact, is striking … geneticists' technical vocabulary … is rife with codes, translations, transcriptions, editing, sense and nonsense, along with comparisons of bases to letters, of genomes to libraries. From cognitive and computer science, meanwhile, engaged in constant conceptual and terminological cross-fertilisation with molecular biology, come information, transmission, representations, programs, and algorithms.[260]

While information has technically been defined in terms of improbability, this is about as far from humanly meaningful notions of information as we can get. In daily experience, information cannot be separated from the way it is *interpreted*, so that it is also dependent on the *context* within which it occurs. If I say "I need some coke", this information would be understood differently depending on whether I am talking to a solid fuel supplier, a waitress, or a drug dealer. Much the same is true in biology, where the significance of a piece of DNA depends on how it is 'interpreted' by an array of enzymes, autocatalytic reactions, and other genes, so that outcomes at an organismic level generally depend on systemic interactions rather than on a sort of linear causation that is traceable directly to genes.[261] As Kay notes, "once the complexities of DNA's context-dependence – genetic, cellular, organismic, and environmental contexts – are taken into account, pure genetic upward causation is an insufficient explanation".[262] Genes in isolation are therefore no more informative than a dictionary; and what is often missing from genetic explanations is the intelligent, embodied context that constitutes the 'reader' who can interpret the 'meaning' of particular configurations of genes.

This question of interpretation is ignored by the view that DNA acts on passive material to create life; and what is denied is the entire interpretive framework which genes contribute to. As Jesper Hoffmeyer points out, for example, birdsong is generally regarded as coded for in the genotype; but we could as well argue that it is coded for in the song of the parents from whom the chick learns.[263] In keeping with this realisation, the 'Developmental Systems Theory' of Susan Oyama and her colleagues suggests that development draws on a range of genetic and non-genetic resources which act together in self-organising systems,[264] rather than being specified in advance through genetic coding. Only in a relatively small proportion of cases can a behaviour, illness, or attribute be reduced to a single, or even a few, genes.

For these reasons, the code-centred understanding of life is clearly inadequate when applied to the natural world; but it is much less inadequate when used to understand the industrialised world, which creates commodities by applying intelligent algorithms to material-made-passive. As Francisco Varela puts it, "To the extent that the engineering field is prescriptive by design, this kind of epistemological blunder is still workable. However, it becomes unbearable and useless when exported from the domain of prescription to that of description of natural systems".[265] In other words, the view that symbolic codes determine material realities is not just used pragmatically to produce specific 'things'; it also becomes a model for the system as a whole, as a symbolic 'language' is applied to and used to reconstruct the entire natural order. Donna Haraway refers to modern biologies as having translated "the world into a problem in coding ... the organism has been translated into problems of genetic coding and read-out ... organisms have ceased to exist as objects of knowledge, giving way to biotic components".[266] This is not *just* a matter of technological development: it is also about *reduction* – the reduction of organic life to passive matter controlled by symbolic 'information'.

It is a reduction which is also perpetrated by other models that reduce embodiment to coding, such as linguistic constructionism, Lacanian psychoanalysis, and structural anthropology; and in this commonality, we can identify a shared ideological vector, Mirowski's 'primal metaphor'.

The same motif is identifiable, for example, in the drive towards centralised bureaucracies which control local behaviours, such as the Andean potato farmers and the British psychotherapists discussed above. Globalisation, with its universalisation of specifications and trade laws, and its homogenisation of cultural values, is clearly consistent with this dominance of localised material and lived realities by distant symbolic authorities. Such schemes, which substitute symbolic algorithms for the emergent ecological interactions of embodiment, will be inadequate until the day dawns when the entire world, along with the vestiges of its humanity and organicity, is determined by symbolic codes quite detached from its organic functioning.

How does this radical transformation affect our human character? This is a question that I will address in the next chapter.

6
The Industrialised Individual

The embedded self

Our great symbolic power does not make our colonisation by industrialism inevitable; but it is nevertheless an Achilles Heel which loosens the constraining power of the real world and thereby makes us vulnerable to a range of ecologically malignant ideological systems. The capacity to focus on alternative realities is perilously close to the capacity to ignore current ones; and the size of the association areas in the human brain could be seen either as our great advantage over other species, or as a deficit in our relation to reality which will eventually prove terminal. *Which* of these views we incline towards will depend on the context; for those traits likely to lead to success in industrial society are not the same ones that would ensure survival or successful reproduction in non-industrial societies. A focus on abstract thought, an individualistic attitude, a casual psychopathy – these traits, although they would be poisonous in most indigenous cultures, would all seem to have short-term benefits for the individual *within industrial society*; so their lack of long-term adaptiveness is due not so much to a mismatch between the individual and the social context as between this social context and *its*, natural, context. In short, what will crucially determine the survival (or not) of *homo sapiens industrialensis* is whether the society that has 'grown out of our brains' is terminally alienated from the natural order, or whether it can rediscover its foundations in the natural world.

But even within industrial society, our 'success' may not be quite what it seems. Consider, for example, some of the implications of our attentional facility. We can direct our attention with a narrow or a broad focus; within the body or towards the outside world; to the past, present, or future; and in ways that are highly oriented towards current

221

realities or in ways that are hypothetical or entirely imaginary. Attention may be consciously articulated or allowed to 'free wheel'; and it varies according to whether we are awake, asleep, or under the influence of a drug. Both *because of* and *despite* this flexibility of focus, subjectivity has a distinctly partial and incomplete view of our life-situation. In addition, individual defence mechanisms employ our capacity to deny, repress, distort, rationalise, and otherwise misperceive our situations to protect our own psychological equanimity. To these misperceptions it is necessary to add all those that result from the intrinsic limitations of our perceptual and cognitive apparatus.

There are further types of misperception which stem from the complexity of the social systems we inhabit. Behaviours may have different effects at different levels of a complex system, and we may be aware of only a few of these effects. As Roy Bhaskar has pointed out, we may not go shopping in order to strengthen the capitalist system nor get married in order to perpetuate the nuclear family; neither do we drive our car to increase global warming nor withdraw our money from the bank in order to contribute to economic collapse; but those may nevertheless be the unintended consequences of our behaviour.[1] Like the bee that unwittingly pollinates the flower, we contribute to systems which we have little awareness of and which partly determine our character and the effects of our actions.

Furthermore, the functioning of the *industrial* system not only does not depend on our awareness of its character, but may actually depend on our *not* being able to understand it.[2] That is, the system may rely on its lack of transparency and inherent complexity in order to enlist our willing participation, so that we misconstrue, say, the motives of advertisers, the importance of social status, or the ecological credentials of corporations. In short, as Katherine Hayles suggests, the comforting view that consciousness has a controlling overview of our situation is mistaken:

> conscious agency has never been 'in control'. In fact the very illusion of control bespeaks a fundamental ignorance about the nature of the emergent processes through which consciousness, the organism, and the environment are constituted. Mastery through the exercise of autonomous will is merely the story consciousness tells itself to explain results that actually come about through chaotic dynamics and emergent structures.[3]

While this may be an overstatement – we do have *some* conscious control over our situation – most people can say little about how they are located either within a natural environment that involves particular watersheds, lunar cycles, ancestry, and local plants and animals, or within an industrial environment that operates in part through the manipulation of awareness, desires, and fears. True, some individuals have more control than others; but these individuals are generally powerful because they have internalised the system's functioning rather than because they can alter it, as many a president or prime minister has found when they try to challenge the political status quo. Real power generally resides in the *system*, not in individuals; and a divide has opened up between the systemic qualities of subjectivity, on the one hand, and those of the industrialist system, on the other. While the latter may originally have been an elaboration of subjectivity, it now assimilates and exploits subjectivity as a resource like any other; and human ignorance is now as important to its operation as human intelligence.

Industrialism therefore largely determines the character of subjectivity in the contemporary world. Certain elements of ideology are part of the landscape as much as the physical geography. These include the 'necessity' of growth; the notion that competition leads to progress and individual freedom; the view that everything can in principle be owned, commodified, and priced; the assumption that intelligent decisions are made on the basis of compromises between economic 'necessity' and ecological health; and the notion that psychological well-being is largely a matter of individual psychological hygiene. Whatever is not at home in this ideological landscape becomes endangered – collective morality, spirit, social interests, wildness; and the dominant form of selfhood within it is not so much the Cartesian *cogito*, with its certainty based in thinking, nor the Kantian knower, imposing structure on a shadowy world, but rather a myopic creature more like a silk worm, living and working within a small sphere of familiarity, yet operating within wider webs that are beyond its comprehension. Those areas within which we appear to have most power are often those in which we are, viewed from a higher level of the industrial hierarchy, most powerless. For example, a car-based transport system gives us an immediate, individual sense of freedom and power which is only indirectly paid for in less obvious ways, as a 1993 advertisement unwittingly admits:

> Illogical as it may seem, the simple act of motoring down the boulevard, exhaust burbling, that's what Viper ownership is all about. Only behind the wheel does it all make perfect sense.[4]

Our immediate sense of control comes at the cost of *being controlled* in other ways, both trivial, such as speed limits, traffic lights, and taxes; and more pervasive, as we are colonised by dreams of power, freedom, and macho autonomy. While car ads often employ images of the carefree individual driving fast and exuberantly down a wild and empty coastal or desert road, the reality is more likely to be one of traffic jams and speed cameras. Furthermore, our alleged 'freedom' is also paid for in other indirect costs such as pollution, noise, and the lack of places children can play safely; so what we experience as 'freedom' predominates only within a small sphere of conscious awareness, outside of which costs come back to haunt us.[5] Motoring strikingly encapsulates the contradictions between fantasy and reality, and between the zealously cultivated zone spotlighted by consciousness and the social, material, and political realities that impinge on our lives in ways we cannot perceive. As the racing driver Emerson Fittipaldi put it: "Outside my car I feel I am nothing; in it, everything changes, and I feel the virile master of the world".[6] An adequate account of human life, then, requires that we recognise the interplay between our limited autonomy and the way that this autonomy is assimilated to wider networks of power that we have little awareness of and even less control over.

Any organic entity, whether a human being, a mollusc, or an ecosystem, is partly defined by the systemic interaction of functional levels and areas; and as Bradford Keeney points out, ecological pathology involves "the maximisation or minimisation of any variable in an ecosystem"[7] – in other words, a loss of systemic balance and integrity. Industrial processes thrive on precisely such losses of systemic balance, tearing apart the natural order as we try to maximise production, profit, and share prices, and minimise cost, waste, and energy use. Consequently, there is a structural conflict in the industrialised world between ecological health and industrial growth; and parallel problems reflecting a similar loss of balance emerge in other arenas such as our daily lives as we try to be the thinnest, the happiest, or the wealthiest, wreaking havoc with complex cultural and natural balances. This maximisation and minimisation appear successful within our symbolic understandings and their technological materialisations so long as we cut out our awareness of the inevitable indirect costs; but they cannot be successful over the longer term or within broader contexts, because as embodied beings in the world, we are multiply connected to the world whether we allow ourselves to be aware of it or not.

Seeing individual experience and behaviour within the context of the interaction between a number of domains, each reflecting 'systemic coalescences' at various levels of functioning, raises the issue of possible conflicts between the functioning of different domains. We cannot assume, for example, that there is some sort of magical synchrony between, say, our biological functioning and the industrialist actualities we exist within, since these are currently embedded within divergent systems. In fact, given that our biological, hormonal, and neurological functioning evolved under the very different conditions that existed thousands of years ago, the question of incongruence between our bodily predispositions and the world we find ourselves in is a real and urgent one, especially as this incongruence is no longer simply an unfortunate by-product of industrial growth, but has become a structural part of the economic system itself through the exploitation of human desire and despair.

This is an issue, however, which theorists have often evaded through a selective attentional focus on disciplinary preoccupations such as biology or culture. In other words, dominant theories veer towards one or other type of reductionism, thereby transforming any possible mismatch between embodiment and social context into a conflict between disciplines. At one extreme, biological or psychological reductionists have methodologically ignored our embeddedness within social and cultural frameworks, taking the autonomous 'thinking self' as a pre-ordained entity to be analysed and dissected. In the world outside academia, legal and corporate orthodoxies also assume the autonomous, responsible agent and consumer who makes decisions and choices, even while they undermine this autonomy. To the extent that they adopt this tactic of reduction to the individual, social scientists are therefore aligned with industrial interests, supporting those structures which operate beyond consciousness and with disregard for human well-being.

But the opposite extreme of cultural reductionism is equally inadequate. By claiming that subjectivity is simply a by-product of cultural, and particularly linguistic, frameworks, any possible tension between the embodied person and his or her social context is eliminated; so it is hard to see why one cultural framework should be preferred over another. Furthermore, the person is disconnected from their embodied sense of agency, since discourses supposedly 'speak through them', and 'constitute' our identities.[8] Foucault is the theorist who most transparently grappled with this issue, turning in his later work from his earlier view that the person was simply an effect of 'power', devoid of independence, through a middle period of tacit confusion in which he seemed to suggest a culturally-constructed self which could also be oppressed by that same culture, towards a more realistic, if implicit, recognition that the person embodies properties that may be independent of, and so may conflict with, cultural and linguistic structures.

Nevertheless, the view that we are discursively or culturally constructed is not without a grain of truth – although it is a different sort of truth from that its authors intend. There *is* a sense in which we *are* partially constructed by social frameworks, particularly visual media, due to our entrainment within industrialism. But this is not a *general* theory of human being: rather, as we have noted before, it is an unwitting account of a pathological process which disrupts our healthy, sensorily-mediated relationship with the world, substituting symbolic interests that are ultimately motivated by capital growth. If, for example, my choice of diet is influenced by advertisements rather than by my embodied sense of what I need, it will indeed be true that this area of my character is 'socially constructed'; but this is a pathological intrusion into my life rather than a healthy form of character development, reflecting my assimilation to webs of meaning that originate in commercial interests and disrupt the relation of my embodied needs to the world. If the starting point of theory is 'the way things are', the result is an *incorporation* of pathology that dangerously excludes any possibility of *critique*. While constructionist approaches are most obviously problematic in this respect, apparently quite different models such as Rational Choice Theory and cognitive-behavioural therapy incorporate ideology more subtly in their emphasis on 'rational' thought, which assumes the two overlapping separations of the person from context and of thinking from embodiment.

A rudimentary systemic understanding is the minimum requirement if we are to act with any degree of sophistication. As we saw in Chapter 5, for example, we can understand recycling, and the 'greening' of industrialism more generally, as enabling us to continue fundamentally the same, destructive way of life, so allowing us to continue up the same industrialist cul de sac for a while longer. In the same way, the psychotherapy and self-help industries often allow people to feel better *despite* the social conditions and huge economic inequalities that are imposed on them; and so the outcome, if not the conscious intent, of these industries is to perpetuate these oppressive conditions. Such ploys have a long history: as Donald Munro points out, the Chinese Communists condemned the Confucian search for tranquillity as an early attempt "to harmonise relations between the rich and poor ... in an attempt to blind the people to the actual contradictions that did exist in society".[9]

Formal logic and the forms of everyday thinking that are modelled on it, which assume a two-dimensional 'playing field' in which something cannot belong to the class of actions that are natural and the class of actions that are unnatural at the same time, are unable to grapple with such complexly nested situations. For example, eating prawns may be 'natural' in a biological sense, but may also embed us within agricultural systems which are destructive of nature. In assessing our actions, it generally makes sense to consider that the more inclusive level trumps lower levels; and ignoring these wider contexts amounts either to blindness or denial. For example, in the film 'The Bridge over the River Kwai', the Colonel decides that because he and his men are forced to build a railway bridge for the Japanese, they will demonstrate the superior standards of British engineering by building the best bridge they can. Only gradually does the Colonel admit that by doing so they will be aiding the Japanese war effort; and after struggling with his conscience, he decides to blow up the bridge they have just built. Only a systemic awareness, giving rise to what Bhaskar calls 'explanatory critiques',[10] can allow us to avoid behaviours that covertly serve industrialism and are therefore destructive of the natural order. In contrast, suggestions that we should 'blur' the boundaries between what is human and what is mechanical, or what is natural and what is industrial, collapse systemic understanding into a single level of the hierarchy, further mystifying the already cognitively challenging task of disentangling the natural from the industrial, and so ultimately hastening the disappearance of the wild.

The shrinkage of experience

The silent assimilation of subjectivity by industrialism, however, does not occur only because we are unable conceptually to grasp the complexities of industrialism's systemic functioning. 'Developed' societies have edged towards living in a symbolic realm that, as we have seen, is materialised as an urban, domesticated environment; and given this contrived consistency between our symbolic models and the landscape, we effectively seal ourselves into a simplified sub-world that is more like a mirror to cognition than a world that resonates with our embodied awareness. Mountain lions, ancient forests, and canyons evoke aesthetic feelings in exactly the way that warehouses, car parks, and refrigerators don't; and urban or agricultural landscapes seldom provoke complex or nuanced emotional reactions. We admire digitally retouched calendar images rather than real landscapes, confining ourselves within the circularity of symbolic reference. The world increasingly accords with our expectations: we are seldom surprised by a scent, and our breath is seldom taken away by a view. Generally, we simply register what is out there. Sensuality, passion, even surprise become experiences that have to be generated *within* us – something *we* supply, rather like the 'personalised' message on a commercial greeting card.

Susan Sontag writes that in "New York sensuality turns completely into sexuality – no objects for the senses to respond to, no beautiful river, houses, peoples. Awful smells of the street, and dirt ... Nothing except eating, if that, and the frenzy of the bed."[11] Sex in uncongenial surroundings becomes *just* sex, conforming more closely to the unromantic psychoanalytic jargon of sexual 'needs' requiring an 'outlet', and reflecting the current disdain for the language of romanticism and sublimation. While sexual techniques are widely promulgated and discussed, the emotional overtones have faded. As Eileen Crist puts it,

> The celebration of frank sexuality strives vainly, almost tragically, to assert Eros in a world where Eros is sorely missing. The contrast is as jarring as images from Brazil's latest Rio Carnival when thirty tons of fish floated on the city's lagoon poisoned by human waste from a ruptured sewage pipeline. (Not that the stench stopped the parades). The more ecosystems are impoverished, the more the senses starve in a world of wounds and eyesores, the more shrill does sexuality become—the quest for carnal intensity striving, it seems, to counterbalance the diminished sensuousness of the world.[12]

This view of sexuality as a residual impulse of sensual connection to a world that no longer exists reminds us of Marcuse's concept of 'repressive desublimation'. We are *allowed* our sexual excesses, perversions, 'freedoms', precisely because these have been drained from the 'real' world of business, work, and politics. In Michael Steinberg's terms, "in a lover's arms we are momentarily released from the stifling and isolating web of words and calculation".[13] But as Marcuse made clear, this does not amount to any sort of 'liberation'. We become like apes imprisoned in a zoo, allowed to copulate compulsively and consume an endless supply of bananas; but we are not permitted to escape from our zoo and interact with the world in a more fulfilling way. Sexuality writ large transcends the egoic self, opening us up to a shared subjectivity; while the pursuit of 'sexual pleasure' is a narcissistic pursuit which isolates us. We have gained in freedom precisely to the extent that the field in which freedom can be exercised has been enclosed and diminished by changes that are almost 'beyond the radar' of human awareness. Thus 'sexual liberation' is industrialism's grudging acquiescence to residual aspects of embodiment – as well as a commercial opportunity.

While the inherently transcendent aspects of sex carry on a sort of rearguard action against encroaching domination by economic rationality, we also adapt to the lack of aesthetic resonances by prioritising cognition. In an un-beautiful place, it is more comfortable to *think* in a detached way about our surroundings than it is to *feel* the absence of beauty. Nobody feels aesthetically stimulated in a multistory car park – although we may *think* about where to park, how to pay, and so on. Lacking an environment that encourages feeling, we become less feeling people. For example, compare picnicking on a quiet, lovely riverbank, and at the side of a motorway. In the first of these, subjectivity escapes from our individual boundaries, reaching out into the landscape, so that the taste of food seems to be enhanced by the surroundings. In the second scenario, however, we recoil from the noise and the traffic fumes, trying to focus on the taste of our lunch and to cut out other sensory experiences. In this case, sensory experience becomes *individualised* as we try to distance ourselves from the noisy environment; and it also becomes *simplified* by this situation. Rather than the total experience of sandwich-and-river-and-sound-of-running-water, we focus just on the taste of the sandwich.[14]

This is an example of 'positive feedback': individualism combined with technology degrades the environment, and a degraded environment drives us further into individualism, so that both subjectivity and the 'environment' conform to a simplistic emotional topography. As feeling is driven 'inside', abandoning its resonance with the world, we construct a world that is more object-ive, unreflective of feeling. All too often, cultural and psychological theory collude with this double imprisonment. An example is SueEllen Campbell's exploration of Edward Abbey's views on the bee and the "soft, lovely, sweet, desirable cactus flower";[15] his need for "the embrace of a friend or lover, the silk of a girl's thigh, the sunlight on rocks and leaves";[16] and his account of "floating down the river in Glen Canyon ... enjoying a very intimate relation with the river ... only a layer of fabric between our bodies and the water".[17] Campbell wonders whether Abbey's enthusiasm for nature can be understood as sublimated sexuality, in conventional Freudian fashion. The implication seems to be that our relation to nature should be *devoid of* eroticism, conforming to a more detached style of interaction. But what if Abbey's writings, far from implying a suspect preoccupation with 'sexuality', rather reflect the *natural* form of our erotic capacity, so that 'sexuality' can be seen as part of something much more pervasive? Is Campbell implying that the *current, narrowed-down* form of sexual expression is its *natural* form? What if the sensuousness of Abbey's experience isn't a 'sublimation' of a fundamentally 'sexual' concern, but is rather the *recognition* of a wider sensuality that potentially exists *in the world* (including in ourselves), and in our relations *to* the world? All too often, we abandon this sort of empathic reaching-out and replace it by an economically-inspired calculus of costs and benefits: 'Does he give me what I need?'; 'Would person X give me more than person Y'? 'Our relationship feels unequal'; and so on. We promise to 'love, honour, and obey', imitating banknotes or business contracts. Passion, as we will shortly see, is reduced to colourless 'preferences'. As Alan Bloom puts it:

> Abstract reason in the service of radically free men and women can discover only contract as the basis of connectedness – the social contract, marriage contract, somehow mostly the business contract as model, with its union of selfish individuals. Legalism takes the place of sentiment.[18]

This contractual model of relation is utterly different to a sensuality that draws one 'out of oneself' *into* the world in an open and generous way; and it is hard to overestimate the emotional havoc wreaked by the assumption that love – and emotionality generally – can be constituted through 'legislative rationality', for unlike house ownership, it must be discovered anew each day. Like other wild things, it cannot be promised in advance, nor can it easily be tracked, understood, or predicted. Like 'falling' asleep, it cannot be egoically willed: rather, it is something one can only *allow* to happen, and it involves a change from habitual modes of perception and awareness. Perhaps Picasso was referring to a similar perceptual change when he said to Gertrude Stein: "When I see you, I do not see you."[19]

Theory, to the extent that it is necessarily codified symbolically, is knowledge *about* rather than knowledge *of*, and can only generate repressive results – as in Campbell's analysis, emasculating Abbey's implicit attempt to rejuvenate the world's lost sensuality and reinstituting a normative repression. The withdrawal of sexuality into the self, complementing a de-sensualised and deromanticised 'environment', is a necessary prerequisite for a liberal individualist economy, just as the confinement of water behind a dam is a prerequisite for its use in power generation; for if sensuality and wildness are all around us, permeating the landscape and other forms of life, how can they be harnessed, disciplined, employed? Our emotional isolation from the natural world leaves us in the position of Del, a character in Alice Munro's *Lives of Girls and Women*, who, resisting social pressures to choose 'feminine' interests, says: "I wanted men to love me, *and* I wanted to think of the universe when I looked at the moon. I felt trapped, stranded; it seemed there had to be a choice when there couldn't be a choice."[20]

Most theory *assumes* this industrialist landscape in which feeling is necessarily enclosed within individual boundaries. The senses become almost redundant because exploration is largely within the *symbolic* realm, a conceptual exploration within a pre-defined world. Consequently, even a leading poet of nature can suggests that a "person with a clear heart and open mind can experience the wilderness anywhere on earth. It is a quality of one's consciousness".[21] What matters is supposedly what happens within the mind, and the material world becomes increasingly irrelevant. This is why reading Freud is about as erotic as a gynaecological examination – because Freud operates not in the realms of the senses, but in a conceptual realm of ideas and rational argument. Like Campbell, he *assumes* the primacy of this conceptual realm over that of raw human experience, so that the sensuality of the Id is replaced by the conceptually ordered bookkeeping of the Ego. Thus we learn that the senses, supposedly, cannot lead us anywhere that is conceptually undiscovered; and as Bloom notes, there "have been hardly any great novelists of love for almost a century … you can be a romantic today if you so choose, but it is a little like being a virgin in a whorehouse".[22]

Subjectivity, trimmed to fit the symbolically reconstituted world, becomes merely a point in social space from which we observe others; and empathic relation is replaced by a conscious – sometimes legally enforced – code. Within this definition of individuality, people are 'naturally' competitive, and so *need* such codes to impose a veneer of consideration for each other. Conscious principle has replaced feeling, instinct, intuition; and *symbolic* relations have replaced *embodied* relations. While it is true that novelists and others can still *use* symbols to express a more complete and embodied world, the trend is for formal symbolic systems to dictate subjectivity, domesticating wildness and emasculating the organic world to fit an economic one. This is a gross diminution of what subjectivity and relatedness can be; and the radical character of the change is concealed by the superficial similarities at the behavioural level. Just as the bluebells in the market appear the same as those I walk past in the wood, so people look, and in some respects behave, as they did before.

The atrophy of memory

Just as we are losing the sense of simultaneous immersion in multiple levels of systemic organisation, so our immersion in natural rhythms and cycles is also slipping away. Christopher Lasch was one of the first to notice our increasingly present-oriented focus, suggesting that we are losing "the sense of belonging to a succession of generations originating in the past and stretching into the future"[23] – one facet of an increasingly narcissistic society. According to Lasch, the past is now viewed as something obsolete, to be superseded by the present, rather than part of a continuing historical tapestry that extends backwards into the lives of our ancestors and forward into the possible futures of our children and grandchildren. Similarly, Rowan Williams suggests that "we're losing the ability to tell stories about ourselves, about a continuous self ... we break up our selfhood into 'bits', into film 'stills' [so that] how we tell the story becomes very problematic".[24] Consciousness tends towards a sequence of disconnected fragments – the mind as computer screen – rather than a continuity of interwoven threads. Walter Benjamin, too, pointed out that the temporal aspect of experience was fading in industrial society, leading to an atemporal, superficial mode of experience that he referred to as 'erlebnis'.[25] Likewise, Richard Terdiman points to the "memory crisis", involving an "uncertainty of relation with the past", threatening "the very coherence of time and of subjectivity" and causing "the disruption of organic connection with the past".[26] And drawing out the implications of these views, Eric Hobsbawm suggests that "it has for the first time become possible to see what a world may be like in which the past, including the past in the present, has lost its role, in which the old maps and charts which guided human beings, singly and collectively, through life no longer represent the landscape through which we move ... we do not know where our journey is taking us, or even ought to take us".[27]

While we think of memory as a *cognitive* function, it is of course also an *embodied* quality, reflected in our evolved physicality, feelings, and awarenesses. The physical makeup of our bodies is a sort of record of our evolutionary past; and as industrial society accelerates according to its own dynamic, so its inconsistency with our embodiment becomes more marked. The ready availability of sugary and fatty foods, and types of travel and work that allow us to become increasingly sedentary are obvious examples of this. Likewise, while we have evolved a fear of heights, we have no such fear of horizontal travel – which statistically is the more dangerous.[28] Our bodies carry silent memories which seek a particular sort of world that has become forgotten.

The reader will by now recognise a recurring pattern when I suggest that this loss of connection with other times is also apparent in the world we create through our actions. If, for example, an area of grassland is ploughed and planted with wheat, this disconnects it from its past history, giving it a meaning that can only be understood within a current, commercial context. The point is not just that changes have occurred in the landscape; but rather that long-term *patterns* of natural change that extended into the distant past and would have extended into the future have been uprooted, so that the meaning of the land is obliterated, ready for the imposition of a new, commercial meaning. This destruction of natural patterns and rhythms is simultaneously an ecological and a human destruction, as Lye Tuck-Po relates in describing the Batek understanding of forest ecology as intrinsically *historical*:

> If the pathways are gone, it becomes harder to keep in touch with the past. Remembering is an act of sociality, connecting an individual to the vast networks of relations that have led to the present. Pathways have no meaning without knowledge of kinship, moments, events, and experiences told and retold over generations. In returning to those old trails, in making remembering a special event that necessitates practical action, the Batek are affirming their bonds with the old people. As long as they do so, the trails will continue to have meaning and to draw the Batek back to times and places past ...
>
> But as landscape deteriorates ... the knowledge becomes cognitive activity alone, bereft of the practical and physical base that gives it continued meaning. The biographical associations, memories, narratives, and conceptual maps are detached from the landscape, becoming history without a place.[29]

Temporality, in other words, is not an abstraction; rather, it is embodied within human physicality and within the world. Destruction of the landscape, or withdrawal from it, therefore destroys our sense of temporal relatedness, so that knowledge becomes 'cognitive activity alone', the past becomes an *idea* preserved only in books, attachments to nature are categorised as 'nostalgia', and the future loses any grounding in history. 'Nostalgia', however, is the yearning for a relationship that has been lost, the vestige of a felt attachment. So from within a symbolically catalogued temporality that abandons meaning, it makes perfect sense to act only according to present needs, without concern for the future.

As we saw in Chapter 2, our understanding of time is not just linear; it is also *discontinuous*, chopping up temporality in much the same way that we conceptually and materially chop up the physical world. This view of time cannot adequately portray natural events, which are often continuous, one situation leading into another. We clear-cut areas of forest to produce agricultural land; and we demolish old buildings to make way for the new. There is no interweaving of old and new here: we clear away the rubble of the old to make way for the new in a sometimes disorienting sequence of sudden changes. The model for these changes is not nature, but cognition, abetted by TV: just as we refocus attention on each rapidly changing image, so we build the same fragmentation into the manufactured world. This contrasts with indigenous cultures' understanding of time as continuous, and also as containing both circular and linear components, as indicated in Robin Ridington's portrayal of the Dunne-za of northern British Columbia:

> Their time is different to ours. The old man and the boy circle around to touch one another, just as the hunter circles around to touch his game. They circle one another as the sun circles around to touch a different place on the horizon with each passing day. During the year, it circles from northern to southern points of rising and setting. It circles like the grouse in their mating dance. It circles like the swans who fly south to a land of flowing water when winter takes the northern forest in its teeth of ice. The sun circles like the mind of a dreamer whose body lies pressed to the earth, head to the east, in anticipation of another day's return. The sun and the dreamer's mind shine on one another. ...
>
> Historical events happen once and are gone forever. Mythic events return like the swans each spring ... [They] are true in a way that is essential and eternal.[30]

Here, human and nonhuman events are related through the rhythms they share, rhythms to which we are mostly blind due to our fixation on the present. At an *embodied* level, however, it may be that we *need* to feel part of the sort of temporally contiguous landscape that is subjectively foregrounded in non-industrial societies. Most variants of environmentalism arise out of an awareness of the contrasts between what *is*, what *should be*, and what *was* – in other words, they reflect the mute, embodied awareness of the lost temporal patterns.[31] If we lose touch with these patterns, we also lose the ethical anchors and sources of 'ontological security' that come with them. If, for example, we accept that wilderness "is a product of ... civilization",[32] or, increasingly, that 'it 'never existed', then theory is constrained to fit present actualities, and 'human nature' becomes identified with *current forms* of human nature, as Alf Hornborg points out:

> the current fashion in anthropology is to dissolve any distinction between the modern and the premodern as a modern fabrication.

Gemeinschaft is now nothing but a fabrication of gesellschaft, and the ecologically sensitive native merely a projection of industrial society. The rather remarkable implication is that, in the course of the emergence of urban-industrial civilisation, no significant changes have been taking place in terms of social relations, knowledge construction, or human-environmental relations. The closely-knit kinship group, locally contextualised ecological knowledge, attachment to place, reciprocity, animism: all of it is suddenly dismissed as myth.[33]

The constriction of theory and experience within a contextless present pushes the past aside as a moral and ontological standpoint. We are encouraged to 'break the mould', to 'challenge conventional wisdom', to engage in 'blue skies thinking' – in other words, to demolish existing structures and to begin again from a pile of conceptual 'raw materials'. However, a tree I have known all my life would be part of a context that is temporally meaningful, locating me within a world that makes sense; and the psychological impact of the destruction of such a tree may be compared to that of bereavement. As one Colorado farmer said, 'it's like losing a kid'.[34] Such embodied feelings of loss, because they do not correspond to a cognitively, legally, or economically recognised loss, are *invisible* to the economic system which, as Marx said, "has made us so stupid and one-sided that an object is only ours when we have it – when it exists for us as capital, or when it is directly possessed, eaten, drunk, worn, inhabited".[35] Such losses, in other words, are already *built into* our economic system, and so seem to reflect 'the way the world is'. Thus industrialism, with its assumptions that everything is owned rather than socially shared, impoverishes us twice over: once when a resource is removed into private ownership, and again when it is physically exploited. Felt relations therefore inhabit a twilight realm of the already half-lost and the cognitively abandoned, which makes their final loss more difficult to recognise or mourn; and in one of those vicious circles that are common to industrialism, we learn to avoid such bereavements by adopting a more emotionally withdrawn, less generous attitude.

Such attitudes become embedded in the infrastructure that surrounds us. For example, a commodity such as a car will be defined not so much by the way it connects past, present, and future as by its success in *distancing* us from a past viewed as obsolete, inefficient, and out of date. It will contain few signs of its origins within iron ore bearing strata of rock in a far off country; and in the not so distant future, its final resting place will be some anonymous breaker's yard. Its meaning, then, is a present meaning, cast loose from its historical contexts, uninformed by either the past that led up to it or the future it may be a step towards. In Rilke's words, it has "no stories to tell".[36]

While contemporary understandings view memory as a purely cognitive process which occurs *within the individual*, indigenous understandings view memory as embodied in the person, in culture, and in the world. But this description is not quite accurate, since it implies a separation between these three realms. For an Australian aboriginal, for example, to remember is to be part of a landscape that is cultural and psychological as well as geographical. Veronica Strang found that visiting a place with an aboriginal person

> almost invariably results in a flood of information about its name, the ancestral beings responsible for its formation, the rituals and responsibilities relating to it, whose 'home' it is, the family or clan it belongs to, the historical events that have taken place at the site, and traditional and current uses of it. Myriad details emerge – who worked on the station nearby and mustered this area; what bush food or craft materials can be found; who was born, dies, or was buried nearby; whose grandparents camped at the waterhole; whether fish were caught, or crocodiles sighted.[37]

For Australian aboriginals, like the Batek, culture and memory are *part of the land*, rather than located in a symbolic realm hovering above the land. In contrast, Strang found that non-aboriginals tended to have highly specialised, often technical, readings of the landscape, as illustrated by a government scientist's description of his approach:

> We've been able to define what we see as broad functional areas ... they have been called biogeographic regions. The finer divisions of those provide a next break up into provinces, and then within provinces there are land types and within the land types we're constructing, we're applying this hierarchical classification for wetlands. ... The hierarchy is one of systems and subsystems and classes and subclasses.[38]

The aboriginal description is detailed and connective, weaving together personal, cultural, and ecological aspects of place; while the scientist's is impersonal and technical, reflecting the imposition of cognitive classifications. And whereas for aboriginals the dangers of the country are pragmatically assessed, for stockmen "the romantic vision of the landscape as dangerous and wild is ... appealing", showing the assimilation of the landscape to a symbolic scenario in which individuals – usually men – heroically do battle with natural hazards. Strang tells us that such romantic evaluations do not "construct a permanent commitment to a specific [part of the] country; the nostalgia is for the experience and not for the land, which remains only a generalised theatre of activity".[39] While the stockmen measure themselves *against* the land, Aboriginals experience themselves as *continuous with* it; and this continuity is temporal as well as ecological in a narrower sense. What is implied here are two quite different forms of subjectivity – one systemically integrated into the natural world, the other defined by its isolation from it.

This leads us back to a recurrent conclusion: industrialised peoples inhabit a symbolic realm that interposes a veil between ourselves and the world rather than relating us to it. In effect, we live in insulating bubbles of constructed experience; but these bubbles are not detached from the world in the sense of being neutral and indifferent to it, but form templates for the transformation of the world into its industrialist form. We too are caught up in this transformation; and while native peoples can be separated from their land only at great psychological cost, we have *already* experienced this loss, so that what we call civilisation is an adaptation to and a defence against an environment that conflicts with our embodied expectations and needs, concealing and misdefining the malaise that results from this mismatch.

The blindness of individualism

The boundaries that industrialism draws around us skew experience and understanding towards individualism; and the belief that our inner character will determine the course of our lives is prioritised in an unbalanced way over social, ideological, and material circumstances. Of course, it is hardly surprising that a society in which everything is owned by someone will need to view individuals as clearly demarcated both from the world and from each other; but our individual isolation is understood not so much as a practical expedient as an assumed aspect of reality. As Michael Steinberg remarks, the "primacy of the individual thinking being ... is not a formal proposition for us in the West. It is the character of experience itself".[40] Thus the favourable tone of concepts such as 'internal locus of evaluation' suggests that it is our *individual* responsibility to *create* meaning, so that the loss of meaning from the outside world is viewed as less of a problem. But as Richard Lichtman asks, while forms of therapy that encourage such individualistic traits may alleviate distress, "is this success ... gained at the expense of accommodating clients to the structure of the society which has so profoundly caused their initial misery?"[41] In other words, while withdrawing into our own refuge of order and meaning may create a temporary haven from social disorder and fragmentation in the world outside, doing so will itself amplify that disorder and fragmentation – a sort of cultural 'tragedy of the commons' in which we all rush to save ourselves from the landslide of social disintegration.

To revisit another recurrent theme of this book, theory all too often incorporates this withdrawal rather than commenting critically on it, so that psychoanalysis, as we saw in Chapter 4, becomes a disembodied theory dealing with disembodied psychological life. There are various explanations for this inward-looking emphasis on psychodynamics. Jeffrey Masson argues that a failure of nerve may have been responsible, and that identifying the relatives who abused his (usually female) patients would have ended Freud's career.[42] Complementarily, David Smail has pointed out that while Freud's theorising drifted ever further from the down-to-earth realities of survival, there is evidence that the character of his work was profoundly motivated by his fear of poverty and his drive towards material success.[43] Ironically, it was Freud himself who recognised that thought can in some cases not merely fail to articulate physical realities and embodied feelings, but can actually provide a web of rationalisation which defends against them; and despite his attempts at self-analysis, this mechanism seems to have operated very effectively to draw a curtain around the factors that influenced his own work. Our headlong flight into symbolism often disguises the material realities of our lives.

Like Freud, psychotherapists of all theoretical persuasions have a material interest in perpetuating the assumption that psychological distress is rooted in causes within the individual and is only distantly related to anything that might be happening outside the domain of psychotherapeutic intervention. This pervasive ideological colouring serves clear political functions: as Smail argues, it is convenient for those who benefit from perpetuating social structures, since they are able to claim that peoples' suffering should be dealt with "not by attacking social injustice, but rather by the *personal* readjustment of the disadvantaged themselves".[44] Similarly, Paul Antze and Michael Lambek note that "the rise of therapeutic culture has gone hand in hand with widespread political disengagement" – for example, in the way that collective guilt among veterans is medicalised as post-traumatic stress disorder. This is not to suggest that therapists are necessarily financially motivated charlatans, as the individualistic emphasis that suffuses therapy is so much a taken-for-granted part of the broader ideological scenery that it requires considerable effort to think outside it. As Antze and Lambek add, "where legal and diagnostic categories become the only legitimate terms in which to remember suffering, then it becomes important to ask what has been forgotten".[45]

Just as therapists focus on individual psychodynamics, so the legal system focuses on individual *blame*. An accident may have many contributory causes; but often the law looks for, and finds, a responsible individual. For example, in 2001, Gary Hart, driving a Land Rover, skidded off the M62 onto the East Coast main line, derailing an express and causing the death of ten people. It was claimed at his trial that he was short of sleep. He was found guilty of causing death through dangerous driving, and jailed for five years.

It was also pointed out, however, that various contributory factors also played a part in the accident. There was a gap in the crash barrier at the point where Hart skidded off. The judge described the crash as a 'freak accident' that was unlikely to reoccur. His defence team claimed that building a motorway just above a railway embankment was inherently dangerous. Furthermore, some commentators criticised government policies which over many years had favoured road transport over the intrinsically safer rail system. Despite these factors which were of a political or infrastructural character, both press coverage and the legal process focused almost exclusively on Hart's state of mind at the time of the crash. As Zygmunt Bauman puts it, in "our 'society of individuals' all the messes into which one can get are assumed to be self-made and all the hot water into which one can fall is proclaimed to have been boiled by the hapless failures who have fallen into it."[46]

Such cases illustrate the 'fundamental attribution error',[47] in which we overemphasise individual, internal causes of behaviour such as personality factors or psychopathology, ignoring the causative factors which lie within the larger social context. For example, Scott Atran points out that "a collective sense of historical injustice, political subservience, and social humiliation"[48] are the most clearly identifiable factors underlying suicide bombing; yet there is a common assumption, both in the media and among researchers, that such 'terrorism' is the result of individual factors such as psychopathology or personality disorders. For years, Jacqueline Rose tells us,

> Israeli secret service analysts and social scientists have been trying to build up a typical profile of the suicide 'assassin', only to conclude

that there isn't one ... burrowing into the psyche of the enemy, far from an attempt to dignify them with understanding, is a form of evasive action designed to blind you to the responsibility for their dilemma that is staring you in the face.[49]

At a more mundane, folk-psychological level, we view qualities such as greed, materialism, and narcissism as existing because of 'human nature', not because we inhabit a political system which encourages us to behave individualistically. To the extent that this stampede towards individualism complements industrialisation, subjectivity is confined within an industrialist 'reality', and becomes incapable of transcending this reality or realigning itself with the natural order – a situation that contrasts with previous eras in which nature, or the gods, or the 'great chain of being', or some similar transcendent entity provided a source of renewal which went beyond what presently existed, so contextualising individuality and social life. For example, Hardt and Negri note that during the decline of the Roman empire, the Christian idea of redemption provided a reference that went beyond the empire.[50] A similar point is made by W. H. Auden: after describing the corruption and decay that permeated the collapsing empire, he ends the poem:

> Altogether elsewhere, vast
> Herds of reindeer move across
> Miles and miles of golden moss,
> Silently and very fast.[51]

Today, however, the 'miles and miles of golden moss' and the 'vast herds of reindeer' are diminishing along with transcendent spiritual awareness; so in the absence of any broader realm into which we could reground ourselves, identity becomes precariously centred around a private island of personal seclusion, defended by our iPods, our social personas, and the constructed boundaries of our cars and houses. Our retreat into increasingly privatistic experience is given added impetus by intrusive economic and ideological structures as corporations and the state maintain ever more sophisticated databases containing our personal preferences, occupational record, shopping habits, and so on. The relatively flimsy and vulnerable carapaces we construct to defend our islands of self-awareness, often involving public personas presented on social networking sites, may in part reflect a natural defensive reaction against the implosion of structures that are at odds with our evolved expectations and needs.

The portrayal of people as 'autonomous individuals' with exaggerated power over their own destinies serves several ideological purposes. Firstly, as noted earlier, it places responsibility for suffering, poverty, and success on individuals rather than on social and economic structures. Secondly, by cutting us off from structures which we might be grounded in, it conceptually reduces the natural world from a realm of shared value and intimate familiarity to an amorphous background. Thirdly, by failing to articulate systemic understandings either of industrialism or the natural order it conceals the differences between them, facilitating the replacement of nature by industrialism, and preventing us from understanding the operation of an economic system which impoverishes vast numbers of people, destroys nature, and hugely enriches a tiny group of relatively powerful individuals. Thus a conscious focus on individual character and self-interest coincides with an actual reduction in the power and autonomy of the individual as industrialism's systemic functioning becomes increasingly pervasive.

Our transmutation from natural to industrial creatures transforms all aspects of life, including our understanding of life itself. Rather than being a defining aspect of all organic entities, life becomes an individual *possession*, something we 'hang on to', 'lose', or fill up with 'experiences' and 'achievements'. Life comes to have an all-or-nothing, digital quality to it, rather than being something that ebbs and flows, strengthens and fades, and moves fluently between individuality and immersion in wider ecologies. In our old age, when most of our life-force has left us, we still cling on to 'life', and view death as a single, tragic event. But just as the 'individual' is an abstraction from the larger, systemic realities we inhabit, so is the individual life or death. This misunderstanding has implications for the way we live – and die; for if life is about interaction with the rest of the world, then we can be more or less alive at any particular moment; and living is more an analogue process of participation in wider life-structures than a digital acknowledgement that our heart is still beating.

The lack of relation between the individual and the world leads to a focus on personality characteristics which are curiously isolated from our activities. 'Self-esteem', for example, is often considered to be an intrinsically desirable characteristic rather than flowing from our own successful activity in the world. As Jean Twenge notes, teachers may be discouraged from correcting children's spelling and grammatical mistakes in case their self-esteem is damaged; and instead, children "should be 'independent spellers' so they can be treated as 'individuals'. Twenge asks us to "imagine reading a nuespaper wyten using that filosofy".[52] Here, self-esteem is viewed as one aspect of an autonomous self which can be specified in the symbolic realm quite independently of our behaviours. All too often, words are used to conceal the relation between self-esteem and our actions in the world; and consequently the truthful use of words is often experienced like a bucket of cold water. Ronald Dworkin relates that a prisoner, incarcerated for robbery and second-degree murder, complained of low self-esteem. The prison psychiatrist responded: "Of course you have low self-esteem. You're a murderer and a thief!"[53]

Twenge shows that the disengagement of self-concept from material reality can lead to some deliriously out-of-touch career choices. She cites the case of UC Berkeley engineering student William Hung, who decided to follow his dream of becoming a singer:

> Hung's singing was tuneless, but it was his jerky, utterly uncoordi-
> nated dancing that caused the American Idol judges to hide discreetly
> behind their ratings sheets as they choked back their laughter ...
> [Later,] Hung said he hopes to make a career out of being a singer. ...
> Rule of the modern world: Doing your best is good enough, even if
> you suck.[54]

Writ large, the consequences of developing a social and political realm defined in symbolic terms that are disengaged from reality may be less amusing. The destructive power of those ideological structures that we fail to perceive reduces democracy to a cosmetic façade for unacknowledged forms of control, so that as Ulrich Beck points out, we are forced to seek "biographical solutions of systemic contradic-tions".[55] In effect, our understanding of individuality is an example of the 'Platonic Backhand': using an industrialist template, we construct a profile of 'the individual' by abstracting certain qualities required by industrialism – intelligence, self-esteem, optimism, and so on – and then judge the adequacy of the person according to how closely they fit this profile. Adrianne Aron and Shawn Corne summarise this 'centripetal' effect of mainstream psychology in describing the writings of Ignacio Martín-Baró:

> Psychology, [Martín-Baró] thought, erases the very real things of
> life that make up what we are as human beings. It acknowledges no
> fundamental differences between a student at MIT and a Nicaraguan
> campesino ... With that denial as its working hypothesis, it simply
> ignores the social and economic conditions that shape the daily

lives of these individuals and their communities; structural problems are reduced to personal problems and factored out as individual differences ... Thus, the final destination of psychology's mission ... is a place where people are befuddled, belittled, homogenised, and left on their own to deal with their social oppression.[56]

Martín-Baró points out that psychological models such as McClelland's 'achievement motivation' and Rotter's 'locus of control' emphasise the individual reasons for 'success' or 'failure', assuming that such concepts realistically portray individuals' power to change their circumstances. However, this is "quite insulting when you look at the continuing efforts of Salvadoran campesinos to rise above conditions of exploitation and misery, only to end up in the same situation or worse than that of their forebears".[57]

In thus sloughing off the material realities that connect us to the world, as well as the systemic powers through which they are ordered, psychology's individual emerges as an abstraction who inhabits an equally abstract symbolic world of individual choice and agency, continuing affluence and progress, of 'green' capitalism and 'sustainable' growth. But once again, the abstraction is a template for a reconstructed reality; and psychology is here normalising an unrecognised *historical* transformation by portraying it as a universal normalcy.

This situation damages personhood and strengthens capitalism – for precisely the same reasons. If the person is needy and damaged, there is a 'market' for fulfilling these needs and providing therapy. If our diet is unhealthy, there is a market for supplements. If we are unhappy, there is a market not only for antidepressants, but for a range of leisure products to divert us – films, magazines, TV programmes. The problems associated with industrialism, then, cannot be realistically regarded as 'collateral damage' that are incidental to 'progress'. Often, they reflect changes which are intrinsic to the violent replacement of one system – the natural order – by another. In order to reconstruct the world, humans must be stripped of their membership within existing ecological and social systems, and must therefore be represented as individuals who 'naturally' make *economic* choices. The feelings of emptiness, the increasing narcissism, and the materialism which accompany industrialisation are simultaneously by-products of consumerism and *necessary* if consumption is to be maximised, demonstrating the systemic character of our entrainment within the industrial order. Similarly, environmental 'problems' such as noise, pollution, and water contamination become business opportunities for makers of sound insulation, air filters, and bottled water companies. From an industrialist viewpoint, a healthy, fulfilled, unmaterialistic person living in an ecologically healthy world represents a current failure and a future marketing opportunity.

The managed self

Beneath the intensifying discourse of individual choice and responsibility there is a powerful undertow in which power ebbs away from the individual and into bureaucratic and technological systems, along with the most significant forms of intelligence, perception and defence. The symbolic forms that were originally derived from human decisions have developed their own systemic momentum, and have returned to colonise us in ways that are both intrusive and cognitively opaque. For example, in Britain today it is now illegal for a pub to allow a singer to perform or a school to invite an author to address groups of schoolchildren without obtaining the appropriate permits. Increasingly, control has become built into the architecture of the modern world – into the intrusive micro-management of our lives through surveillance and centralised databases, through the increasingly sophisticated advertising and public relations industries, and through the built realities of the modern world.

Any individual who has tried to challenge bureaucratic decisions will quickly become aware that bureaucracies tend to constitute an effective means of dissipating the power of individuals or groups into the sands of administrative structure; and complementarily, the disorienting diversity of consumer choice serves an important *political* function in distracting us from our inability to affect significant decisions. Thus the assumption of 'individual freedom' as it is applied in modern industrial society mystifies rather than emancipates us, covertly confining us within a trivial consumerist space, while the important decisions are made outside this space, in much the same way that the Chinese Room insulates the person from defining meanings. Just as the deeply fissured landscape of the Colorado Plateau channels water into the eroded gulches and canyons which earlier floods have created, so social architectures that originally resulted from human choices now channel thought and behaviour in clearly demarcated ways, so that 'freedom' is freedom to perform within closely constrained limits, and behaviour which might materially affect the direction of industrialism becomes almost inconceivable.

Travel, for example, is closely surveilled and controlled by government bureaucracies and passport controls. How we travel is determined by car licensing authorities and train companies; how fast we travel is enforced by speed cameras and road design; which areas we can travel through is determined by the laws of private property, trespass laws, and other regulations; and where we travel will be recorded by automatic number plate recognition technology and mobile phone companies. Our internet browsing history, which can be regarded as a sort of updated Rorschach test of our interests, concerns, and fantasies will also be monitored, potentially giving governments and corporations unprecedented access to our private worlds. Richard Thaler and Cass Sunstein suggest what is, in fact, already happening – that we should be 'nudged' by a 'choice architecture' which is designed to guide us in particular directions.[58] And the word 'architecture' suggests that such 'nudges' are often physical; for there are resonances between Bentham's panoptical prison and the layout of supermarkets and malls, designed to channel us in particular directions so that we make particular choices. In particular, the character of our overwhelmingly car-oriented transport system 'nudges' us in the direction of individualism in a way that, say, a well-developed rail system would not. Symbolic structure has developed, expanded, and returned to reconstruct the world ideologically and physically.

Under these circumstances, 'truth' is often skewed away from consistency with the natural order towards consistency with ideology. This is the sort of divergence encountered in 1995 by a radio weatherman in California who (correctly) predicted rain for a large Republican rally, and was fired for doing so.[59] Similarly, George Monbiot relates that Sir David King,

> the British government's chief scientist, proposed that a 'reasonable' target for stabilizing carbon dioxide in the atmosphere was 550ppm CO2 … It would be 'politically unrealistic', he said, to demand anything lower. Simon Retallack from the Institute for Public Policy Research reminded Sir David that his duty is not to convey political reality but to represent scientific reality. King replied that if he recommended a lower limit, he would lose credibility with the government. It seemed to me that his credibility as a scientific adviser had just disappeared without trace.[60]

When the world is rebuilt in the image of ideology, we are sealed within an ideological sphere that is both symbolic and material; and the 'end of ideology' is the end, not of ideology, but of our power to recognise it. This, as we have seen, is the culmination of a long history which can be traced most obviously to Descartes and before him, Plato, during which mind was split from body, initiating what would, more widely, become the industrialist domination of nature. But it was not enough that mind and body be separated into opposing factions; for just as the propensities of wild nature insist on 'inconvenient truths' which may eventually cripple industrial society, so body erupts into mind, especially through sexual fantasy which, as Augustine put it, "paralyses all powers of deliberate thought".[61] Hence, as we saw earlier, the movement of the confession from an admission of sexual *acts* to an admission of sexual *thoughts*; and this exposure of what is internal was later generalised more widely in various publications, autobiographical narratives, letters, and so on.[62] It is not enough for our overt behaviour to accord with social expectations; our minds, too, are colonised by expectations which come from outside the body. So as Foucault argues, the "spiritual struggle against libido … consists … in turning our eyes continuously downwards or inwards to decipher, among the movements of the soul, which ones come from the libido" – in effect, a distancing of the self from our own embodiment, which 'we' then observe and control. Sexual ethics, Foucault tells us, "do not only consist in learning the rules of a moral sexual behaviour, but also in constantly scrutinising ourselves as libidinal beings".[63]

Just as our needs become orchestrated through influences outside ourselves, so the mechanisms of control that penetrate our inner spaces can best be understood not as an extension of self-control, but rather as reflecting the greater reach of external controls. Whereas in the past, behaviour could be said to be significantly limited by individual conscience, today we are observed by batteries of CCTV cameras as society becomes increasingly 'panoptical', arguably causing an assimilation of the self by the observing systems and a shrinkage of genuine autonomy, so that willingness to take socially responsible action shrivels. As Hille Koskela remarks, "electronic means have more and more often [been] used to replace informal social control in an urban environment: the eyes of the people on the street are replaced by the eyes of surveillance cameras".[64] In the natural world being watched is often experienced as a threat; so being observed this way, although consciously we largely ignore it, is experienced at an embodied level as a chronic unease, as an intrusion into our private space. As William Safire puts it, "To be watched at all times, especially when doing nothing seriously wrong, is to be afflicted with a creepy feeling. … It is the pervasive, inescapable feeling of being unfree."[65] Thus a system which claims to reinforce our social consciences and powers of observation is in embodied practice destructive of those elusive but crucial attributes of shared space which ensure social cohesion and congeniality.

The repression and expression of embodiment

In a natural environment, we are subject to controls in the form of predators, diseases, availability of food, shelter, and water, and so on; but none of these greatly threatens our individuality and agency. In the industrial world, on the other hand, we are immersed in a symbolic environment that is in some respects continuous with our thought processes, so that the point at which *thinking* morphs into *being thought* is unclear. Control enters individuality itself, and our thought processes are subject to influences such as education, propaganda, radio, advertising, and TV. Our relation to industrialism, therefore, is not that of a separate creature to its environment; rather, we are parasitised so that our lives are co-opted to serve purposes quite outside our awareness. Habermas's notion of the 'colonisation of the lifeworld' comes closest to describing this phenomenon. The discrepancy between our supposed individual autonomy and our actual lack of autonomy is veiled behind the constant references to 'democracy', the individualistic explanations of successes and failures, and the media websites that encourage us to 'have our say'; and the same messages are conveyed by films, our heroes being cool, in-control Bond-figures. Thus one of the most significant ideological elements we are colonised by is the individualistic belief that we are not colonised; and this divergence between our limited understanding of industrialism and the actual functioning of industrialism is essential to that functioning itself.

Under these circumstances, repression has become more subtle and more insidious as it is 'built into' the ideological and physical scenery of the industrial world, becoming one aspect of a social unconscious that has both institutional and physical dimensions. Desires are narrowed-down and isolated within particular niches, like wild water, where they can often be harnessed for commercial purposes. Thus eroticism is stimulated, used to sell commodities and lifestyles, and extinguished from some areas of life such as bureaucracy. Psychological life becomes part of a managed landscape; and it is appropriate that Freud used the draining of the Zuider Zee as a model for the ego's appropriation of portions of the id. But in reassigning nature's predispositions we create symptoms, whether ecological or psychological; and as Anne Spirn shows in her study of the Mill Creek area of West Philadelphia, denying the existence of natural features of the landscape and trying to build over them results in the reassertion of these features, often in ways that are socially and psychologically damaging. Mill Creek, imprisoned and buried within a 20-foot diameter pipe, has destroyed the neighbourhood it runs through:

> For more than 60 years, the ground has fallen in, here and there, along the line of the sewer. The creek has undermined buildings and streets and slashed meandering diagonals of shifting foundations, and vacant land across the urban landscape. Local newspapers have chronicled the long series of broken pipes and cave-ins. In the 1940s, 47 homes were demolished because they were 'plagued with rats and filled with sewer vapor'. In 1945, a neighbourhood of small row homes built above the sewer was destroyed when the sewer collapsed. In 1952, a 35-foot-deep cave-in on Sansom Street swallowed two cars, and the porches of three homes crumbled into the crater. On 17 July 1961, the sewer caved in beneath Funston Street near 50th. Initially, four houses were destroyed and three people killed; ultimately 111 homes were condemned and demolished, leaving hundreds homeless and many others fearful of further collapse. 'We haven't been ordered to leave. We're just too frightened to stay here', one person told a reporter. Months later, Philadelphia's Evening Bulletin described residents' complaints of sewer odours and their frustration at the city's slow response in repairing the 30-foot chasm. Entire city blocks are now open within the buried floodplain. Young woodlands of ailanthus, sumac and ash have grown up on older lots, urban meadows on lots vacated more recently. Many community gardens in this part of West Philadelphia lie within the old floodplain of Mill Creek; older gardeners remember when buildings sank, their foundations eroded by high groundwater, undermined by subsiding fill.[66]

Despite the drive to wipe the slate clean and impose an industrialist landscape, nature insists that memory is part of the present, as Toni Morrison points out in drawing parallels between our own memories and those that are embodied in the landscape:

> You know, they straightened out the Mississippi River in places, to make room for houses and livable acreage. Occasionally the river floods these places. 'Floods' is the word they use, but in fact it is not flooding; it is remembering. Remembering where it used to be. All water has a perfect memory and is forever trying to get back to where it was. Writers are like that: remembering where we were, what valley

we ran through, what the banks were like, the light that was there and the route back to our original place. It is emotional memory – what the nerves and skin remember as well as how it appeared. And a rush of imagination is our 'flooding'.[67]

Water appears to be the most pliable and yielding of materials; but when it is dammed and channelled without regard for its original place within a natural topology, it can become a fearsome and destructive force. Generally, nature is flexible in allowing us to clear domains within which we impose symbolic patterns; but exploiting this flexibility without regard for natural structure invites a sort of nemesis as the dissonance between industrialism's demands and the systemic predispositions of nature become unsupportable. The direct and indirect effects of climate change are only the most obvious ways that nature will outflank and undermine human symbolic and physical practices that ignore the natural architecture of nature.

Our contravention of natural realities includes those that while not easily expressible in scientific terms, are nevertheless essential to our quality of life. As Jack Manno notes, for example, nomads and hunter-gatherers who make their own clothing and grow or collect their own food will appear statistically as 'poor' because they have little measurable income. As their lands become 'developed', logged, and farmed, and they become sucked into the industrial economy, their income will tend to increase (for example, as they become factory workers); so an apparent alleviation of poverty will be accompanied by an actual *increase* in experienced poverty as communally shared benefits such as common land and unpolluted water disappear.[68] The divergence between experienced realities and dominant types of symbolic representation is also illustrated by the account given by an aid worker who moved to a slum area of Glasgow after working in India for ten years. She reports that the Adivasi, who live in the Nilgiri mountains in Tamil Nadu, "were clear about their own notions of wealth – 'our community, our children, our unity, our culture, the forest'. Money was not mentioned at all." The Adivasi, although they had to walk half a kilometer to find water, and were grateful if they had a kilo of rice to feed their families, "didn't see themselves as poor. They saw themselves as people without money". In contrast, the (objectively more wealthy) Glaswegians were "dispirited, depressed, often alcoholic. Their self-esteem had gone. Emotionally and mentally they were far worse off than [the Indian poor]".[69]

Many qualities that are overlooked by bureaucratic and scientific codes involve the higher-order emergent properties of nature; but writers of fiction have sometimes found more inclusive and complete ways of expressing such qualities. This is a very particular form of fiction that is constrained and disciplined by its anchorage to other levels of reality; for while the elemental details are 'made up', it conveys truths at higher levels. Morrison, for example, argues that since we haven't yet found a way to preserve subjectivity, it will necessarily be the first casualty of historical recording. We record the physical remains of civilisations – the bones, the crumbling walls, the broken pots – but not the subjective states of the inhabitants. Morrison therefore refers to her writing as "a kind of literary archaeology: on the basis of some information and a little bit of guesswork you journey to a site to see what remains were left behind and to reconstruct the world that these remains imply. What makes it fiction is the nature of the imaginative act: my reliance on the image ... on the remains ... in addition to recollection, to yield up a kind of truth". She continues:

> the crucial distinction ... is not the difference between fact and fiction, but the distinction between fact and truth. Because facts can exist without human intelligence, but truth cannot. So ... I'm looking to find and expose a truth about the interior life of people who didn't write it (which doesn't mean that they didn't have it).[70]

A reliance on concrete facts, then, cannot by itself offer an adequate description of reality, and is itself, as Bernard Williams has pointed out, "an offence against truthfulness".[71] As Morrison implies in her statements above, fiction – or at least great fiction – can convey emergent truths more fully and accurately than a purely factual account, because one can go some way towards restoring the subjective qualities omitted by the latter. As David James Duncan suggests,

> fiction-making and lying are two different things. To write *War and Peace* required imaginative effort. To embezzle money from a bank does, too. [This] does not make Tolstoy a bank robber. *War and Peace* is an imaginative invention but also, from beginning to end, a truth-telling and a gift-giving. We know before reading a sentence that Tolstoy "made it all up", but this making is as altruistic and disciplined as the engineering of a cathedral. It uses mastery of language, spectacular acts of empathy, and meticulous insight into a web of individuals and a world to present a man's vast, haunted love for his Russian people. And we as readers get to recreate this love in ourselves. We get to reenter the cathedral.[72]

There is a parallel here with a point made by Robert Ulanowicz: the constituent species of ecosystems can change while the emergent properties of the whole system remain constant, so that the changing 'cast' of the ecological drama doesn't affect the message conveyed by the play as a whole.[73] By the subversive act of pointing to those elusive human and worldly qualities that are slipping away partly because they are hard to objectify and measure, fiction can enrich subjectivity towards our full humanness. But such qualities are also part of our everyday lives, as Holmes Rolston implies:

> [H]iking with my father, we had just left the car at a stranger's place in the mountains … realising that I had forgotten to lock the car, I turned to go back. 'Never mind", my father said, 'Did you see that man's tomatoes?'[74]

As Rolston suggests, there is something about contact with the natural world that ensures a certain sort of integrity; and the growing divide between nature and industrialism's symbolic templates is an unacknowledged source of psychological problems. While we may try to redefine ourselves as primarily symbolic beings, we still need the touch of others and the sights, sensations, and smells of the natural world in order to remain healthy, as we noted above. The categorisation and compartmentalisation that separates the psycho-disciplines from research on the environment has concealed an important relation between well-being and the accessibility of natural areas, as an example of David Smail's illustrates:

> Mrs Arkwright moved from a bleak public housing estate – open, asphalt, streaked with graffiti, litter and dog shit – where her existence anywhere beyond her own front door had been one of perpetual tension if not outright fear, to a new house she and her husband had bought in a pleasant residential suburb a few miles away. From the moment of her move her 'agoraphobia' started to diminish until after a few months it became most of the time absent from her awareness. At first, though, she was puzzled and mistrustful of herself, almost ashamed. 'Do you think there's something peculiar about me?' she asked soon after moving house. 'It sounds funny to say it, I know, but it's so nice to walk down roads lined with trees.'[75]

Although there is now plentiful evidence that interaction with natural landscapes fosters mental and physical health,[76] there is a curious reluctance to admit the complementary possibility – that the absence of such interaction fosters ill-health and social problems. Psychopathology, we are led to believe, is largely an *internal* problem, so that the distressed person often feels that their unhappiness is 'their fault' – which, as Smail points out, is a "great and often very damaging mistake".[77]

'Incongruence', then, is not merely an individual property; it is also a structural attribute of the industrial world, including academia, which often legitimises dissociations between natural and symbolic forms. For example, we attempt to address climate change not by eliminating the burning of fossil fuels, but through carbon trading; and the widespread destruction of wildness is seen through the lens of 'biodiversity reduction'.[78] The problems and the 'solutions' are thus safely kept within that portion of the symbolic realm which we can access, often having only a tenuous relation to what is happening in material reality. Andrew Vayda has noted the chasm between human discourse and behavioural realities; and as he notes in relation to the spread of the water hyacinth in the southeastern USA, in "seeking satisfactory explanations of why all this happened, we must look first not to the basic values or ideals which seed-sellers or seed-buyers held about nature but rather to the contexts in which they acted".[79] Similarly, Yi-Fu Tuan has pointed out that the reverent Chinese attitudes towards nature did not prevent them from extensive deforestation.[80] Our symbolic understandings and attitudes often seem to form a sort of parallel universe to the physical world rather than one which is functionally engaged with it; and to the extent that such engagement occurs, human symbolic activity is as likely to generate camouflage or rationalisation as it is to express material realities. Often the workings of industrialism seem to pass conscious symbolic processes by, proceeding under the cover provided by their distracting power.

The repression of repression

In the divergence of symbolism from embodiment, Freud was clearly an advocate of the dominance of the symbolic, arguing for a 'dictatorship of the intellect'; but he did at least recognise the existence of embodiment in the form of the id, acknowledging that "our own psychical constitution" includes "a piece of unconquerable nature".[81] Later theorists often prefer to exclude the realm of embodiment altogether, as we saw in Chapter 4, claiming that both 'human nature' and the natural world are social creations. This epistemological tactic completes the conquest of nature, which is no longer seen as an enemy to be vanquished, but simply as an outgrowth of the larger, social realm. Just as an empire's battle to assimilate a culturally distinct people may be presented as 'a matter of internal security', so the conflict between nature and industrialism, viewed bureaucratically, is transformed into questions of the 'terms of reference' of an inquiry, the 'externalities' of corporate operations, or the practicalities of policing an 'illegal protest'. The role of social constructionism in the epistemological realm is therefore analogous to that played by the US Department of Homeland Security or the British Home Office in drawing the boundaries of empire, excluding understandings and activities perceived as unwelcome.

For example, the concept of repression has recently been understood not as acting on memories and fantasies deriving from embodied, and especially sexual, aspects of the person, but rather as existing within an entirely symbolic universe. Michael Billig, for example, views repression as a *linguistic* phenomenon; and he claims that in contrast to "Freud's vision of the battle between the id and the ego, between biology and civilisation", "the roots of repression do not necessarily lie in biologically inborn desires". Rather,

> The objects of repression … would themselves be formed in language. More than this, the psychic battleground may not be dominated by the conflict between language and wordless instinct. Instead, the battleground may be within language … the objects of repression … would themselves be formed in language."[82]

Desires, we are told, are not primary, embodied human qualities, but epiphenomena of language: "[b]ecause we speak, we have desires which must be expressed".[83] Here is no conflict between embodiment and symbolism: rather, repression becomes a purely symbolic activity, pushing into the background an earlier, unmentioned, and much more radical repression. Examples of repression given by Billig include what is left unsaid; not shouting out at a lecture; and not thinking about the origins of our inexpensive clothes:

> If my clothes and other possessions are to be mine, then my imagination needs to be habitually curtailed. … I should not imagine those strange hands, which once touched my possessions. Indeed, my possessions would cease to feel mine – and I would cease to be my good consuming self – if I took seriously the dark, busy fingers, working in conditions of oppression far removed from my life-world. Those anonymous fingers, no matter how many hours they labour, will

never be able to own the sorts of possessions that I take for granted. This is not a pleasurable line of thinking. It is more comfortable to act habitually as if there can be consumption without production.[84]

Billig's version of repression does not involve dark fantasies of erotic perversions or the murder of rivals, but mild social embarrassment or fleeting consumerist guilt. It is a peculiarly cerebral, passionless, disembodied sort of repression – which is hardly surprising, given its linguistic origins. Rather than explaining repression, Billig exemplifies it in his choice of subject matter; for the image of the individual conveyed here is not that of a passionate, physically vital creature, but of a thoroughly domesticated, almost virtual being for whom action can only be *symbolic* action. The theory of repression, ironically, has itself become repressive, and now constitutes a means of denying the existence of a deeper, embodied knowledge. As Paul Stoller remarks, "sensuous awakening is a very tall order in an academy where mind has long been separated from body."[85]

Similar points can be made about Foucault's famous critique of the 'repressive hypothesis'. Rather than being repressed, Foucault tells us – that is, "condemned to prohibition, nonexistence, and silence"[86] – the form of sexuality is changed, so that the

> most important elements of an erotic art linked to our knowledge about sexuality are not to be sought in the ideal … of a healthy sexuality, nor in the humanist dream of a complete and flourishing sexuality, and certainly not in the lyricism of orgasm and the good feelings of bio-energy … but in [the] multiplication and intensification of pleasures connected to the production of truth about sex.[87]

This 'production of truth about sex' involves all the "learned volumes", the "consultations and examinations", "all the stories told to oneself and to others", and so on – all this, according to Foucault, "constitutes something like the errant fragments of an erotic art". We are dealing here, he continues, "not nearly so much with a negative mechanism of exclusion as with the operation of a subtle network of discourses, special knowledges, pleasures, and powers", a process that "spreads [sex] over the surface of things and bodies, arouses it, draws it out and bids it speak, implants it in reality and enjoins it to tell the truth".[88] As we noted in Chapter 4, the agenda here is a transformation of embodied realities into virtual ones, so that the human being becomes a purely symbolic creature. *This* is the great repression that in its academic variants is accomplished most influentially by Freud and Foucault. This is clear in Foucault's description of sex as being "driven out of hiding and constrained to lead a discursive existence"[89] – a description curiously reminiscent of Bacon's earlier intention to "follow and as it were hound nature in her wanderings".[90] Beneath the much-scrutinised differences between the writings of Freud, Foucault, and Bacon, then, there lies a common aim of converting embodiment into discourse.

Like many discursively inclined theorists, Foucault conflates sexuality with its discursive *representation*. In denying that what we experience is a "censorship of sex", and arguing for the flourishing of "an apparatus for producing an even greater quantity of discourse about sex",[91] he overlooks the fact that no amount of discursive representation can compensate for the repression of *embodied*, *lived* sex. Thus sex becomes a 'construct':

> Sexuality must not be thought of as a kind of natural given which power tries to hold in check, or as an obscure domain which knowledge tries gradually to uncover. It is the name that can be given to a historical construct: not a furtive reality that is difficult to grasp.[92]

Furthermore, beginning in the eighteenth century a sexual discourse emerged, Foucault tells us, that was derived "not from morality alone but from *rationality as well*", one that involved "analysis, stocktaking, classification, [and] quantitative and causal studies".[93] Thus sex is no longer a slippery, sweaty, intensely erotic activity, but one that is closer to accountancy; and 'passions' have indeed become 'rational choices'. This is somewhat reminiscent of those 'conservation' studies of wild creatures that involve similar techniques – for example, the use of radiocollars, the reference to taxonomies, the measurement of population dynamics, the fencing-off of 'wildlife preserves', and so on. In both these situations, what we see is the assimilation of the wild and the organic to the discursive. Just as wilderness becomes 'socially constructed' within these discourses, so sex, according to Foucault, is "a rather complex *idea* that was formed inside the deployment of sexuality".[94]

Freud envisaged repression as an essentially *individual* defence mechanism; but what seems to have happened is that as individuals' symbolic capacities have become extended into a widely shared *system*, the locus of repression has moved *outside* the individual, redefining us as essentially symbolic beings. This redefinition constitutes a massive repression that is all the more powerful for being unnoticed. As we noted earlier, Silverstein has highlighted the extent to which some topics such as the situation in Palestine have become taboo areas about which discussion is pre-emptively avoided. Thus plays, debates, and lectures have all been cancelled in order to avoid controversy and the dissemination of awareness.[95] It is not that knowledge is being repressed within the individual, or that protests are occurring against particular views; rather, a decision is being made that particular issues will not be *considered*; and so the issue is architecturally expunged from the communicational arena. Such 'pre-emptive censorship', as Silverstein terms it, occurs prior to any individual defence mechanisms; and while there has always been a social component to repression, the balance seems to have shifted in recent years. Behind the relatively trivial defence mechanisms that are the subject matter of coffee table discussions and earnest academic debates lurks a deeper form of repression that engulfs and colours our taken-for-granted ways of living, speaking, and theorising; and the resonances of this deeper repression extend far beyond the consulting room into every aspect of our lives.

Fragmenting subjectivity

The decontextualisation of psychological distress and its arrangement into separate categories ensure that psychiatric classifications mystify and reinforce the underlying split between symbolism and nature. As Kleinman, Das, and Lock have suggested, "experience has been splintered off into measurable attributes ... [which] are then managed by bureaucratic institutions and expert cultures that reify the fragmentation while casting a veil of misrecognition over the whole.[96] Thus we are guided into assessing our actions only by reference to the particular 'fragment' we are currently embedded within. Consistently with this idea, David Healy has suggested that market ideology leads to "the atomisation of distress". For example, a drug used to treat hypertension is found to be "100% effective" in reducing blood pressure while the psychological side-effects are ignored, so that the treatment has "disconnected individuals from their social milieu".[97] Such splits and compartmentalisations are part of a cognitive style that begins in school, where we are taught that knowledge is divided up into maths, physics, languages, history, and so on; and, of course, what we learn at school is itself separated from what we learn at home. Later, most of us learn to specialise in a single subject, which often becomes the basis of our careers. We are, in effect, systematically steered away from any holistic understanding of how society functions, so that we become knowledgeable about detailed issues while remaining blind to the larger picture.

This lack of holistic awareness also leads to the sort of *moral* deficiencies described by Ulfried Geuter in his discussion of the professionalisation of German psychology before and during the war. He found that "individuals began to see themselves as psychologists, members of a group ... In the pursuit of their goals they were relatively blind to, when they did not actively affirm, the social and political context in which professionalisation took place." Wehrmacht psychologists often referred to themselves as operating within an 'island' or 'oasis'; and even after the war had ended, their self-examinations tended to be limited to whether they had fulfilled their professional tasks dutifully – indicating, as Geuter suggests, a "certain self-deception".[98] While it is easy to see this situation as reflecting Sartrean 'bad faith' – that is, self-deceptively viewing our behaviour as unavoidable when it is in fact optional – it is important to recognise that this is not simply an *individual* failing, but rather a structural property of the entire system of specialisation on which modern society is based. Indeed, science itself, with its emphasis on objectivity, its reductionism, and its blindness to the uses to which it is put may, as Geuter implies, be open to accusations of promiscuously serving any ends, be they worthy or not. Thus what Stanley Cohen calls 'inner emigration' – that is, a retreat into a particular symbolic niche, away from the challenging realities of an integrated world – may be understood as a culturally approved cognitive orientation rather than as an individual defence mechanism.[99] Here we can how the two meanings of the term 'integrity' converge in the relation between wholeness and truth.

While the *underlying* disconnection here is that of symbolism from the natural order, this is manifested in a variety of more directly experienced splits such as those between the mind and the body, bureaucratic and folk knowledge, thought and feeling, and the urban world and wild nature. This symbolic compartmentalisation also allows a further split, divorcing our actions from their consequences and allowing narcissism, anthropocentrism, and delusion to thrive. The dramatic increases in narcissism during the past half-century[100] do not only reflect personality changes: they are also increasingly embodied in the media, the law, and the physical structure of the built world, which celebrate and legitimate consumption while distancing us from its effects. The contrasts between the shop front and the delivery gate, or the 'beauty strip' and the clearcut it hides, or the lifestyles of consumers and geographically distant workers are matters not only of psychological compartmentalisation, but also of infrastructural and geographic configurations. Such contrasts are papered over with a façade of pseudo-intimacy, for example in ads that use frolicking children or happy-looking lovers to 'humanise' the underlying commercial message, or in the use of first names by call centre workers who have never met us and who have only been talking to us for a few seconds. Aidan Davison points out that care, friendship and love have become "emotional tools for ensuring the passive loyalty of the subjects of technological culture to the reign of the master rationality".[101] Likewise, as Sut Jhally puts it, advertising

> takes the *images* of the life people really want – a life of meaning, of connection, of socialising, of friendship, of family, of sexuality ... and links them to objects. So advertising is both true and false at the same time. If it was simply false, it wouldn't work; but advertising is true to the extent that it reflects our real desires.[102]

Sensing this prostitution of the realm of intimacy, we become more distrustful of personal relationships; and the vicious privatistic circle is given another twist.

The orchestration of experience

In our 'environment of evolutionary adaptation', sensory intimacy with the landscape is likely to have fostered an awareness of ecological realities such as the dependence of animal and vegetable life on the presence of water, the complex interactions of predator and prey, and the immediate human consequences of changes in the environment; and the absence of this intimacy would have threatened one's survival. While a capacity to focus on details such as the movement of a prey animal would also have enhanced one's survival prospects, one would also need to be aware that one might oneself become prey; and cognitively mapping one's location would have been necessary if one was to return home in a world without signposts, maps, or compasses. Thus survival was dependent on moving flexibly between two complementary abilities: the ability to focus closely on details, and the ability to be aware of the overall context of these details.

Today, many of us inhabit a quite different landscape – one built to accord with our conceptual assumptions, and presented to us through electronic media which convey the fragmentations I described above. The attractive appearance of consumer goods is foregrounded, while their origins in resource extraction and processes of production are hidden from us, so that the *wholeness* of the world is veiled; and in the short term we can survive without a broader contextual awareness. Predators do not, generally, lurk in the shadows: the things we pay attention to, such as road signs, advertising hoardings, and brake lights require little careful observation, for they are *designed* to grab our attention rather than remain hidden. In a shopping mall, we are captivated by the sequence of shopfronts, our mood cemented by the feel-good muzak, oblivious to anything that disturbs the focus on consumption. While in an Ivy League university town recently, I noticed the bag women, often pushing their few possessions along the pavement on supermarket trolleys, ignored by the bright-eyed, affluent young coeds who seemed to swim around them without actually noticing them. It was as if they were invisible, living in a parallel universe that existed beyond the sensory powers of the happy shoppers.

Consciousness, then, has become more like a slide show than an intimacy with the landscape. Our shifting attention has become a way of *not* engaging with the world as we skim lightly and unquestioningly over its seductive surface, neither focusing intensely on details nor retaining a clear sense of our place in the overall scheme of things. As Jonathan Crary argues,

> we are in a dimension of contemporary experience that requires that we effectively cancel out or exclude from consciousness much of our immediate environment. ... Western modernity since the nineteenth century has demanded that individuals define and shape themselves in terms of a capacity for 'paying attention', that is, for a disengagement from a broader field of attraction, whether visual or auditory, for the sake of isolating or focusing on a reduced number of stimuli. That our lives are so thoroughly a patchwork of such disconnected states is not a 'natural' condition but rather a product of a dense and powerful remaking of human subjectivity in the West over the past 150 years.[103]

Any culture, of course, directs its participants toward certain aspects of the world rather than others; but industrialism extends this tendency to create what is effectively an alternative, symbolically constructed world only selectively connected to embodied realities, so that an apparatus of sensory information is used not so much to *inform* us about the character of the world as to camouflage it. While fourteenth century natural philosophers, as we saw in Chapter 5, had to withdraw into a world of abstraction in order to maintain the conceptual order, today this ploy has become perfected by the electronic media and embodied within the built environment. Our 'culture', if such it be, is as notable for what it conceals as for what it reveals.

In this situation, the growth of a grounded, coherent identity is inhibited as the self lurches between scattered islands of isolated meaning. Television programming constructs a series of mostly unrelated scenes, interspersed with advertisements, all under the guise of 'entertainment', so that one minute we are watching multiple shootings or car crashes, the next we see a happy couple strolling in the countryside. These fragments are then reintegrated into new industrialist collages in which products are linked with desirable emotional states, status, or sexual success, reconstructing selfhood in ways that makes us passengers riding towards an unknowable destination. Just as industrialism obliterates natural structure and reconstitutes the resulting 'natural resources' into commodities, so it splinters subjectivity and reformulates it to produce the materially and emotionally needy consumer/worker. Despite – or, more accurately, because of – the meaninglessness of our diet of tasty informational titbits, it is curiously addictive as we continue to search for what is not being provided. Just as we tend to eat more if food lacks nourishment, so we turn up the volume to compensate for the absence of meaning; and as we saw earlier, "stimulation becomes a substitute structure".[104]

The bombardment by music, advertising, and rapidly changing televisual images affects our ability to focus attention. Just as contemporary ways of getting about require no physical discipline or fitness, so the contemporary media require no subjective effort on the part of the individual, who simply has to watch the screen and soak up the sequence of rapidly changing images. This is unlikely to foster strong, autonomous attentional processes. It is not so much that we disentangle those things that we need to attend to through careful observation and selection; rather, our attentional capacities are entrained within sensory arrays that are contrived for exactly that purpose. The locus of agency thus moves somewhat away from the attending individual towards the hidden, remote designers of our sensory environment, as our evolved *capacity* for orienting attention is engaged by often hidden structures and organisations *outside* ourselves; and with the advent of television, the orientation of attention becomes less an individual choice than a product of technology and the producer's whim. As Crary notes, even while corporations spend millions of dollars on such manipulations, they have to simultaneously deny their effectiveness in order to maintain the myth of the "rational and volitional human subject".[105] There is a certain irony in the fact that, in Crary's words, "one of the ways an immense social crisis of subjective dis-integration is metaphorically diagnosed is as a deficiency of 'attention'".[106]

In effect, then, attentional decisions are now 'outsourced' to commercial structures over which the self has no control. This becomes a vicious circle as the person's diminished power to make attentional decisions makes them more vulnerable to media influences which will further reduce attentional autonomy. In effect, as Crary argues, attention becomes the "means by which a perceiver becomes open to control and annexation by external agencies."[107] As Neil Postman summarises the commercially desirable qualities of this style of consciousness, "bite-sized is best, complexity must be avoided ... nuances are dispensable ... qualifications impede the simple message ... visual stimulation is a substitute for thought ... verbal precision is an anachronism".[108] In effect, one is held captive within symbolic systems that are external to the self and often conceptually opaque, disabling our capacity to monitor and assess them. It is not that these symbolic systems are irrational: rather, they are perfectly rational in following the 'logic of the market' which leads to capital growth, but this logic expresses itself in a mystifying *illogic* at the subjective level, feeding off and exacerbating a lack of individual coherence and integrity. Commercial 'logic' requires that the person should be disoriented and mystified, so that order exists not at the individual level, but in the hidden workings of industrialism.

Given these conditions, it is entirely realistic that as Jean Twenge has documented, children and college students increasingly believe that their lives are controlled by external forces. In one study by Twenge and her colleagues, the correlation between scores on an 'Externality Scale' – which measures the belief that "larger forces control your fate" – and the year the data were collected was a "stunning" .70.[109] As the veteran British politician Tony Benn puts it, "increasingly, people are being managed rather than represented".[110] In turn, the resulting feelings of powerlessness, as Twenge documents, lead to cynicism and political quietism, tightening the vicious circle towards even greater impotence, creating a reservoir of compliant, suggestible consumers who are utterly unable to comprehend the forces through which they are manipulated.

Under these circumstances, it is not surprising that embodiment becomes further dissociated from our symbolic investment in the systems that colonise us. We can usefully remind ourselves here of DeGrandpre's distinction between the 'technological' self "in which the interface with technology synchronises our conscious minds to the rhythm of the microchip and gives us increasingly realistic virtual worlds in which to live and dream", and the 'social' self, "which moves within human relationships that operate at relatively slower speeds and with old-fashioned rules of engagement." To DeGrandpre, "it seems that until the … technological self takes over the … social self – a time when we will live only among virtual beings of our own design – the divided self will continue to be deeply conflicted".[111] Although I would characterise this conflict as one between embodiment and symbolism rather than between 'social' and 'technological' selves, it is clear that the thrust of De Grandpre's insight parallels my own suggestions. Without the stupefying and superficially reassuring beat of electronically mediated 'entertainment', which serves much the same function as Prozac, the implications of such conflicts would come crashing in on us.

The exclusiveness of rationality

Since the fourteenth century, as we saw in the previous chapter, order has increasingly been sought in the symbolic realm rather than in the raw materiality of existence, becoming a quality that is *applied to* reality rather than *found in* it. While it is true that science maintains a firm foothold on reality, this is only after a good deal of selective measurement, abstraction, and calculation; and the sort of 'reality' generated by science is distantly observed rather than sensorily immediate and emotionally involving. One does not find the valency of carbon attached to a diamond; nor are latent heats observable by the naked eye. Among industrialised peoples, who have mostly lost the capacity to make spiritual and mythological sense of a world that is vastly more complex than our cognitive abilities, rational order becomes a sort of psychological liferaft that we cling to avoid the implosion of meaninglessness; and for our own psychological equanimity and physical ease, we rebuild the world to fit our needs, pushing the unpredictable and the random to the periphery of our awareness. We prefer our trains to follow timetables; the floors of buildings are reassuringly horizontal; every plant has a name; and words have the same meaning in Liverpool as in Littlehampton.

Anchoring subjectivity to rationality in this way exemplifies the wider reduction of materiality to fit industrialist symbolic schemes, exterminating anything discordant with these schemes. As Mary Midgley has commented, however, if "it is a bad idea to exterminate the natural fauna of the human gut … trying to exterminate the natural fauna and flora of the human imagination is perhaps no more sensible".[112] This insight has important consequences for the systems we allow to develop, since apparently inessential, duplicated, and redundant processes often stabilise a system, and conversely too much order and efficiency typically lead to brittleness and catastrophe.[113] Monocultures of the mind,[114] like agricultural monocultures, are vulnerable to parasitisation or sudden collapses to a degree that biodiverse systems are not; and as Robert Ulanowicz argues, apparently 'irrational' associations and even 'mistakes' seem to underly much human creativity.[115] In both technological and natural systems, as well as in those involving human meanings, redundancy and diversity are essential for health.

Mainstream psychology has waged a long war to convert the untidy diversity of feeling into something more conceptual, as any student who has struggled through some of the emotionally anaemic texts dealing with 'maternal deprivation' will recognise. Indeed, the primacy of emotional bonds in human life and in the lives of other creatures has often only reluctantly been admitted. In the first half of the twentieth century, in an era before the corporate world had learned that feelings could be commercially exploited, John Watson was one of many experts who warned of the dangers of mother love and the difficulties that lay ahead for the 'over-kissed baby'. The children's hospitals of the time were driven, as Robert Sapolsky puts it, by a "worship of sterile, aseptic conditions at all costs, and a belief among the (overwhelmingly male) pediatric establishment that touching, holding and nurturing infants was sentimental maternal foolishness'.[116] The implication of such views is that the skin – exquisitely equipped as it is with sense receptors – is not a medium of connection to a sensuous world, but is rather a separative boundary that distinguishes us from an alien environment. As Erica Burman notes, there is a masculinist ideology at work here, in that the course of development was – and to a considerable extent, is – one of movement "from *attachment*, a stereotypically feminine quality, to a stereotypically masculine *detachment*."[117] Even here, the language is mechanistic, the terms attachment and detachment being more redolent of the connection between railway wagons than of affection between human beings. Language, psychology, and social ideology converge systemically to mechanise human relations.

Psychoanalysis also carries a lightly veiled unease about the intimate relation between mother and child. Deborah Blum notes that for Freud, both adult and child may be led astray from reality by "dreams and fantasies of a better life", impeding ego development and adjustment to reality.[118] But what is 'reality' here? Is it constituted by the social actualities of Freud's era, which we now recognise to be grossly deficient in the case of childrearing; or is it a healthier situation we fantasise about? Does the relation between mother and child imply the possibility, in microcosm, of a wider relationality that could transcend the mechanical and the rational? Could it be that the intense, embodied experiences of childhood raise expectations that are stifled during socialisation, being incapable of fulfillment under current social conditions? By locating theory *within* a preconceived web of social assumptions about 'reality', we tacitly support these assumptions, reinforcing the 'centripetal' characteristics of industrialism.

Given the low esteem in which feelings are generally held, those who feel strongly about the natural world are likely to look around for scientific support. Usually, scientific 'findings' are portrayed as isolated from feelings; but in many cases it may be that science is used to rationalise pre-existing feelings. Eugene Hargrove, for example, points to Ian Douglas-Hamilton's study of the attitudes of rangers in a Tanzanian national park as an example of such rationalisation. In this case, elephants were demolishing most of the trees, and one obvious course of action was to cull the elephants. But none of the rangers wanted to shoot the elephants, feeling that they had great intrinsic value. Nevertheless "they did not believe that their feelings could be part of a professional justification for not shooting the elephants. Given that such justifications were closed off for him, Douglas-Hamilton concluded that he was supposed to find some facts that would independently justify this position so that aesthetic considerations would not have to be mentioned"[119] – a conclusion that is consistent with Donald Worster's observation that ecologists "picked out their values first and only afterwards came to science for its stamp of approval."[120] Here, despite the overt emphasis on rationality, ecologists are sensing that feeling can recognise qualities of the world that are otherwise excluded – something that David Hume was aware of when he suggested that "Reason is, and ought only to be the slave of the passions, and can never pretend to any other office than to serve and obey them".[121] But for many psychologists since Hume, passions have been at best a diversion from the central, cognitive, character of persons; and the struggle by reason to subdue and dominate emotion is one of the dominant forms taken by the more general campaign of symbolism to dominate embodiment.

Margaret Archer has cogently summarised the progress of this long-standing campaign, charting the utilitarian's shift from 'passions' to 'pleasures', which was followed by the economists' reconceptualisation of 'pleasures' as 'preferences', which in turn were reconstituted by the rational choice theorists as, well, 'rational choices'.[122] Intelligent individuals, according to Gary Becker, do not base their behaviour on "fickle" emotions such as love, but make decisions on the basis of "more durable characteristics". Furthermore, because "maturation, knowledge that comes from living together, and inevitable frictions due to the many difficulties of life often transform love toward a spouse into indifference, loathing, disdain, and other unpleasant feelings", "[f]orward-looking persons, however deeply in love, force themselves to recognise that their love may not last, [and] expensive presents, dowries, and other gifts are commonly demanded up front while a spouse's love shines brightly".[123] Nobody could accuse Becker of being a dewy-eyed romantic. While Archer may well be right to argue that rational choice theory doesn't fit the reality of life today, a materialistic focus is not only an economic assumption but also a well-researched social trend. For example, Jean Twenge tells us that "1990s high school students were twice as likely as their 1970s counterparts to say that 'having a lot of money' was 'very important'".[124]

Rational choice theory is one of the most blatant instances of the colonisation of the social sciences by a fundamentally economic model. Becker has argued that most human behaviours, including personal habits, tastes, and child-rearing behaviours, as well as love and sympathy, can be understood as reflecting a drive to 'maximise utility' – 'utility' being understood as dependent on the person's consumption of goods and their stock of "personal and social capital". For example, encountering a beggar "may lower utility because a person feels guilty or uncomfortable", so we should "outsmart beggars and avoid contact with them."[125] Likewise, he argues that children can be regarded as consumer goods:

> As consumer durables, children are assumed to provide 'utility'. The utility from children is compared with that from other goods via a utility function or a set of indifference curves. The shape of the indifference curves is determined by the relative preference for children ...
>
> I will call more expensive children 'higher quality' children, just as Cadillacs are called higher quality cars than Chevrolets ...
>
> By and large, children cannot be purchased on the open market but must be produced at home.[126]

Although this is an extreme example, the narrowing of human being towards pure rationality and the equation of psychological well-being with the possession of 'utility' are diffusing outwards from economics to colonise everyday thought and discourse. The president of the Intelligence Factory, for example, is quoted in an airline magazine article as saying that "parents always have to be managing their assets, including their children".[127] The economists' assumption of the rational individual, stemming from Adam Smith's belief in the human propensity to "truck, barter, and exchange",[128] has fatefully narrowed our understanding of humanity towards the 'ideal consumer'; and as Karl Polanyi remarked, "no misreading of the past ever proved more prophetic of the future".[129] Unfortunately, basing decision-making on 'maximising utilities', far from guaranteeing satisfaction and fulfillment, has been found to be associated with unhappiness, relationship difficulties, lack of empathy, manipulativeness, and other generally undesirable characteristics;[130] and as we saw earlier, when this sort of 'rational' approach has been applied to child-rearing, the result has been emotional devastation. Similarly, the contemporary trend towards 'freeing' mothers from 'oppressive' childcare duties, while in some respects positive, is not one that will enhance the emotional welfare of the child. As Sarah Hrdy suggests, "[d]enial of infant need runs like an invisible and insidious counter current through publications purporting to correct the 'river of mother-blame' coursing through our society."[131]

In adulthood, too, the attempt to fit human psychology to the needs of the economic system are damaging. As Robert E. Lane summarises the effects of prioritising material wealth,

> The materialist is economic man, endowed with all the qualities that are said to be necessary for market economies to do their work. If he were also endowed with rationality (which is itself negatively related to well-being), the market would then maximise his well-being – would it not? Consider the following paradox: market rationality leads the materialist to pursue wealth; more wealth (beyond the poverty level) has little effect on well-being. Furthermore, something about the kind of people who choose (if that is the verb) to be materialists is associated with low life satisfaction. Materialists, economic men, endowed with the qualities that economists assume are the characteristics of winners, tend, in fact, to be losers from the start.[132]

As Jon Elster has pointed out, desires often reflect prior adaptations to what is viewed as possible ('sour grapes'); and so the notion that 'rational' choices can be made by 'free' individuals is a problematic one – especially when the individual's ideas about what is desirable have been shaped by advertising, the media, and socialisation.[133] In other words, as Archer argues, 'rational' choice tends to be aligned with prevailing social actualities and to discourage utopian or revolutionary thinking which *transcends* these actualities,[134] so contributing to the 'centripetal' tendencies of industrial society and the tendency towards 'pre-emptive repression'. As such, the 'rational' analysis of costs and benefits is problematic not only for individual well-being, but also in relation to the health of the natural order. Garrett Hardin's famous account of the 'tragedy of the commons', in which each herdsman gains from pushing the common grazing resources to the point of collapse, elegantly illustrates how the ideal of the rational person and the objective of maximising individual utilities leads directly to environmental disaster. In Hardin's words, "ruin is the destination toward which all men rush, each pursuing his own best interest in a society that believes in the freedom of the commons."[135] This basic model applies in varying forms to many aspects of the way we live, illustrating the logical endpoint of Gellner's 'single-stranded' society in which the balance between multiple considerations is lost.

My concerns about such economically inspired constructions of the person are twofold – firstly, that they inaccurately represent human being, and secondly, that they increasingly accurately represent human being. To start with the first objection, Becker's approach, as Archer argues, misrepresents people as economic agents rather than complete human beings, with only lip service paid to characteristics that cannot easily be expressed in economic terms. But to turn to the second concern, models such as Becker's which elaborate the perspective of the rational individual are also intended as models of the way we *should* behave and, increasingly, *are* behaving; and so these models work alongside other aspects of social 'reality' to channel us into economically-consistent habits.

The economic metaphor, therefore, pervades virtually the entire sphere of life in the industrial world; and this metaphor is not simply an *addition* to the repertoire of human possibilities: it *replaces* other modes that are ultimately inconsistent with it, in much the same way as commercial agriculture replaces indigenous ecologies. In James Buchan's terms, "as money enters the system of values, and then displaces all other values like a cuckoo the eggs in a nest, reality as the world of things perceived and wanted and believed must be completely revised".[136] This induces a cascade of follow-on effects as reality is rewritten to fit the economic model. Tim Kasser, for example, found that materialistic values are associated with low concern for the environment; and that extrinsic (e.g. economic) motivation decreases intrinsic motivations such as enjoyment. Related to this is the finding that over the three decades following the 1960s, the "percentage of students believing that it is very important or essential to develop a meaningful philosophy of life decreased from 89% to around 40%, [whereas] the percentage of those who believed that it is very important or essential to be very well off financially rose from just over 40% to over 70%".[137]

In fact, the view that we own possessions is probably no more accurate than the view that *they* own *us*, making a nonsense of the notion of the rationally choosing individual. At the individual level, we seem to make the choices, while at the corporate level, products are 'rationally' designed and sold in order to stimulate desire; so the 'rationality' of the consumer complements the 'rationality' of the manufacturer as elements within a larger, and largely unsuspected, system. At the corporate and governmental level, however, it was recognised at an early stage that consumer goods constituted a means of political and social control, a way of fostering dependency on the capitalist system. For example, in regard to the British occupation of India, Felix Padel notes that the Konds of what is now Orissa were subdued partly through trade. As one report of 1836 relates, making coloured cloth available to the Konds "will tend greatly to promote their intercourse with us, and by giving them new tastes and new wants will, in time, afford as the best hold we can have on their fidelity as subjects, by rendering them dependent upon us for what will, in time, become necessities of life".[138] Indeed, even in the nineteenth century it was recognised that the key to lasting domination of people lay in the moulding of human nature towards the individualism which regarded the land and what was on it as in 'private ownership' – alien as this was to the Konds who regarded the forest as owned by no-one and shared in common by everyone. As one British administrator recognised in 1872, colonisation is not primarily a *physical* act:

> It is only by what we have implanted in the living people, rather than what we have built upon the dead earth, that our name will survive. The permanent aspect of British rule in India is the growth of Private Rights.[139]

The land itself was equally rigidly controlled; and Padel compares the Indian situation to the Highland Clearances:

> The animals and birds they shot in such numbers, with such blood-lust, are a potent symbol of the habits of control they imposed in their working life over the forest and its human inhabitants. If one considers the huge tracts of forest which the British felled for profit, as well as the countless animals which they shot for fun, one could say that they were engaged in a sacrifice of wild-life, or of nature itself.[140]

Just as wild species were cleared from the forest, so wild thinking was cleared from the people's minds in order that the rational clarity of commercial logic and Western categories could be implanted. Padel notes that while botanists' "classification of forest species is very impressive in one way … for subtler levels of understanding 'adivasis' such as the Konds possess a much vaster body of knowledge". This knowledge, however, was never recognised as such by the British:

> the hierarchy which the British established over Konds comes down to a hierarchy of knowledge. British ideas about nature or about tribal people were seen as knowledge; tribal ideas as ignorant superstition. Colonial discourse allowed no relationship between the two, only a one-way relationship of power. At the heart of this hierarchy there is thus a sacrifice of mutual learning, or of mutual relationship.[141]

Indeed, this 'sacrifice of mutual relationship' is at the heart of the entire industrialist quest for rational domination; for rather than the multiply-connected systemic relations that characterise the natural order and the cultural systems that grow out of it, industrial colonisation involves a one-way relationship in which the dominating power does not learn from the dominated, but simply imposes its own structure, sweeping aside whatever is inconsistent with it.

The fading of embodied meaning

That is not to imply, of course, that there are no continuities between the embodied and symbolic realms. Words can sometimes express truths sensed through embodiment, reminding us of the potential to create a symbolic realm which extends rather than replaces natural realities and ensuring a synchrony between thought and nature. And as Lakoff and Johnson have shown, embodiment is reflected in language, so that categories and metaphors are often derived from physical experience.[142] We 'jump to a conclusion', 'reach out towards' someone, or 'grasp' a situation; and we may be 'gobsmacked', 'overwhelmed', or 'gutted' when a realisation 'strikes' us. But the connection between embodiment and symbolism should not be overstated. As we noted in Chapter 2, language is often used in a merely nominalistic way; and the industrialised mind uses metaphors of embodiment as a sort of temporary scaffolding in the construction of a nature-free symbolic realm, so that thought, as Piaget put it, eventually "becomes free from the real world".[143] Secondly, as I argued in Chapter 4, there are many meanings which go beyond words, capable of being sensed only when we sweep away words to clear a space of silence, as Carlos Fuentes indicates in his story "Confederacy of shadows":

> Silence. Tranquillity. Solitude. That's what unites us, thought Laura, holding Santiago's burning hand in her own. There is no greater respect or tenderness than that of being together and silent, living together but living the one for the other without ever saying so. With no need to say so. Being explicit might betray that deep tenderness which was only revealed in a tapestry of complicity, intuition, and acts of grace.[144]

Language's power to represent embodied experience is limited and often clumsy; and the insistence on the primacy of words betrays the world, forcing it to fit a symbolic measure. As Eugene Gendlin says, the silence that goes beyond words is "both vague and more precise" than language.[145] Such silently embodied knowledge, because it is difficult to communicate, tends to hide in the shadows of industrial society, remaining mostly tacit, driven into the hidden corners of our lives by its inconsistency with the clamour of words and the sensory assault of the media.

Some of the most profound varieties of meaning are rooted in embodiment. For example, we derive our sense of security from our mother's touch – itself nested within broader family, community, and natural contexts. Our memories of such meanings become more distant as we are educated to 'grow up' into symbolic understandings; so as we noted earlier, they are often denigrated as 'nostalgic'. Because of this increasing absorption of experience by words, argues David Smail, "our memory of bliss is not something we can get to grips with in our thoughts, but remains in our bodies as a kind of aching absence".[146]

Furthermore, to jettison our links with the natural world is to abandon a certain type of thought; for as we noted earlier, thought may be *en-worlded* as well as *embodied*. Lévi-Strauss's claim that 'animals are good to think with' is also applicable to the rest of the world. Not only can we think *about* the world; we can also think *through* it – as has occurred both historically and in our own childhoods[147] – ensuring a style of thought that is consistent with the world's natural forms. The developmental trajectory from 'sensori-motor' to a type of 'operational' thought which celebrates its isolation from the material world is thus not only a triumph of symbolic power, but also potentially an alienation which sets us against both the world and our own bodies, both of which have evolutionary histories which ensure their complementarity and communication. For example, if we feel 'bogged down', or that we've 'come to the end of the road', or that a task 'is a big mountain to climb', we are saying something about our *relation to the world*, not just our own bodies. Childhood learning does not take place only in the arena of one's own mind; it is also about climbing trees and tripping over rocks, watching a heron fly and reaching out to an injured friend. Such experiences, ideally, ground us within a world that makes sense – emotionally and experientially – in a pre-cognitive way, as we grow into the landscape, much as the roots of a tree reach into the crevices of the surrounding rock.

For peoples who live in a way that is attuned to the natural order, the continuity between experience and the landscape is an essential source of strength and emotional well-being, ensuring what Carl Rogers referred to as 'congruence'. The anthropologist Robert Williamson, for example, who has lived and worked extensively among the Inuit of northern Canada who were often so badly damaged by forced acculturation in mission schools, tells how

> The most restorative factor was the habitat. It has always been there, waiting for the soul-drained need of the hurt Inuit to move again into its ambit, returning to the places where the old souls were also waiting to be invoked. I remember an aeroplane landing on the ice near a hunting camp, returning some people from hospital. Among them was a young hunter, his hospital pallor contrasting with the bronze spring-time glow of the faces greeting the new arrivals. He asked about the whereabouts and situation of his family. Sadly, the people told him his wife had died during his two-year absence in the sanatorium, and their two children separately sent to a missionary residential school.
>
> He spoke earnestly to the owner of a dog-team that had come out to meet the plane, a kinsman's concern and understanding written on his features. And right there and then, straight from the aircraft onto the ice, he drove the borrowed team off into the surrounding country, promising to return in a few days. He drove that team non-stop, except for pauses to rest the dogs and hunt for them – for two days and two nights, and did not sleep until the third day. As we watched him heading over the ice and up the coast until he was out of sight, we understood.[148]

The absence of such a restorative world is demoralising, as is an environment that is being steadily degraded; and although at a conscious, conceptual level we seem to be enriching ourselves as we degrade the natural world, at a deeper level the enrichment is illusory as we withdraw into a 'self' defined through its symbolic isolation both from our bodies and from the world. Lacking embodied meanings, the need for meaning has become an insistent psychological drive, and we fill it by any available means – compulsive sex, drugs, food, instantly available music, and a constant diet of superficially entertaining pap. 'Liberated' – as some would see it – from the embodied foundations of meaning, we grasp at whatever is on offer like a drowning man grasps at a piece of floating wreckage. As Donald Spence remarks, "the search after meaning is especially insidious because it always succeeds."[149]

The absence of a permanent world within which we could ground our meanings is particularly apparent when we compare our own social assumptions with those of non-industrial peoples. In an insightful paper, William Bevis points out that the natural world that is described by most native American novelists is neither the rationally ordered world of the scientist nor the world of everyday understanding: it is just *the world*, sometimes comprehensible, but often recognised as extending beyond human understandings. In James Welch's writing, for example, the natural world "is strangely (to whites) various, objective, unsymbolic, *as if it had not yet been taken over by the human mind*". There is, as Bevis puts it, an "apparent fragmentation of the natural world into a huge cast of individual 'micro-characters', a fragmentation that has not been properly noted because it does not fit white formulas ... Cows, bats, mosquitoes, blackbirds, coyotes, magpies act in their individual, peculiar ways".[150] Like the mediaeval cosmos described in Chapter 5, this is a world which cannot be forced into the mould of a single theoretical framework: its meanings are multiple, various, often species specific; and only some are accessible to human awareness. For the Cree, for example, 'bear reality' and 'caribou reality' are as real and valid as 'human reality'.[151] One thing does not 'symbolise' another, conforming to any anthropocentric scheme of comprehension: things and creatures just *are*, in their own peculiar ways. Bevis illustrates this by referring to the writing of D'Arcy McNickle. Archilde, McNickel tells us, is at Mission School, and one afternoon a cloud

> by curious coincidence ... assumed the form of a cross – in the reflection of the setting sun, a flaming cross. The prefect was the first to observe the curiosity and it put him into a sort of ecstasy ...

'The Sign! The Sign!' he shouted. His face was flushed and his eyes gave off flashing lights – Archilde did not forget them.

'The Sign! Kneel and pray!'

The boys knelt and prayed, some of them frightened and on the point of crying. They knew what the sign signified ... the second coming of Christ, when the world was to perish in flames.[152]

The cloud, of course, melts away, as Bevis tells us; "but curiously Archilde does not need this empirical proof to reject Christianity's symbolic use of nature":[153]

It was not the disappearance of the threatening symbol which freed him from the priest's dark mood, but something else. At the very instant that the cross seemed to burn most brightly, a bird flew across it ... It flew past and returned several times before finally disappearing – and what seized Archilde's imagination was the bird's unconcernedness. It recognised no 'sign'. His spirit lightened. He felt himself fly with the bird.[154]

For Archilde, nature's meaning is intrinsic rather than dependent on symbolic interpretation; and in his reaction, we glimpse a groundedness that we often lack. For Archilde, meanings are embodied within the world. The world comes first.

This attitude of letting the world be, rather than retreating into symbolic meaning, is a sort of truth-telling which places accurate perception of the world above the contrived coherence of any abstracted symbolic realm. We find similar attitudes in other non-industrial societies. For example, Lye Tuck-Po relates that Batek mythology doesn't add up to a coherent narrative. Rather, she "ended up with loose ends and details that didn't quite fit".[155] And she quotes Alan Campbell's view that we have to "accept the narratives as they are, in all their crazy, fragmented incoherence. ... Those with a purist itch will keep looking for some complete, definitive version, but that kind of search is a blatant result of literacy".[156] In contrast to the scientific drive towards internal consistency, non-industrial societies are happy to accommodate a degree of symbolic incoherence, since they recognise that symbolism is secondary, a way of representing the world – which remains *one* world regardless of how we construe it. Similarly, Tim Ingold points out that the "Ojibwa way of dealing with perception is ... fundamentally anti-taxonomic, reducing to a shambles any attempt to bring it within the bounds of a neatly ordered system of classificatory divisions."[157]

This openness to the world is almost universal outside industrialism, and raises important questions about our need to interpose a quasi-world of discourse and concepts between ourselves and the real world. As Robert Desjarlais puts it, "any theoretical orientation that values representation over presentation, or discourse over felt experience, can only take us so far in explaining what life (and death) is like".[158] Furthermore, knowledge that is 'antitaxonomic' is not necessarily *structureless*: in fact, peoples such as the Ojibwa are well attuned to their environment, and possess a tacit confidence in its intrinsic coherence. Embodied knowledge and folk biology may overlap with Western science in certain respects; but they may also contain a good deal that is not expressible in scientific terms, mainly because they reflect a quite different order of knowledge. As Eugene Gendlin argues, most post-Kantian philosophy has overstated the influence of 'forms', so that "nothing is considered to have an order of its own. Everything is taken as ordered by imposed forms, patterns, and rules. Most modern philosophers have utterly lost an order of nature, human nature, the person, practice, the body. ... All order is assumed to be entirely imposed by a history, a culture, or a conceptual interpretation".[159]

Given our retreat from embodiment, feeling often lacks the opportunity for its expression; and so our emotional reactions, when they do occur, can be out of proportion to the events that trigger them. For example, the outpourings of grief following the death of Princess Diana, even among those who barely knew of her existence or work when she was alive, suggests hidden reservoirs of untapped emotion. Furthermore, our emotions are groomed by advertising, Hollywood, and television soaps, although our personal lives offer a partial opportunity to escape the ordered and controlled social realm. As Morris Berman argues, "falling in love is literally the one ecstatic or mystical experience left open, and it serves as a haven from the culture of repression and control".[160] Today, even the resonances shared by wild love and wild landscapes are endangered, as lovemaking becomes sex and immersing oneself in a landscape becomes tourism, and the feelings involved no longer *transcend* individualism and its social context, but are instead ordered *by* them. To the extent that such individualised experiences fall short of the transcendence our embodiment expects and hopes for, our relationships often seem curiously unsatisfying in ways that evade conscious understanding. Love and ecstasy cannot be fitted within social or economic templates; neither can they be designed, predicted, or manufactured. In her dystonic tale of the future world, *The Handmaid's Tale*, Margaret Atwood captures the poignant sense of an embodied experience that is always just out of reach as we confine ourselves within a symbolic pseudo-life:

> Falling in love ... was the central thing; it was the way you under-stood yourself; if it never happened to you ... you would be like a mutant, a creature from outer space ...
>
> Falling in love, we said ... We believed in it, this downward motion: so lovely, like flying, and yet at the same time so dire, so extreme, so unlikely. God is love, they once said, but we reversed that, and love, like Heaven, was always just around the corner. The more difficult it was to love the particular man beside us, the more we believed in Love, abstract and total. We were waiting, always, for the incarna-tion. That word, made flesh.[161]

Atwood is referring here to the same phenomenon that Margaret Archer is pointing to when she says that the "sense is absorbed into the concept";[162] and this absorption is a social phenomenon as well as an academic fashion. We are captured by ideas, fantasies, and delusions – of love, freedom, wealth, beauty – which exist only within a symbolic universe that has abandoned its material foundations.

Liberating ourselves from reality

One of the dangers of denying embodiment is that rather than ground-ing symbolism in the world in order to realistically appraise situations, we act according to our preferences, wishes, and fantasies, thereby becoming seriously out of touch with reality. There is now a substantial literature which demonstrates, in the words of a seminal review article, that in the industrial world "overly positive self-evaluations, exaggerated perceptions of control or mastery, and unrealistic optimism are characteristic of normal human thought";[163] and most significantly, these are *collective*, not merely individual, problems. Indeed, an 'opti-mistic' outlook is well on the way to becoming a social imperative. While these flights of fantasy may stave off the alarm and despondency that would flow from a more realistic recognition of our situation, they also encourage us to postpone effective action. Consider, for example, the views of the late Julian Simon, an influential economist who for many years suffered from what he described as 'a deep depression', and who became famous as a scourge of 'doomsaying' environmentalists. Although Simon claimed that the origins of his depression "had nothing to do with population growth or the world's predicament", he notes on the same page that

> As I studied the economics of population and worked my way to the views I now hold – that population growth, along with the lengthening of human life, is a moral and material triumph, my outlook for myself, for my family, and for the future of humanity became increasingly more optimistic. Eventually I was able to pull myself out of my depression.[164]

Applying the principles of cognitive therapy to resolve his own low mood, Simon became so enamoured with this approach that he later wrote a book about it. Summarising the principles involved, he observed that "everyone knows the old saw about seeing the glass half empty or half full. Even truer is that you can often choose which glass to look at, a glass which is full or one which is empty. Sadness and depression usually are optional".[165] We can, he continued, "will [our] attention away from depressing thoughts", adding that "we have some choice over what we pay attention to, just as we choose one television programme over another".[166]

Reality, unfortunately, does not consist of television channels that we can switch on and off; and one of the functions of industrialist symbolism is to defend us against whatever threatens our comforts and certainties. George Gilder, for example, like Simon, isolates thinking from reality, talking of "the strange effort to subject the mind to the laws of matter – life to the laws of death", and seeing the "limits to growth as merely its frontiers". Likewise, intellectuals'

> morbid anxieties about 'nonrenewable' resources, 'finite' reserves, limits of growth ... all bespeak the predicament of any mortal worshipper of matter and flesh. Matter is 'nonrenewable', flesh is finite and exhaustible, co-eds flee the withered touch, youth is fleeting and beset by natural laws and depletions of energy.[167]

In the same vein, Gilder claims that "the energy crisis is most essentially a religious disorder, a failure of faith. It can be overcome chiefly by worship".[168] There is enough here to keep a psychoanalyst employed for years; and these brief extracts from one of Gilder's books are perhaps sufficient to indicate the intensity of the denial involved. This has its roots in the same fear which drove the witch-hunts of the seventeenth century or the destruction of wildlife in the Americas; it is, in other words, fear of the natural order, of mortality, of the embodiment we share with all creatures, to be counteracted by the insistence on a symbolic Heaven here on earth, a divine breast that will always provide regardless of our greed and destructiveness.

If mood is viewed ecologically – that is, as part of an interactive web that connects the physical world with our own needs, senses, and experiences – then sadness is often understandable given the deteriorating state of the natural world. But if we lose sight of the interplay between experience and what exists outside us, splitting these two aspects of reality, sadness becomes 'depression', an inexplicable individual 'pathology'; and a potentially sadness-inducing situation becomes an emotionally neutered actuality such as a 'housing development', 'timber sale' or whatever. Psychiatric understandings of 'depression', therefore, while necessary and appropriate in some cases, also function to repress painful or politically inconvenient meanings by splitting experience from reality. Like any defence mechanism this ploy can only be partly successful; so industrialism carries with it a hidden burden of loss and despair, the sources of which are concealed by psychiatric categorisation.

If we choose to make ourselves feel good by 'cherry picking' those aspects of experience and those thoughts that lead us to believe that everything is rosy in the garden, we are blinding ourselves to vital feedback. While cognitive therapy may be useful in overcoming unrealistically negative interpretations, it is equally clear that it can reinforce socially-embedded delusions, permitting us a precarious psychological comfort at the price of a deeper and more dangerous alienation from reality, and promoting a consensual withdrawal into idealism. This becomes clear when Simon justifies his view that resources are "not ... finite in any operational sense" by arguing that "the resource system [is] as unlimited as the number of thoughts a person might have".[169]

As Simon himself admits, "a solid body of research in recent years suggests that depressives are *more accurate* in their assessments of the facts ... than are non-depressives, who tend to have an optimistic bias".[170] In fact, although this is still a hotly debated topic, the evidence supporting the 'depressive realism' hypothesis is now very substantial; and complementarily, there is a burgeoning literature demonstrating that "unrealistically optimistic beliefs about the future are held by normal individuals with respect to a wide variety of events".[171] As Alloy, Albright, Abramson, and Dykman summarise this evidence in a review, the "findings of depressive realism and non-depressive optimistic distortions suggest that the primary active ingredient in cognitive therapy may not be the enhancement of realistic self-appraisal ... but rather the training of depressed clients to engage in the sort of optimistic biases and illusions that non-depressives typically construct for themselves."[172] Simon's own writings seem to illustrate exactly this problem, predicting (in 1981) that there are enough resources to last the human race for "seven billion years",[173] and that

> Mining of the moon will begin in 1990. The material from 50 million tons of moon rocks can be used to make solar-powered satellites that will provide all the earth's energy needs by 2000 ... space is an ideal location for many types of manufacturing, including the making of electronics equipment. Space manufacturing can begin in the 1980's, becoming a multi-billion dollar business in a few decades.[174]

Like Gilder, Simon seems to have entirely abandoned the realm of physical reality, seeking refuge within a more malleable symbolic sphere – as indicated by his claim that "the ultimate constraint is not energy but information. Because we can increase the stock of information without limit, there is no need to consider our existence finite."[175] Such views exist in a symbolic sphere liberated from all physical limitations – the permanently blessed, hypomanic world of the advertiser. Such ungrounded fantasies can today be recognised as wishful thinking; but the popularity and influence of Simon's books suggest that there is a hunger for this sort of good news, not merely among the public as a whole but, more worryingly, among those in powerful political positions. As Stanley Cohen suggests, the prospect of entrusting our future to these "optimists, with their positive illusions and creative self-deceptions, is not reassuring. ... People highly endowed with positive illusions – notably about their own omnipotence – commit the most appalling atrocities. The admired qualities of high self-esteem, a sense of mastery, faith in their capacity to bring about desired events, and unrealistic optimism were possessed in abundance by Mussolini, Pol Pot, Ceausescu, Idi Amin and Mobutu"[176] – and, one might add, by George W. Bush and Tony Blair.

   In an era of widespread political disengagement, the capacity of psychological distress to spur us into action is particularly valuable. Movements such as 'positive psychology', by fostering short-term happiness in the absence of an adequate political analysis, therefore build up problems for the future since they short-circuit this relationship between mood and necessary action. Furthermore powerlessness – a leading cause of depression – is often reframed by psychotherapists as a *belief* that one is powerless: in other words, it transfers the problem from *political reality* to one's *belief system*. Of course, effective action may not always be possible, and I would not claim that such symbolic decontextualisations of psychological distress are always counterproductive. But if social and political structures are such that people *are* increasingly powerless – as I believe is the case – then that needs to be recognised and addressed. As Christopher Lasch argued in response to critics who suggested that his *The Culture of Narcissism* was overly 'pessimistic', blindly encouraging 'optimism' is a "failure of nerve" by theorists who prefer to inhabit theoretical webs of symbolic meaning rather than engaging with real problems.[177] 'Learned helplessness' is not just a psychological problem, but can often be an index of political oppression that follows quite logically from our positioning as individuals within political structures. For example, as Martín-Baró notes in his analysis of Latin American politics,

> social relations are structured in such a way that they deprive most people of the minimal resources necessary for shaping their lives. As one of the supreme principles of life in society, private property enshrines the continuing plunder of the majority, who have no real chance to control their own destiny. One's birthplace becomes one's destination. Hence fatalism is a social, external, and objective reality before it becomes an internal and subjective personal attitude.[178]

Depression and anxiety are sometimes necessary feedback mechanisms; and the attempt to deal with them simply by adjusting the way we think is itself symptomatic of a chronic disengagement from reality that undermines democracy. As David Smail has argued,

> It does not suit the interests of ... power that the hard realities of [the] world are too well understood by those ... who profit from it least. For us there needs to be – and has been – created other forms of world, not real, where we may lead disembodied lives, detached from ... the levers of power. It is a world of make-believe.[179]

As Smail adds, this sort of ideological defence "lends itself wonderfully well to a society which seeks ideologically to detach its citizens from their embodied relation to a material world".[180]

This detachment of mood from the realities we live within has also been encouraged by the emergence of a *technology* of mood control. Especially since the development of SSRIs (Selective Serotonin Reuptake

Inhibitors), the pharmaceutical industry has encouraged us to view mood as a matter of neurochemical balance rather than as a reflection of our relationship to the world. SSRIs, according to Ronald Dworkin, create a mood that is "stuporous and purposely unknowing", allowing us to become "totally self-contained organisms" with weak consciences. As one of Dworkin's patients remarked about the effects of his medication, "I see the same things as before, but I don't care so much. I still feel good no matter what happens." As Dworkin summarises his argument:

> Medical science should confine itself to the treatment of clinical depression, rather than extend itself into the realm of everyday unhappiness. Medical science "helps" unhappy people by clouding their thoughts, by making them less aware of the world, and by sapping their urge to see themselves in a true light. People medicated for everyday unhappiness gain inner peace, but they do so through a real decrement in consciousness.[181]

Drugs, as Luc Sante remarks, are resorted to when "the more commonplace illusions fail",[182] reinforcing what Dworkin refers to as the "disconnect between the inner and outer life". Likewise, John Gray suggests that drug use "is a tacit admission of a forbidden truth. For most people, happiness is beyond reach. Fulfilment is found not in daily life but in escaping from it. Since happiness is unavailable, the mass of mankind seeks pleasure".[183] It is questionable whether such disconnections are desirable even as individual defences; but in an age when apathy is allowing governments and corporations to destroy the prospects for future life on earth, it is certainly undesirable that they become sedimented into human character. As John Rodman pointed out, "engineering away the needs that are frustrated by the conditions of oppression" is to cause the oppression to be embodied in character structure,[184] so that we 'freely choose' to live constricted lives within a ravaged environment. This is a form of human extinction: one that exploits our adaptability and capacity for self-deception, using the myth of conscious control as the means by which we are controlled.

Having lost his grounding in the real world, the industrialised individual lives within webs of symbolic delusion derived largely from economic imperatives and a narcissistic preoccupation with the self. The cognitive powers that gave rise to industrialism are ill-prepared to grapple with the unintended consequences of those same powers. Externalising cognition into the world, we have also externalised control over technology; and the symbolic systems that were originally enhancements of human intellectual abilities have now become autonomous forms of organisation that have returned to restructure human character, fracturing our relations to the rest of the natural world and tearing apart previously complementary aspects of our own nature. Our intoxication by the symbolic forms that our ancestors first set in motion has allowed us to ignore the embodied realities of our existence and to base our ways of living on the commercial exploitation and destruction of almost all nonhuman forms of life.

Today, this disconnection from foundational realities has become an unexamined part of the way we live. What counts as truth and reality have become clouded, often referring to situations, events, and entities that exist entirely within the symbolic realm rather than expressing embodied life more fully; and even fundamental aspects of life such as mortality are veiled by these evolving confabulations. As Terrence Deacon remarks, we try to forget our inevitable fate by "submerging the constant angst with innumerable distractions, or trying to convince ourselves that the end isn't really what it seems by weaving marvellous alternative interpretations of what will happen in 'the undiscovered country' on the other side of death".[185] Furthermore, "the dark side of religious belief and powerful ideology is that they so often provide twisted justifications for arbitrarily sparing or destroying lives", trapping us within

> a web of oppression, as we try through ritual action and obsessive devotion to a cause to maintain a psychic safety net that protects us from our fears of purposelessness. The interaction of symbolic cultural evolution and unprepared biology has created some of the most influential and virulent systems of symbols the world has ever known. Few if any societies have ever escaped the grip of powerful beliefs that cloak the impenetrable mystery of human life and death in a cocoon of symbolism and meaning. The history of the twentieth century, like all those recorded before it, is sadly written in the blood that irreconcilable symbol systems have spilt between them … Symbols are subject to being rendered meaningless by contradiction, and this makes alternative models of the world direct threats to existence.[186]

Such scenarios are the result of our captivation by symbolic systems that override our embodied sensings, natural compassion, and the empathic reach of subjectivity. Losing their grounding in nature, such systems become self-sufficient and self-serving, developing their own dynamics and views of what constitutes truth, seldom troubling themselves to anchor truth to the underlying propensities of the living world. The first step in our search for solutions to this predicament is to realistically assess our current situation – a task that has until now generally been evaded as we generate illusory 'solutions' that are really ways of staving off the consequences of our destructive ways of being for a little longer.

Notes

Preface

1. Bennett, 'De rerum natura', 9.
2. Luke, 'Cyborg enchantments', 58.
3. Hillman, *The Essential James Hillman*, 28.

1 Symbolism Breaks Free

1. Donald, *The Origins of the Modern Mind*, 114–5.
2. Donald, *The Origins of the Modern Mind*, 313–5.
3. Tomasello, *The Cultural Origins of Human Cognition*.
4. Donald, *The Origins of the Modern Mind*, 316.
5. Hutchins, *Cognition in the Wild*.
6. Tooby and Cosmides, 'The psychological foundations of culture', 47.
7. Donald, *A Mind So Rare*, 316.
8. Donald, *A Mind So Rare*, 315.
9. Hayles, *How We Became Posthuman*, 48–9.
10. Dawkins, 'Viruses of the mind'.
11. Wynn, *The Evolution of Spatial Competence*.
12. Feist, *The Psychology of Science and the Origins of the Scientific Mind*, 183.
13. Deacon, *The Symbolic Species*, 345.
14. Renfrew, 'Mind and matter: Cognitive archaeology and external symbolic storage', 2.
15. Miller, 'A production of amino acids under possible primitive earth conditions'.
16. Boyd and Richerson, *The Origin and Evolution of Culture*, 66–82.
17. Hutchins, *Cognition in the Wild*.
18. Minsky, 'Virtual molecular reality'.
19. Gibson, *Neuromancer*.
20. As in Sherman and Judkins' suggestion that virtual reality "is the hope for the next [i.e. this] century. It may indeed afford glimpses of Heaven." Quoted by Robins, 'Cyberspace and the world we live in'.
21. Lakoff and Johnson, *Philosophy in the Flesh*.
22. Marais, *The Soul of the White Ant*.
23. Bateson, *Steps to an Ecology of Mind*, 434.
24. Elkin, *Aboriginal Men of High Degree*; Ingold, *The Perception of the Environment*.
25. Ingold, *The Perception of the Environment*, 14.
26. Cf. Ingold, *The Perception of the Environment*, 19.
27. Richards, *The Romantic Conception of Life*, 471.
28. Barron, *Creativity and Personal Freedom*.
29. Bateson and Bateson, *Angels Fear*, 30.

30. Hoffmeyer, *Biosemiotics*, 175.
31. McGilchrist, *The Master and His Emissary*, 116.
32. Richards, *The Romantic Conception of Life*, 13.
33. The scare quotes are intended to draw attention to the parallels with form that evolves 'at the edge of chaos', that area between order and randomness where complexity is maximal.
34. Katherine Hayles, in *How We Became Posthuman*, 148, points out that Maturana's notion of 'autopoesis' also produces this sort of cognitive 'autism'.
35. Turner, *The Abstract Wild*, 79–80.
36. Deacon, *The Symbolic Species*, 140.
37. Gergen, *The Saturated Self*.
38. Deacon, *The Symbolic Species*, 416.
39. Boyle, *The Tyranny of Numbers*, xiii.
40. Martin, 'The contexts of environmental decision-making'.
41. Stewart, 'Drax trial held in a climate of injustice'.
42. O'Flaherty, *Highways*.
43. Twenge, *Generation Me*.
44. Deacon, *The Symbolic Species*, 336.
45. James, *Some Problems in Philosophy*, 51.
46. Deacon, *The Symbolic Species*, 111.
47. Deacon, *The Symbolic Species*, 112.
48. Lane, *The Loss of Happiness in Market Democracies*.
49. Joy, "Why the future doesn't need us".
50. Deacon, *The Symbolic Species*, 453.
51. Odling-Smee et al., *Niche Construction*, 341.
52. Henneberg, 'Evolution of the human brain'.
53. Gibbons, 'Paleoanthropology: Bone Sizes Trace the Decline of Man (and Woman)'.
54. Deacon, *The Symbolic Species*, 162.
55. Donald, *A Mind So Rare*, 285.
56. Donald, *Origins of the Modern Mind*, 324.
57. Donald, *A Mind So Rare*, 285.
58. Borgmann, *Holding On to Reality*, 79.
59. Borgmann, 'The destitution of space', 14.
60. Laing, *The Divided Self*.
61. Searle, *The Rediscovery of the Mind*.
62. Donald, *A Mind So Rare*, 316.
63. Donald, *A Mind So Rare*, 149.
64. Hutchins, *Cognition in the Wild*, 362.
65. Hutchins, *Cognition in the Wild*, 363. (Emphasis in original.)
66. Hutchins, *Cognition in the Wild*, 363. (Emphasis in original.)
67. Hutchins, *Cognition in the Wild*, 365.
68. Hutchins, *Cognition in the Wild*, 365.
69. Deacon, *The Symbolic Species*, 335–6.
70. Deacon, *The Symbolic Species*, 339.
71. Deacon, *The Symbolic Species*, 335–6.
72. Deacon, *The Symbolic Species*, 372.
73. Deacon, *The Symbolic Species*, 375.

74. Geertz, *The Interpretation of Cultures*, 49.
75. Deacon, *The Symbolic Species*, 345.
76. Deacon, *The Symbolic Species*, 349.
77. Ingold, *The Perception of the Environment*, 20.
78. Tuck-Po, *Changing Pathways*, 30.
79. Ryan, 'A people-centered approach to designing and managing restoration projects', 215.
80. Ryan, 'A people-centered approach to designing and managing restoration projects', 213.
81. Epstein, 'Integration of the cognitive and the psychodynamic unconscious', 710.
82. Bruner, *Actual Minds, Possible Worlds*.
83. Rogers, *Client Centered Therapy*.
84. Tversky and Kahneman, 'Extensional versus intuitive reasoning: The conjunction fallacy in probability judgement'.
85. McGilchrist, *The Master and His Emissary*.
86. Keep and Mayhew, 'The assessment: Knowledge, skills, and competitiveness'.
87. Ingold, *The Perception of the Environment*, 22.
88. Ingold, *The Perception of the Environment*, 22.
89. Drengson, 'The Life and Work of Arne Naess: An appreciative overview', 39.
90. Ingold, *The Perception of the Environment*, 54.
91. Argyrou, *Anthropology and the Will to Meaning*, 62.
92. Claude Bernard, quoted by Evernden, *The Natural Alien*, 16–17.
93. Deacon, *The Symbolic Species*, 416.
94. Deacon, *The Symbolic Species*, 435.
95. Deacon, *The Symbolic Species*, 436.
96. See, for example, Legler, 'Body politics in American nature writing'.
97. Turner, *Beyond Geography*.

2 The Natural and the Industrial

1. Gray, *Straw Dogs*, 16.
2. Callicott, 'The wilderness idea revisited', 241.
3. Vogel, 'The nature of artifacts', 164.
4. Bateson, *Steps to an Ecology of Mind*, 469. (Italics in original.)
5. Rolston, *Conserving Natural Value*, 86.
6. Soper, 'Nature and culture: The mythic register', 69, 70.
7. Franklin, *Nature and Social Theory*, 21.
8. The reference is to Clifford Geertz's 'Common sense as a cultural system', in his *Local Knowledge*.
9. Jacoby, *The End of Utopia*, 38.
10. Hornborg, *The Power of the Machine*, 49.
11. Casey, *Getting Back Into Place*.
12. Nabhan, *Cultures of Habitat*, 182.
13. Nabhan, 'Cultural perceptions of ecological interactions', 146.
14. Kretch, *The Ecological Indian: Myth and History*, 149.
15. Ingold, *The Perception of the Environment*, 314.
16. Cushman, 'Why the self is empty', 605.

17. Nabhan, *The Desert Smells Like Rain*, 96.
18. Nabhan, *Cultures of Habitat*, 162.
19. Nabhan, *Cultures of Habitat*, 163.
20. Nabhan, *Cultures of Habitat*, 11.
21. Nabhan, *Cultures of Habitat*, 11.
22. See, for example, the papers in Maffi, *On Biocultural Diversity*.
23. Smith, *From the Land of Shadows*.
24. Kahn, *The Human Relationship with Nature*.
25. Walker et al., 'Properties of ecotones'.
26. Manno, *Privileged Goods*, 41.
27. Manno, *Privileged Goods*, 49, 51.
28. Ingold, *The Perception of the Environment* , 217. (My italics).
29. Atran, 'Modes of thinking about living kinds', 225.
30. Atran, 'Modes of thinking about living kinds', 253.
31. Atran, 'Modes of thinking about living kinds', 253.
32. Atran, 'The vanishing landscape of the Petén Maya lowlands'. In Maffi, *On Biocultural Diversity*.
33. Chiu, 'A cross-cultural comparison of cognitive styles in Chinese and American children', 241.
34. Thoreau, *Walden*, 106–14.
35. Collier, 'Explanation and Emancipation'. In Archer, Bhaskar, Collier, Lawson, and Norrie, *Critical Realism*, 470.
36. Taussig, *The Devil and Commodity Fetishism in South America*, 36.
37. On 'top-down' influences on systemic behaviour, see Ulanowicz, *A Third Window*, Chapter 5.
38. Lave, *Cognition in Practice*, 1.
39. Lave, *Cognition in Practice*, 82.
40. Lave, *Cognition in Practice*, 82.
41. Lovelock, *Gaia*.
42. Maturana and Varela, *Autopoiesis and Cognition*.
43. Hayles, *Chaos and Order*, 57.
44. Ingold, *The Perception of the Environment*, 321.
45. Kauffman, *Reinventing the Sacred*, 150, 162.
46. For example, see Hern, 'Why are there so many of us?'; Lovelock, *Gaia*.
47. Soto and Sonnenschein, 'Emergentism by default'.
48. Buchmann and Nabhan, *The Forgotten Pollinators*, 174.
49. Buchmann and Nabhan, *The Forgotten Pollinators*, 44–5.
50. Buchmann and Nabhan, *The Forgotten Pollinators*, 181–2.
51. Buchmann and Nabhan, *The Forgotten Pollinators*, 82.
52. Kauffman, *The Origins of Order*.
53. See, for example, Allen and Starr, *Hierarchy*, 31.
54. Pimm, *The Balance of Nature?*
55. Budiansky, *Nature's Keepers*, 17.
56. Luke, 'Green consumerism'.
57. Ulanowicz, *Ecology, the Ascendant Perspective*, 64.
58. Ulanowicz, *Ecology, the Ascendant Perspective*, 64.
59. Worster, 'The ecology of order and chaos', 14.
60. Seed, *Ceremonies of Possession in Europe's Conquest of the New World 1492-1640*, 175.
61. Basso, *Wisdom Sits in Places*, 8, 34.

62. Basso, *Wisdom Sits in Places*, 108.
63. Spirn, *The Language of Landscape*, 11, 18.
64. Nettle and Romaine, *Vanishing Voices*, 16.
65. Padel, *The Sacrifice of Human Being*, 264.
66. Padel, *The Sacrifice of Human Being*, 264–5.
67. Bowker and Stoll, 'Use of dichotomous choice nonmarket methods to value the whooping crane resource'.
68. Whitmire, 'Blue Ridge Parkway pegs value of vistas'.
69. Enwegbara, 'Toxic colonialism'.
70. Manno, *Privileged Goods*, 53.
71. Manno, *Privileged Goods*, 53, 133.
72. Smail, *Taking Care*, 143.
73. Rappaport, *Ritual and Religion in the Making of Humanity*, 454.
74. Rappaport, *Ritual and Religion in the Making of Humanity*, 455.
75. Pomeroy, *Marx and Whitehead*, 171.
76. Shiva, *Biopiracy*.
77. Buell, *From Apocalypse to Way of Life*, 227.
78. Zbigniew Herbert, quoted by Buchan, *Frozen Desire*, 108.
79. Colin Blakemore, speaking in the National Portrait Gallery, 22 April 2004.
80. Fox, *Toward a Transpersonal Ecology*, 214–5. It is perhaps no coincidence that at the time of writing this, torture seems once again to have become a part of Western foreign policy.
81. Ewen, *All Consuming Images*, 34.
82. Terdiman, *Present Past*, 129.
83. Terdiman, *Present Past*, 129.
84. A skeuomorph is an architectural feature that once served a structural purpose, but has been retained for merely aesthetic reasons.
85. Monbiot, *Captive State*.
86. Kenrick, 'Contrast effects and judgements of physical attractiveness'.
87. Farrelly, *Blubberland*, 111.
88. Farrelly, *Blubberland*, 115.
89. Farrelly, *Blubberland*, 113.
90. Steinberg, *The Fiction of a Thinkable World*, 144.
91. Mingers, *Self-Producing Systems*; Maturana and Varela, *Autopoiesis and Cognition*.
92. Kauffman, *The Origins of Order*; Ulanowicz, *A Third Window*.
93. Schaffer, 'Stretching and folding in lynx fur returns'; Schaffer and Kot, 'Do strange attractors govern ecological systems?'. For a more recent review of studies of this much-analysed data, see Gamarra and Solé, 'Bifurcations and chaos in ecology'.

    Likewise, avalanches are not *predictable*; but neither are they *random*. See, for example, Kauffman, *At Home in the Universe*, 29.
94. Emily Martin, 'Fluid bodies, managed nature'. In Braun and Castree, *Remaking Reality*, 68.
95. Stuart Ewen, interviewed in the *Century of the Self* (Produced by Adam Curtis), BBC 4. Broadcast 29 April–2 May 2002.
96. Quoted by Pullman, *The Atom in the History of Human Thought*, 147.
97. Boden, *The Creative Mind*, 280.
98. Giddens, *Modernity and Self-Identity*, 20–1.
99. Holling and Sanderson, 'Dynamics of (dis)harmony in ecological and social systems', 63.

100. Johannes, *Words of the Lagoon*.
101. Mithen, *The Prehistory of the Mind*, 196–7.
102. Crist, *Images of Animals*, 170.
103. Crist, *Images of Animals*, 132.
104. Buchan, *Frozen Desire*, 110–11.
105. Hayles, *Chaos Bound*, 279.
106. Bhaskar, *A Realist Theory of Science*.
107. Ulanowicz, *A Third Window*.
108. Cushman, 'Why the self is empty'.
109. Ulanowicz, *A Third Window*, 145.
110. Hayles, *Chaos Bound*, 280.
111. Botkin, *Discordant Harmonies*, 159.
112. Pimm, *The Balance of Nature?*
113. Childs, *The Secret Knowledge of Water*, 65.
114. Vogel, 'Environmental philosophy after the end of nature', 34–5.
115. Hansen, Sato, Kharecha, Russell, Lea, and Siddall, 'Climate change and trace gases'.
116. Bhaskar, *From Science to Emancipation*, 28.
117. Cronon (ed.), *Uncommon Ground*, 36.
118. Botkin, *Discordant Harmonies*, 193.
119. Braun and Castree (eds), *Remaking Reality*, 9.
120. Smith, *Uneven Development*, 11.
121. Braun and Castree, *Remaking Reality*, 13.
122. Connerton, *How Modernity Forgets*, 40.
123. Connerton, *How Modernity Forgets*, 43.
124. Connerton, *How Modernity Forgets*, 46.
125. Connerton, *How Modernity Forgets*, 46.
126. Connerton, *How Modernity Forgets*, 2.
127. Connerton, *How Modernity Forgets*, 47.
128. Barthes, *Mythologies*, 78.
129. Schumaker, *The Age of Insanity*.
130. Silverstein, 'Land of the free?'
131. Holmwood, 'BBC Gaza appeal row: unions protest'.
132. Berman, *The Twilight of American Culture*, 96.
133. Kahn, *The Human Relationship with Nature*.
134. For example, Oliver James reports that "despite substantially increased affluence since 1950, a 25 year old American today is between three (Kessler et al., 1994) and ten (Wickramratne et al., 1989) times more likely to be suffering from major depression than then. Two meta-analyses by Twenge (2000) produced the startling conclusion that that the average American child in the 1980's reported more anxiety than child psychiatric patients in the 1950's." See James, 'They muck you up'.

3 Growing Out of the World

1. Edward O. Wilson, 'Biophilia and the Conservation Ethic'. In Kellert and Wilson, *The Biophilia Hypothesis*, 32.
2. Kahn and Friedman, 'Environmental views and values of children in an inner-city Black community'.

3. Kahn, *The Human Relationship with Nature*, 114.
4. Atran and Medin, *The Native Mind and the Cultural Construction of Nature*, 128–9.
5. Atran and Medin, *The Native Mind and the Cultural Construction of Nature*, 132.
6. Atran and Medin, *The Native Mind and the Cultural Construction of Nature*, 138.
7. Atran and Medin, *The Native Mind and the Cultural Construction of Nature*, 139, 140.
8. Mithen, *The Prehistory of the Mind*, 51.
9. Myers, *Children and Animals*.
10. Nabhan, *Cultures of Habitat*, 75.
11. Winnicott, *Playing and Reality*.
12. Buck-Morss, 'Socioeconomic bias in Piaget's theory', 40.
13. Atran and Medin, *The Native Mind and the Cultural Construction of Nature*, 44.
14. Stephen Kellert, 'Experiencing nature: Affective, cognitive, and evaluative development in children'. In Kahn and Kellert (eds), *Children and Nature*, 118.
15. Schumaker, *The Age of Insanity*, 157.
16. Myers, *Children and Animals*, 149.
17. Myers, *Children and Animals*, 150.
18. Eaton, *Human and Animal*.
19. Myers, *Children and Animals*, 57.
20. Myers, *Children and Animals*, 57.
21. Myers, *Children and Animals*, 153.
22. Myers, *Children and Animals*, 153.
23. Kohlberg, *The Psychology of Moral Development*.
24. See, for example, Scott and Willits, 'Environmental attitudes and behavior'.
25. Cynthia Thomashow, 'Adolescents and ecological identity: attending to wild nature'. In Kahn and Kellert (eds), *Children and Nature*, 265.
26. Barrett, *Irrational Man*, 198.
27. Marin, 'The new narcissism', 55.
28. Ingold, *The Perception of the Environment*, 43.
29. Laura Rival, 'The growth of family trees: Understanding Huaorani perceptions of the forest'. *Man*, 28(4) (1993), 649.
30. Ingold, *The Perception of the Environment*, 47.
31. Gedo, *The Artist and the Emotional World*, 8.
32. Bernstein, *Living in the Borderland*, 92.
33. Stern, *The Interpersonal World of the Infant*, 162–3.
34. Berne, *Games People Play*, 158.
35. Mead, *Mind, Self, and Society from the Standpoint of a Social Behaviorist*, 37.
36. Myers, *Children and Animals*, 4.
37. Myers, *Children and Animals*, 48.
38. Nabhan, 'Cultural parallax in viewing North American habitats', 98.
39. Piaget, *The Psychology of Intelligence*, 151.
40. Searles, *The Nonhuman Environment in Normal Development and Schizophrenia*, 114.
41. Smail, *Power, Interest, and Psychology*, 91.
42. Frumkin, 'Beyond toxicity'.

43. See, for example, Russell, 'What is wilderness therapy?'
44. Goodman, *Growing Up Absurd*, 12.
45. Goodman, *Growing Up Absurd*, 13.
46. Robert Bly, quoted by Scheibe, 'Mirrors, masks, lies, and secrets', 61.
47. Cushman, 'Why the self is empty'.
48. DeGrandpre, *Ritalin Nation*, 49.
49. DeGrandpre, *Ritalin Nation*, 31.
50. Laura Rival, 'The growth of family trees'.
51. Scheper-Hughes and Sargent, *Small Wars*, 12.
52. See Sluzki and Ransom, *Double Bind*.
53. Scheper-Hughes and Sargent, *Small Wars*, 12.

4   Lost in (Symbolic) Space

1. Stoller, *Sensuous Scholarship*, 85.
2. Kleinman, 'Everything that really matters', 318.
3. Desjarlais, *Body and Emotion*, 29.
4. Desjarlais, *Body and Emotion*, 247.
5. Connerton, *How Societies Remember*, 101.
6. Connerton, *How Societies Remember*, 94.
7. Connerton, *How Societies Remember*, 104.
8. Leder, *The Absent Body*, 3.
9. Ingold, *The Perception of the Environment*, 169–70.
10. Smail, *Power, Interest, and Psychology*, 90.
11. Edwards, *Discourse and Cognition*, 45.
12. Potter, *Representing Reality*, 97.
13. Collier, *Critical Realism*, 86.
14. Deacon, *The Symbolic Species*, 451.
15. Lyng and Franks, *Sociology and the Real World*, 69.
16. Shweder, ' Menstrual pollution, soul loss, and the comparative study of emotions', 184–5.
17. Thomas, 'Some problems with the notion of external storage', 152.
18. Reiss, *The Discourse of Modernism*, 36–7.
19. Romanyshyn, *Technology as Symptom and Dream*, 82.
20. Nadeau, *The Environmental Endgame*, 66.
21. http://news.bbc.co.uk/1/hi/health/2819595.stm. Accessed 16 October 2010.
22. Heather Blears, interviewed on the BBC Radio 4 one o'clock news, 2 August 2005.
23. Monbiot, *Captive State*.
24. Steinberg, *The Fiction of a Thinkable World*, 165.
25. Steinberg, *The Fiction of a Thinkable World*, 147.
26. Austin, *How To Do Things With Words*, 94.
27. Johnson, 'The emergence of meaning in bodily experience'. In Den Outen and Moen, *The Presence of Feeling in Thought*, 153.
28. Mark Fettes, 'Critical Realism and Ecological Psychology: Foundations for a Naturalist Theory of Language Acquisition', on the Website for Critical Realism, at www.raggedclaws.com/criticalrealism/. Accessed 24 April 2009.
29. Brown, *Life Against Death*, 167.

30. Miller, 'Free the Media', 10.
31. Jacqueline Rose, *The Last Resistance* (London: Verso, 2007), 126.
32. McQueen, *The Essence of Capitalism*, 259.
33. Goux, *Symbolic Economies*, 131.
34. Goux, *Symbolic Economies*, 132.
35. Scheper-Hughes and Lock, 'The mindful body', 23.
36. Scheper-Hughes and Lock, 'The mindful body', 23.
37. Goux, *Symbolic Economies*, 132.
38. D. Bickerton, quoted by Jeffrey Wollock, 'Linguistic diversity and biodiversity', in Maffi (ed.), *On Biocultural Diversity*, 254.
39. Miller and Philo, 'Silencing dissent in academia', 244–5.
40. Gendlin, 'How philosophy cannot appeal to experience', 39.
41. Roger Keil and John Graham, 'Reasserting nature: Constructing urban environments after Fordism'. In Braun and Castree, *Remaking Reality*, 102, 122.
42. Hannigan, *Environmental Sociology*, 187.
43. Soper, 'Nature and culture: The mythic register', 70.
44. Monbiot, 'Leave it in the ground'.
45. Sass, *Madness and Modernism*, 209.
46. Noel Castree and Bruce Braun, 'The construction of nature and the nature of construction: analytical and political tools for building survivable futures'. In Braun and Castree, *Remaking Reality*, 7.
47. Michael, *Reconnecting Culture, Technology, and Nature*, 20.
48. Michael, *Reconnecting Culture, Technology, and Nature*, 25.
49. Tonnies, *Community and Society*.
50. Bauman, *Intimations of Postmodernity*.
51. Jeffrey Wollock, 'Linguistic diversity and biodiversity'. In Maffi, *On Biocultural Diversity*, 254.
52. Smail, *The Nature of Unhappiness*, 218.
53. Masson, *The Assault on Truth*.
54. Freud, 'Fragment of an analysis of a case of hysteria', 64.
55. Hrdy, *Mother Nature*, 396.
56. Quoted by Hrdy, *Mother Nature*, 396.
57. Hrdy, *Mother Nature*, 396.
58. Winnicott, *The Maturational Process and the Facilitating Environment*, 133.
59. Brown, *Life Against Death*, 151.
60. Brown, *Life Against Death*, 151.
61. Mackay, 'Psychotherapy and the idea of meaning', 365.
62. Rogers, *Solving History*, 12.
63. Hillman and Ventura, *We've Had 100 Years of Psychotherapy and the World's Getting Worse*, 11–12.
64. David Smail, 'Psychotherapy and tragedy'. In House and Totton, *Implausible Professions*, 161.
65. Mitchell, *Relational Concepts in Psychoanalysis*, 18.
66. Eagleton, *Literary Theory*, 60.
67. Steinberg, *The Fiction of a Thinkable World*, 53.
68. Mitchell, *Relational Concepts*, 3.
69. Mitchell, *Relational Concepts*, 19.
70. Fairbairn, *Psychoanalytic Studies of the Personality*; Guntrip, *Schizoid Phenomena, Object Relations, and the Self*.

71. Frank Sulloway, *Freud: Biologist of the Mind* (London: Burnett Books, 1979), 4.
72. Smail, *Power, Interest, and Psychology*, 7.
73. Smail, *Power, Interest, and Psychology*, 7.
74. Mackay, 'Psychotherapy and the idea of meaning', 359.
75. Arthur Kleinman, quoted by Crossley, *Rethinking Health Psychology*, 18.
76. Kleinman, *Rethinking Psychiatry*, 136–7.
77. Kleinman, *Rethinking Psychiatry*, 87.
78. Laing and Esterson, *Sanity, Madness, and the Family*.
79. Billig, *Freudian Repression*, 7.
80. Lacan, *The Seminar of Jacques Lacan*, 9.
81. Sokal and Bricmont, *Intellectual Impostures*, 19.
82. Storr, *The Dynamics of Creation*, 63.
83. Vallentin, *Einstein*.
84. Derrida, *Limited Inc.*, 148.
85. Derrida, *Memoirs of the Blind*, 2.
86. Descartes, *Meditations on First Philosophy*, 48.
87. Derrida, *Memoirs of the Blind*, 36–7.
88. Derrida, *Memoirs of the Blind*, 37.
89. Chomsky, *Keeping the Rabble in Line*, 163–4.
90. Powell, *Jacques Derrida*, 176.
91. Evernden, *The Natural Alien*, 16–17.
92. Williams, *Truth and Truthfulness*, 11.
93. Gitlin, 'The politics of communication and the communication of politics', 336.
94. Scheper-Hughes and Sargent, *Small Wars*, 29.
95. Jacoby, *The End of Utopia*, 141.
96. Bennett, *Cultural Pessimism*.
97. Not that I am suggesting that the 'raw data' are in some way free of cultural and linguistic influences – only that they are subjected to a further colonisation as they are assimilated into the favoured discourses of each discipline.
98. Barnes and Duncan, *Writing Worlds*, 2–3.
99. Hare, 'Why fabulate?'
100. Donald, *A Mind So Rare*, 276–7.
101. Orwell, *A Collection of Essays*, 175.
102. Raine, *Farewell Happy Fields*, 25.
103. For example, Rozin and Kalat, 'Specific hungers and poison avoidance as adaptive specializations of learning.'
104. Steinberg, *The Fiction of a Thinkable World*, 94.
105. Rival, 'The growth of family trees', 636.
106. Rival, 'The growth of family trees', 636. (Italics in original.)
107. Williamson, 'The Arctic habitat and the integrated self'.
108. Berman, *Wandering God*, 168.
109. Berman, *Wandering God*, 169.
110. Abbey, *Desert Solitaire*, xi.
111. Steinberg, *The Fiction of a Thinkable World*, 77.
112. Rogers, *Nature and the Crisis of Modernity*, 174.
113. Phillips, *Contested Knowledge*, 73.

114. Borgmann, *Holding On to Reality*, 120.
115. Braver, *A Thing of This World*, 425.
116. Carrette, *Foucault and Religion*, 41–2.
117. Carrette, *Foucault and Religion*, 42.
118. Whitehead, *Modes of Thought*, 21.
119. Rogers, *Nature and the Crisis of Modernity*, 98.
120. Archer, *Being Human*, 125.
121. Hornborg, *The Power of the Machine*, 231.
122. Hacking, *The Social Construction of What?*, 11.
123. Kalland, 'Indigenous knowledge', 323.
124. Kalland, 'Indigenous knowledge', 323.
125. Peterson, 'Toward a materialist environmental ethic', 376.
126. Medin and Atran, *Folkbiology*, 6.
127. Sokal and Bricmont, *Intellectual Impostures*. See also Ehrenreich, 'Farewell to a fad'.
128. Pinch and Collins, 'Private science and public knowledge'.
129. Latour and Woolgar, *Laboratory Life*, 235.
130. John Snow, 'On the mode of communication of cholera'. In Snow, *Snow on Cholera*, 31–2.
131. Devaney, *Since At Least Plato … and Other Postmodernist Myths*, 171.
132. Shapiro, *Origins*, 261–2.
133. Latour, Bruno. 'On the partial existence of existing and nonexisting objects'. In Daston, *Biographies of Scientific Objects*.
134. Latour, 'On the partial existence of existing and nonexisting objects', 249.
135. Latour, 'On the partial existence of existing and nonexisting objects', 249–250.
136. Collier, *Critical Realism*, 76.
137. Latour and Woolgar, *Laboratory Life*, 45.
138. Latour and Woolgar, *Laboratory Life*, 75.
139. Feynman, 'What is science?'
140. Veena Das, 'Transactions in the construction of pain'. In Kleinman, Das, and Lock, *Social Suffering*, 74.
141. Brody, *The Other Side of Eden*, 194.
142. Brody, *The Other Side of Eden*, 194–5.
143. Carpenter, *Oh, What a Blow That Phantom Gave Me!*, 84.

5 How the Mind Took Over the World

1. Hay, *You Can Heal Your Life*, 7.
2. Simon, *Models of My Life*, 306, 308–9.
3. Simon, *Models of My Life*, 35, 86.
4. Stoller, *Sensuous Scholarship*, 53.
5. Nora, 'Between memory and history: Les lieux de memoire', 8, 16.
6. Steinberg, *The Fiction of a Thinkable World*, 109.
7. De Shazer, *Keys to Solution in Brief Therapy*.
8. Steinberg, *The Fiction of a Thinkable World*, 147.
9. Clarke, *Descartes*, 62.

294 *Notes*

10. Reiss, *The Discourse of Modernism*, 31–2.
11. Reiss, *The Discourse of Modernism*, 35.
12. Reiss, *The Discourse of Modernism*, 142.
13. 'The chymistry of Isaac Newton'. Dibner Collection MS. 1031B, Dibner Library for the History of Science and Technology, Smithsonian Institution. http://webapp1.dlib.indiana.edu/newton/mss/norm/ALCH00081/Accessed 13/6/08.
14. Bhaskar, *A Realist Theory of Science*, 55.
15. Thomas Kuhn, quoted by Mirowski, *More Heat Than Light*, 102–3.
16. Hoffmeyer, *Signs of Meaning in the Universe*, 38.
17. Mayr, *Authority, Liberty, and Automatic Machinery in Early Modern Europe*, 117.
18. Ulanowicz, *A Third Window*.
19. Fichte, *The Vocation of Man*, 91.
20. Kant, *Critique of Pure Reason*, xvi.
21. Braver, *A Thing of This World*, 35.
22. Hosinski, *Stubborn Fact and Creative Advance: An Introduction to the Metaphysics of Alfred North Whitehead*.
23. Midgley, *Science and Poetry*, 61.
24. Buchan, *Frozen Desire*.
25. Borgmann, *Holding on to Reality*, 10.
26. Adorno, *Negative Dialectics*.
27. Ingold, *The Perception of the Environment*, 242.
28. Cohen, *States of Denial*, 86.
29. Galilei, *Dialogue Concerning the Two Chief World Systems*, 207.
30. Steffens, *James Prescott Joule and the Concept of Energy*.
31. Reiss, *The Discourse of Modernism*, 169.
32. Quoted by Reiss, *The Discourse of Modernism*, 169.
33. Quoted by Reiss, *The Discourse of Modernism*, 174.
34. Reiss p. 174. But we should note here that Reiss – and Campanella – are using the term 'signs' in the way that most linguists and philosophers understand it: that is, as a component of the mind-constructed realm of thought that is then applied to the world, rather than in the way a biosemiotican would use it, as a bridge between the realm of thought and entities in the 'external' world.
35. Hayles, *How We Became Posthuman*, 12.
36. Borgmann, *Holding on to Reality*, 2.
37. Dupuy, *The Mechanisation of the Mind*, 29–30.
38. Sahlins, *Islands of History*, 149.
39. Dupuy, *The Mechanisation of the Mind*, 31.
40. Locke, *An Essay Concerning Human Understanding*, 2; 6; 9.
41. Boyle, *Schizophrenia: A Scientific Delusion?*
42. O'Hanlon, 'Not strategic, not systemic'.
43. Martín-Baró, *Writings for a Liberation Psychology*, 111.
44. Desjarlais, Eisenberg, Good, and Kleinman, *World Mental Health*, 55.
45. Prigogine and Stengers, *Order out of Chaos*, 46.
46. Temperton and Hobbs, 'The search for ecological assembly rules and its relevance to restoration ecology'.
47. Worster, 'The ecology of order and chaos', 12.
48. Pauly, Review of *Medicine, Mind, and the Double Brain*.
49. Hall, *Philosophers at War*.

50. 'A Basis for Choice' (London: Further Education Unit, 1979).
51. Payne, 'All things to all people'.
52. Sampson, 'Cognitive psychology as ideology', 731.
53. Sampson, 'Cognitive psychology as ideology', 731.
54. The concealed interplay between 'individual' subjectivity and ideological systems that we touched on in the previous chapter produces erroneous claims both of 'objectivity' and autonomous 'subjectivity'. This is a problem that goes some way towards undermining democracy, since it erodes the possibility of 'individual choice'.
55. Crary, *Suspensions of Perception*, 309.
56. E.g. Braun and Castree, *Remaking Reality*, 7.
57. Braun and Castree, *Remaking Reality*, xiii.
58. Braun and Castree, *Remaking Reality*, xiii.
59. Gade, *Nature and Culture in the Andes*, 6.
60. Mann, 'Three trees', 32.
61. Hay, *A Companion to Environmental Thought*, 22.
62. Budiansky, *Nature's Keepers*, 5.
63. Ulanowicz, *Ecology*.
64. A. S. Romer was a biologist who pointed out that evolutionary changes often have the effect of enabling organisms to continue living in the same basic way. In other words, superficial changes facilitate a deeper *constancy*.
65. Crary, *Suspensions of Perception*, 12.
66. Sass, *Madness and Modernism*, 279.
67. Sass, *Madness and Modernism*, 279–80.
68. Reported by Hare, 'Why fabulate?'
69. 'The New Shock of the New', BBC Television, 1 July 2004.
70. Goux, *Symbolic Economies*, 181.
71. Goux, *Symbolic Economies*, 181.
72. Taylor, *The Picture in Question*, 6.
73. Budiansky, *Nature's Keepers*, 86.
74. Budiansky, *Nature's Keepers*, 79.
75. Hull and Robertson, 'The Language of Nature Matters', 106.
76. Vogel, 'Environmental philosophy after the end of nature', 32.
77. As Eileen Crist shows in her 'Can an insect speak?'
78. Ingold, 'The optimal forager and economic man', 26.
79. Worster, 'The ecology of order and chaos', 13.
80. Borgmann, *Holding on to Reality*, 105.
81. Kaye, *Economy and Nature in the Fourteenth Century*, 215.
82. Kaye, *Economy and Nature in the Fourtenth Century*, 176–7. Kaye's book is an excellent review of the beginnings of natural science in the thirteenth and fourteenth centuries; and I will draw extensively on it in this section.
83. Murdoch, 'The analytic character of late medieval learning', 198.
84. Kaye, *Economy and Nature in the Fourteenth Century*, 8.
85. Le Breton, 'Dualism and Renaissance'.
86. Le Breton, 'Dualism and Renaissance', 60–1.
87. Romanyshyn, *Technology as Symptom and Dream*, 114–5.
88. Le Breton, 'Dualism and Renaissance', 53.
89. Olson, *Scottish Philosophy and British Physics, 1750–1880*, 123.
90. Murdoch, 'The analytic character of late medieval learning', 174. (My italics.)

91. Kaye, *Economy and Nature in the Fourteenth Century*, 67.
92. Kaye, *Economy and Nature in the Fourteenth Century*, 53.
93. Aristotle, *The Politics*, Book 1, Chapter 9.
94. Kaye, *Economy and Nature in the Fourteenth Century*, 67.
95. Kaye, *Economy and Nature in the Fourteenth Century*, 233.
96. Kaye, *Economy and Nature in the Fourteenth Century*, 129.
97. Kaye, *Economy and Nature in the Fourteenth Century*, 229.
98. Kaye, *Economy and Nature in the Fourteenth Century*, 133.
99. Kaye, *Economy and Nature in the Fourteenth Century*, 166.
100. Kaye, *Economy and Nature in the Fourteenth Century*, 14, 17.
101. Kaye, *Economy and Nature in the Fourteenth Century*, 17.
102. Kaye, *Economy and Nature in the Fourteenth Century*, 168.
103. Kaye, *Economy and Nature in the Fourteenth Century*, 18.
104. Kaye, *Economy and Nature in the Fourteenth Century*, 115.
105. Reiss, *The Discourse of Modernism*, 35, 36.
106. Romanyshyn, *Psychological Life*, 30.
107. On the corpse as a model for the person, see Romanyshyn, *Technology as Symptom and Dream*, Chapters 4 and 5.
108. Nadeau, *Environmental Endgame*, 104.
109. Mirowski, *More Heat than Light*, 224–5. As Fisher was to discover, such a frivolous 'science' may come home to roost; for it was he who suggested, a few weeks before the 1929 stock market crash, that "stock prices seem to have reached what looks like a permanently high plateau". See Coleman, 'The age of inexpertise'.
110. William Jevons, quoted by Nadeau, *The Environmental Endgame*, 105.
111. Nadeau, *The Wealth of Nature*, 23.
112. Nadeau, *Environmental Endgame*, 110.
113. Webley, Burgoyne, Lea, and Young, *The Economic Psychology of Everyday Life*, 2.
114. Nadeau, *The Wealth of Nature*, 77.
115. Goux, *Symbolic Economies*, 97.
116. Goux, *Symbolic Economies*, 98.
117. Nadeau, *Environmental Endgame* 36–7.
118. Brightman, *Grateful Prey*, 287–8.
119. Goux, *Symbolic Economies*, 98.
120. Goux, *Symbolic Economies*, 99.
121. Mirowski, *More Heat Than Light*, 4.
122. Maturana and Varela, *Autopoiesis and Cognition*.
123. Goux, *Symbolic Economies*, 113.
124. Goux, *Symbolic Economies*, 109.
125. Buchan, *Frozen Desire*, 71.
126. Nadeau, *Environmental Endgame*, 141.
127. Barbara Crossette, 'Kofi Annan's astonishing facts!' *New York Times*, 27 September 1998.
128. Leder, *The Absent Body*, 34–5.
129. Butler, *Luck, or Cunning, as the Main Means of Organic Modification?*, 49.
130. Buchan, *Frozen Desire*, 59.
131. Buchan, *Frozen Desire*, 191.
132. Vera, *Grazing Ecology and Forest History*.

133. Gillson, 'Testing non-equilibrium theories in savannas'.
134. Crist, 'Beyond the climate crisis'.
135. Alf Hornborg, 'Ecological embeddedness and personhood: Have we always been capitalists?' In Messer and Lambek (eds), *Ecology and the Sacred*, 90.
136. Kahn, *The Human Relationship with Nature*.
137. Daniel Pauly, quoted by Vera, 'The Shifting Baseline Syndrome in restoration ecology', 107.
138. Vera, 'The Shifting Baseline Syndrome in restoration ecology', 98.
139. Frank R. Wilson, quoted by Pallasmaa, *The Thinking Hand*, 33.
140. Hutchins, *Cognition in the Wild*, 356, 370.
141. Brody, *The Other Side of Eden*, 255.
142. Long, Tecle, and Burnette, 'Cultural foundations for ecological restoration on the White Mountain Apache reservation'.
143. Scott Atran, 'The vanishing landscape of the Petén Maya Lowlands.' In Maffi, *On Biocultural Diversity*, 166–7.
144. Rogers, *Nature and the Crisis of Modernity*, 92.
145. Atran, *The Cognitive Foundations of Natural History*.
146. Rodman, 'The liberation of nature?' 104.
147. Geary, *The Origin of Mind*, 195.
148. Brody, *The Other Side of Eden*, 277.
149. Hoffmeyer, *Signs of Meaning in the Universe*, 55.
150. Pallasmaa, *The Thinking Hand*, 20.
151. Brody, *The Other Side of Eden*, 254.
152. Lave, *Cognition in Practice*, 69–70, 82, 171.
153. Lave, *Cognition in Practice*, 150.
154. Lave, *Cognition in Practice*, 171.
155. Lave, *Cognition in Practice*, 82.
156. Holling and Sanderson, 'Dynamics of (dis)harmony in ecological and social systems', 65.
157. Brody, *The Other Side of Eden*, 254.
158. Brody, *The Other Side of Eden*, 269.
159. Ingold, *The Perception of the Environment*, 25.
160. Ingold, *The Perception of the Environment*, 55.
161. Richard Nelson, quoted by Ingold, *The Perception of the Environment*, 55.
162. Ingold, *The Perception of the Environment*, 55.
163. Ingold, *The Perception of the Environment*, 55.
164. Ingold, *The Perception of the Environment*, 47.
165. Cheney, 'Truth, knowledge, and the wild world', 125.
166. Ingold, *The Perception of the Environment*, 25.
167. Gellner, *Plough, Sword, and Book*, 44.
168. Gellner, *Plough, Sword, and Book*, 64–65.
169. Holling, 'The resilience of terrestrial ecosystems'.
170. Dijksterhuis and Nordgren, 'A Theory of Unconscious Thought'.
171. See, for example, Beck et al., 'Probabilistic Population Codes for Bayesian Decision Making'; Soon, Brass, Heinze, and Haynes, 'Unconscious determinants of free decisions in the human brain'.
172. Lewicki, Hill, and Czyzewska, 'Nonconscious acquisition of information'.
173. Bechara and Damasio, 'The somatic marker hypothesis'.
174. Kasser, *The High Price of Materialism*.

175. Plumwood, *Feminism and the Mastery of Nature*.
176. Cushman, 'Why the self is empty'.
177. Hallowell, *The Ojibwa of Berens River, Manitoba*, 61.
178. Derrida, *Memoirs of the Blind*, 109.
179. Haraway, 'A Cyborg Manifesto'.
180. In *Gender Trouble*, Butler writes: "Gender is not to culture as sex is to nature; gender is also the discursive/cultural means by which 'sexed nature' or a 'natural sex' is produced and established as 'prediscursive,' prior to culture, a politically neutral surface *on which* culture acts."(p. 7).
181. Callicott, 'The wilderness idea revisited'.
182. Ulanowicz, *A Third Window*, 118.
183. Watzlawick, Bavelas, and Jackson, *Pragmatics of Human Communication*.
184. Vogel, 'The nature of artifacts', 153.
185. Descartes, *Meditations on First Philosophy*, 15.
186. Cathcart and Klein, *Plato and a Platypus Walk into a Bar*, 57.
187. Clarke, *Descartes*, 52.
188. Clarke, *Descartes*, 191.
189. Plato, *Phaedo*, 30.
190. Deely, *Descartes and Poinsot*.
191. Quoted by Lorraine Daston, 'The coming into being of scientific objects'. In Daston, *Biographies of Scientific Objects*, 2.
192. Deely, *Descartes and Poinsot*, 14.
193. Bhaskar, *A Realist Theory of Science*, 45.
194. Hosinski, *Stubborn Fact and Creative Advance*, 38.
195. Deely, *Descartes and Poinsot*, 71–2.
196. Deely, *Descartes and Poinsot*, 71, 58.
197. Mirowski, *More Heat Than Light*, 4.
198. Todorov, *The Conquest of America*, 16.
199. Todorov, *The Conquest of America*, 21.
200. Todorov, *The Conquest of America*, 19.
201. Todorov, *The Conquest of America*, 19.
202. Todorov, *The Conquest of America*, 21.
203. Todorov, *The Conquest of America*, 22.
204. Todorov, *The Conquest of America*, 89.
205. Todorov, *The Conquest of America*, 111.
206. Todorov, *The Conquest of America*, 116.
207. Todorov, *The Conquest of America*, 175.
208. Todorov, *The Conquest of America*, 171.
209. Todorov, *The Conquest of America*, 169.
210. Todorov, *The Conquest of America*, 174.
211. Todorov, *The Conquest of America*, 97.
212. Locke, *An Essay Concerning Human Understanding*, 102–4.
213. Deacon, *The Symbolic Species*, 448.
214. Borgmann, *Holding On to Reality*, 1.
215. Borgmann, *Holding On to Reality*, 88.
216. Ingold, *Key Debates in Anthropology*, 115.
217. Brooke, *Jung and Phenomenology*, 60–1.
218. Naess, *Ecology, Community, and Lifestyle*, 60–1.
219. Miles, 'Psychological benefits of volunteering for restoration projects', 223.

220. Bateson, 'A theory of alcoholism'. In *Steps to an Ecology of Mind*.
221. Jackson, *Paths Toward a Clearing*, 174.
222. Wollock, 'Linguistic diversity and biodiversity'. In Maffi, *On Biocultural Diversity*, 254.
223. Cheney, 'Truth, knowledge, and the wild world', 109.
224. Favareau, 'The evolutionary history of biosemiotics', 7.
225. Quoted by Deely, *Descartes and Poinsot*, 15.
226. Wetherick, 'Against cognitive psychology', 22. (My italics.)
227. Midgley, *Science and Poetry*, 116.
228. Deacon, *The Symbolic Species*, 447.
229. Favareau, 'The evolutionary history of biosemiotics', 10.
230. Totton, 'The baby and the bathwater'.
231. Van der Ploeg, 'Potatoes and knowledge', 210–12.
232. Van der Ploeg, 'Potatoes and knowledge', 221.
233. Shiva, *Monocultures of the Mind*.
234. Denis Postle, 'Counselling in the UK: jungle, garden, or monoculture?' In House and Totton, *Implausible Professions*.
235. Midgley, *Science as Salvation*, 13.
236. Hornborg, *The Power of the Machine*, 231.
237. Havel, 'The power of the powerless', 29–30, 33.
238. Havel, 'The power of the powerless', 27, 39.
239. Turner, *The Abstract Wild*.
240. Tuck-Po, *Changing Pathways*, 21–2, 20.
241. Eidelson and Eidelson, 'Dangerous ideas', 185.
242. Arasteh, 'Denying history', 182.
243. Humphrey, 'Reconciliation and the therapeutic state'.
244. Arasteh, 'Denying history', 182.
245. Sampson, 'Cognitive psychology as ideology', 737.
246. Sampson, 'Cognitive psychology as ideology', 735–6.
247. Sampson, 'Cognitive psychology as ideology', 735.
248. Hayles, *How We Became Posthuman*, 36–7.
249. Sampson, 'Cognitive psychology as ideology', 736.
250. Öhman and Mineka, 'The malicious serpent', 5–9.
251. Rosch, 'Universals and cultural specifics in human categorization', 183.
252. Rosch, 'Universals and cultural specifics in human categorization', 190.
253. Rosch, 'Universals and cultural specifics in human categorization', 178.
254. Hugh Brody, *The Other Side of Eden*, 246.
255. Hoffmeyer, *Biosemiotics*, 25.
256. See Richard Woods and Matthew Campbell, 'Air France 447: The computer crash', *The Sunday Times*, 7 June 2009.
257. Graeme Dalling, 'I suffered from anorexia'. *The Guardian*, 11 July 2009.
258. Lily Kay, 'A book of life? How the genome became an information system and DNA a language'. *Perspectives in Biology and Medicine* 41 (1998), 507.
259. Dawkins, *The Extended Phenotype*.
260. Oyama, 'Speaking of Nature', 53.
261. Kauffman, *At Home in the Universe*.
262. Kay, 'A book of life?', 523–4.
263. Hoffmeyer, *Biosemiotics*, 104.

264. Oyama, *The Ontogeny of Information*.
265. Francisco Varela, quoted by Hayles, *How We Became Posthuman*, 155.
266. Haraway, 'A cyborg manifesto', 164.

6 The Industrialised Individual

1. Bhaskar, *The Possibility of Naturalism*, 35.
2. Margaret Archer. 'Realism and morphogenesis', in Archer, Bhaskar, Collier, Lawson, and Norrie, *Critical Realism*, 365–6.
3. Hayles, *How We Became Posthuman*, 288.
4. Quoted by Freund and Martin, *The Ecology of the Automobile*, 3.
5. Even, apparently, if we are one of the relatively rich inhabitants of a society that unequally diverts the costs towards the poor, as Wilkinson and Pickett argue in *The Spirit Level*.
6. Emerson Fittipaldi, quoted by Freund and Martin, *The Ecology of the Automobile*, 87.
7. Keeney, *Aesthetics of Change*, 125–6.
8. Philo and Miller, *Market Killing*, 59–70.
9. Munro, *The Concept of Man in Early China*, 160.
10. Bhaskar, *Scientific Realism and Human Emancipation*.
11. Susan Sontag, Journals: 'Early 1959, New York City': Guardian, 14 September 2006.
12. Eileen Crist, personal communication.
13. Steinberg, *The Fiction of a Thinkable World*, 145.
14. Cf. Marcuse, *One-Dimensional Man*, 73.
15. Abbey, *Desert Solitaire*, 25.
16. Abbey, *Desert Solitaire*, xiii.
17. Abbey, *Desert Solitaire*, 176.
18. Bloom, *Love and Friendship*, 27.
19. Derrida, *Memoirs of the Blind*, 68.
20. Munro, *Lives of Girls and Women*, 150.
21. Gary Snyder, quoted by William Cronon, 'The trouble with wilderness; or, getting back to the wrong nature'. In Cronon, *Uncommon Ground*, 89.
22. Bloom, *Love and Friendship*, 25.
23. Lasch, *The Culture of Narcissism*, 5.
24. Rowan Williams, 'Conversations with Rowan Williams', Channel 4 television, 17 October 2003.
25. Benjamin, *Illuminations*.
26. Terdiman, *Present Past*.
27. Hobsbawm, *The Age of Extremes*, 16.
28. Evans, *Traffic Safety and the Driver*, 103.
29. Tuck-Po, *Changing Pathways*, 100.
30. Ridington, *Trail to Heaven*, 70, 72.
31. See my earlier remarks, in Chapter 1, on the 'naturalistic fallacy'.
32. Cronon, 'The trouble with wilderness', 69.
33. Hornborg, 'Ecological embeddedness and personhood'. In Messer and Lambek, *Ecology and the Sacred*, 90.
34. Sommer, 'Trees and human identity'.

35. Marx, 'Private property and communism', 91–2.
36. Gross, *Lost Time*, 178.
37. Strang, *Uncommon Ground*, 200.
38. Strang, *Uncommon Ground*, 203.
39. Strang, *Uncommon Ground*, 215.
40. Steinberg, *The Fiction of a Thinkable World*, 162.
41. Lichtman, 'The illusion of the true self', 127.
42. Masson, *The Assault on Truth*.
43. Smail, *Power, Interest, and Psychology*, 1–4.
44. Smail, *Power, Interest, and Psychology*, 9.
45. Antze and Lambek, *Tense Past*, xxiv.
46. Bauman, *The Individualized Society*, 9.
47. Ross, 'The intuitive psychologist and his shortcomings.'
48. Atran, 'The genesis of suicide terrorism', 1536.
49. Rose, *The Last Resistance*, 135.
50. Hardt and Negri, *Empire*, 21.
51. Auden, 'The Fall of Rome', 218.
52. Twenge, *Generation Me*, 61.
53. Dworkin, *Artificial Happiness*, 253.
54. Twenge, *Generation Me*, 86.
55. Beck, *Risk Society*, 137.
56. Martín-Baró, *Writings for a Liberation Psychology*, 5.
57. Martín-Baró, *Writings for a Liberation Psychology*, 92–3.
58. Thaler and Sunstein, *Nudge*.
59. Jan Null, 'What happens when forecasters come under a cloud'. *San Jose Mercury News*, 24 February 2004.
60. Monbiot, 'Environmental feedback', 107.
61. Michel Foucault and Richard Sennett, 'Sexuality and solitude'. *London Review of Books* 3(9), 21 May 1981, 5.
62. Foucault, *The History of Sexuality: Volume 1: An Introduction*, 63.
63. Foucault and Sennett, 'Sexuality and solitude', 5.
64. Quoted by Gray, 'Urban surveillance and panopticism', 318.
65. Quoted by Gray, 'Urban surveillance and panopticism', 318.
66. Spirn, 'Restoring Mill Creek', 398.
67. Morrison, 'The Site of Memory', 119.
68. Manno, *Privileged Goods*, 132.
69. Mari Marcel-Thekaekara, 'Poor relations'. *The Guardian*, Saturday February 27th 1999, Review, 3.
70. Morrison, 'The Site of Memory', 112–13.
71. Williams, *Truth and Truthfulness*, 12.
72. Duncan, 'When compassion becomes dissent'.
73. Ulanowicz, *A Third Window*, 73–5.
74. Rolston, *Environmental Ethics*, 16–17.
75. Smail, *The Nature of Unhappiness*, 85.
76. Frumkin, 'Beyond toxicity'.
77. Smail, *The Nature of Unhappiness*, 87.
78. Turner, *The Abstract Wild*.
79. Vayda, *Methods and Explanations*, 7.
80. Tuan, 'Discrepancies between environmental attitudes and behaviour'.

81. Freud, 'Civilisation and Its Discontents', 86.
82. Billig, *Freudian Repression*, 52, 71.
83. Billig, *Freudian Repression*, 71.
84. Billig, *Freudian Repression*, 257.
85. Stoller, *Sensuous Scholarship*, xii.
86. Foucault, *The History of Sexuality Vol. 1: An Introduction*, 6.
87. Foucault, *The History of Sexuality*, 71.
88. Foucault, *The History of Sexuality*, 71–2.
89. Foucault, *The History of Sexuality*, 33.
90. Francis Bacon, quoted by Merchant, *The Death of Nature*, 168.
91. Foucault, *The History of Sexuality*, 23.
92. Foucault, *The History of Sexuality*, 105.
93. Foucault, *The History of Sexuality*, 23–4. (My italics.)
94. Foucault, *The History of Sexuality*, 152. (My italics.)
95. Richard Silverstein, 'Land of the Free?' *The Guardian*, 10 October 2007.
96. Kleinman, Das, and Lock. 'Introduction to "social suffering"'.
97. Healy, *The Antidepressant Era*, 254.
98. Geuter, *The Professionalisation of Psychology in Nazi Germany*, 259, 268.
99. Cohen, *States of Denial*.
100. Twenge, *Generation Me*, 68–71.
101. Davison, *Technology and the Contested Meanings of Sustainability*, 85.
102. Sut Jhally, interviewed in *Psywar* (produced by Scott Noble; Metanoia Films, 2010).
103. Crary, *Suspensions of Perception*, 1.
104. De Grandpre, *Ritalin Nation*, 49.
105. Crary, *Suspensions of Perception*, 72.
106. Crary, *Suspensions of Perception*, 1.
107. Crary, *Suspensions of Perception*, 5.
108. Postman, *Amusing Ourselves to Death*, 107.
109. Twenge, *Generation Me*, 139.
110. Tony Benn, 'Big ideas that changed the world'. Channel 5 Television, 21 June 2005.
111. DeGrandpre, *Ritalin Nation*, 31.
112. Midgley, *Science as Salvation*, 13.
113. As seems to have occurred in the case of the Challenger disaster. See Ulanowicz, *Ecology*, 92, 152–3.
114. Shiva, *Monocultures of the Mind*.
115. Ulanowicz, *Ecology*, 92–3.
116. Quoted by Blum, *Love at Goon Park*, 36–7.
117. Burman, *Deconstructing Developmental Psychology*, 142.
118. Blum, *Love at Goon Park*, 55.
119. Hargrove, 'Taking environmental ethics seriously', 18.
120. Worster, *Nature's Economy*, 336.
121. Hume, *A Treatise of Human Nature*, 156.
122. Archer, *Being Human*, Chapter 2.
123. Becker, *Accounting for Tastes*, 236–7.
124. Twenge, *Generation Me*, 99.
125. Becker, *Accounting for Tastes*, 5–6, 232–3.
126. Becker, *The Economic Approach to Human Behaviour*, 173, 178.

127. Quoted by Kasser, *The High Price of Materialism*, 67.
128. Smith, *An Inquiry into the Nature and Causes of the Wealth of Nations*, 25.
129. Polanyi, *The Great Transformation*, 43.
130. Kasser, *The High Price of Materialism*.
131. Hrdy, *Mother Nature*, 494.
132. Lane, *The Loss of Happiness in Market Democracies*, 158.
133. Elster, *Sour Grapes*.
134. Archer, *Being Human*, 59.
135. Hardin, 'The tragedy of the commons', 1245.
136. Buchan, *Frozen Desire*, 94.
137. Kasser, *The High Price of Materialism*, 92, 77–8, 104.
138. Padel, *The Sacrifice of Human Being*, 178–9.
139. Sir W. W. Hunter, quoted by Padel, *The Sacrifice of Human Being*, 180.
140. Padel, *The Sacrifice of Human Being*, 181.
141. Padel, *The Sacrifice of Human Being*, 263–4.
142. Lakoff and Johnson, *Philosophy in the Flesh*.
143. Piaget, *The Psychology of Intelligence*, 151.
144. Carlos Fuentes, 'Confederacy of shadows'. *The Guardian*, 4 September 2004.
145. Eugene Gendlin, 'Thinking beyond patterns: Body, language, and situations'. In Den Outen and Moen, *The Presence of Feeling in Thought*, 49.
146. Smail, *Taking Care*, 28.
147. Piaget, *The Psychology of Intelligence*.
148. Williamson, 'The Arctic habitat and the integrated self', 189–90.
149. Spence, *Narrative Truth and Historical Truth*, 108.
150. Bevis, 'Native American novels', 599, 602–3.
151. Ingold, *The Perception of the Environment*, 51.
152. McNickle, *The Surrounded*, 102.
153. Bevis, 'Native American novels', 603.
154. McNickle, *The Surrounded*, 102–3.
155. Tuck-Po, *Changing Pathways*, 79.
156. Alan T. Campbell, quoted by Tuck-Po, *Changing Pathways*, 79.
157. Ingold, *The Perception of the Environment*, 97.
158. Desjarlais, *Body and Emotion*, 249.
159. Gendlin, 'Thinking beyond patterns'. In Den Outen and Moen, *The Presence of Feeling in Thought*, 24.
160. Berman, *Coming to our Senses*, 205.
161. Atwood, *The Handmaid's Tale*, 237.
162. Archer, *Being Human*, 125.
163. Taylor and Brown, 'Illusion and well-being', 193.
164. Simon, *The Ultimate Resource*, 9.
165. Simon, *Good Mood*, 93.
166. Simon, *Good Mood*, 182, 168.
167. Gilder, *The Spirit of Enterprise*, 69.
168. Gilder, *The Spirit of Enterprise*, 70.
169. Simon, *The Ultimate Resource*, 346–7.
170. Simon, *Good Mood*, 142.
171. Taylor and Brown, 'Positive illusions and well being revisited', 24.
172. Alloy, Albright, Abramson and Dykman, 'Depressive realism and nondepressive optimistic illusions', 72.

173. Julian Simon, 'Pre-debate statement'. In Myers and Simon, *Scarcity or Abundance?*
174. Simon, *The Ultimate Resource*, 89.
175. Simon, *The Ultimate Resource 2*, 82–3.
176. Cohen, *States of Denial*, 58.
177. Lasch, 'A response to my critics'. Theories such as that of Giddens which disregard the growing asymmetry between individual agency and the power of the industrial system are often regarded as preferable because they are more 'optimistic'. Thus wishful thinking becomes a criterion by which social theories are assessed.
178. Martín-Baró, *Writings for a Liberation Psychology*, 215.
179. Smail, *Power, Interest, and Psychology*, 54.
180. Smail, *Power, Interest, and Psychology*, 56.
181. Dworkin, 'The medicalisation of unhappiness', 85, 99.
182. Quoted by Rogers, *Nature and the Crisis of Modernity*, 130.
183. Gray, *Straw Dogs*, 141.
184. Rodman, 'The liberation of nature?', 100.
185. Deacon, *The Symbolic Species*, 437.
186. Deacon, *The Symbolic Species*, 437.

Bibliography

Abbey, Edward. *Desert Solitaire: A Season in the Wilderness* (New York: Ballantine, 1968).

Adorno, Theodor. *Negative Dialectics* (London: Routledge and Kegan Paul, 1973).

Allen, T. F. H. and Starr, Thomas B. *Hierarchy: Perspectives for Ecological Complexity* (Chicago: Chicago University Press, 1982).

Alloy, Lauren, J. S. Albright, L. Y. Abramson and B. M. Dykman. 'Depressive realism and nondepressive optimistic illusions: The role of the self'. In R. E. Ingram (ed.), *Contemporary Psychological Approaches to Depression: Theory, Research, and Treatment* (New York: Plenum, 1990).

Antze, Paul and Michael Lambek (eds), *Tense Past: Cultural Essays in Trauma and Memory* (New York: Routledge, 1996).

Arasteh, Kamyar. 'Denying history'. *American Psychologist* 59(3), April 2004, 182.

Archer, Margaret. *Being Human: The Problem of Agency* (Cambridge: Cambridge University Press, 2000).

Archer, Margaret, Roy Bhaskar, Andrew Collier, Tony Lawson, and Alan Norrie (eds), *Critical Realism: Essential Readings* (London: Routledge, 1998).

Argyrou, Vassos. *Anthropology and the Will to Meaning* (London: Pluto Press, 2002).

Aristotle. *The Politics*, Trans. T. A. Sinclair (Harmondsworth: Penguin, 1962).

Atran, Scott. *The Cognitive Foundations of Natural History: Towards an Anthropology of Science* (Cambridge: Cambridge University Press, 1990).

Atran, Scott. 'Modes of thinking about living kinds'. In David R. Olson and Nancy Torrance (eds), *Explorations in Culture and Cognition* (Cambridge: Cambridge University Press, 1996).

Atran, Scott. 'The genesis of suicide terrorism'. *Science* 299, March 2003, 1534–9.

Atran, Scott and Douglas Medin. *The Native Mind and the Cultural Construction of Nature* (Cambridge, MA: MIT Press, 2008).

Atwood, Margaret. *The Handmaid's Tale* (London: Vintage, 1996).

Auden, W. H. 'The Fall of Rome'. In *Shorter Collected Poems* (London: Faber and Faber, 1996).

Austin, J. L. *How To Do Things With Words* (Oxford: Oxford University Press, 1962).

Barnes, Trevor J. and James S. Duncan. *Writing Worlds: Discourse, Text and Metaphor in the Representation of Landscape* (London: Routledge, 1992).

Barrett, William. *Irrational Man: A Study in Existential Philosophy* (Westport: Greenwood Press, 1977).

Barron, Frank. *Creativity and Personal Freedom* (Princeton: Van Nostrand, 1968).

Barthes, Roland. *Mythologies* (London: Cape, 1972).

Basso, Keith. *Wisdom Sits in Places* (Albuquerque: University of New Mexico Press, 1996).

Bateson, Gregory. *Steps to an Ecology of Mind* (St Albans: Paladin, 1973).

Bateson, Gregory and Mary Catherine Bateson. *Angels Fear: Towards an Epistemology of the Sacred* (New York: Macmillan, 1987).

Bauman, Zygmunt. *Intimations of Postmodernity* (London: Routledge, 1991).

Bauman, Zygmunt. *The Individualized Society* (Cambridge: Polity, 2001).

Bechara, Antoine and Antonio Damasio, 'The somatic marker hypothesis: A neural theory of economic decision'. *Games and Economic Behavior* 52 (2005), 336–72.

Beck, Jeffrey M., Wei Ji Ma, Roozbeh Kiani, Tim Hanks, Anne K. Churchland, Jamie Roitman, Michael N. Shadlen, Peter E. Latham, and Alexandre Pouget, "Probabilistic population codes for Bayesian decision making". *Neuron* 60(6), December 26th 2008, 1142–52.

Beck, Ulrich. *Risk Society: Towards a New Modernity* (London: Sage, 1992).

Becker, Gary S. *The Economic Approach to Human Behaviour* (Chicago: University of Chicago Press, 1976).

Becker, Gary S. *Accounting for Tastes* (Cambridge, Mass: Harvard University Press, 1996).

Benjamin, Walter. *Illuminations* (London: Fontana, 1992).

Bennett, Jane. 'De rerum natura'. *Strategies: Journal of Theory, Culture, and Politics* 13(1), May 2000, 9–22.

Bennett, Oliver. *Cultural Pessimism: Narratives of Decline in the Postmodern World* (Edinburgh: Edinburgh University Press, 2001).

Berman, Morris. *Coming to our Senses: Body and Spirit in the Hidden History of the West* (New York: Simon and Schuster, 1989).

Berman, Morris. *Wandering God: A Study in Nomadic Spirituality* (Albany: State University of New York Press, 2000).

Berman, Morris. *The Twilight of American Culture* (New York: Norton, 2001).

Berne, Eric. *Games People Play* (Harmondsworth: Penguin, 1968).

Bernstein, Jerome S. *Living in the Borderland: The Evolution of Consciousness and the Challenge of Healing Trauma* (London: Routledge, 2005).

Bevis, William. 'Native American novels: homing in'. In Brian Swann and Arnold Krupat (eds), *Recovering the Word: Essays on Native American Literature* (Berkeley: University of California Press, 1987).

Bhaskar, Roy. *The Possibility of Naturalism* (London: Harvester Press, 1979).

Bhaskar, Roy. *A Realist Theory of Science* (London: Verso, 1997).

Bhaskar, Roy. *From Science to Emancipation* (New Delhi: Sage, 2002).

Bhaskar, Roy. *Scientific Realism and Human Emancipation* (London: Routledge, 2009).

Billig, Michael. *Freudian Repression: Conversation Creating the Unconscious* (Cambridge: Cambridge University Press, 1999).

Bloom, Alan. *Love and Friendship* (New York: Simon and Schuster, 1993).

Blum, Deborah. *Love at Goon Park: Harry Harlow and the Science of Affection* (New York: Berkley Books, 2002).

Boden, Margaret A. *The Creative Mind: Myths and Mechanisms* (London: Routledge, 2004).

Borgmann, Albert. *Holding On to Reality: The Nature of Information at the Turn of the Millennium* (Chicago: University of Chicago Press, 1999).

Borgmann, Albert. 'The destitution of space: from cosmic order to cyber disorientation'. *Harvard Design Magazine* 10, Winter/Spring 2000, 12–17.

Botkin, Daniel B. *Discordant Harmonies: A New Ecology for the Twenty-First Century* (New York: Oxford University Press, 1990).

Bowker, James and Stoll, John R. 'Use of dichotomous choice nonmarket methods to value the whooping crane resource'. *American Journal of Agricultural Economy* 23(5), 1987, 372–81.

Boyd, Robert and Peter J. Richerson. *The Origin and Evolution of Cultures* (Oxford: Oxford University Press, 2005).

Boyle, David. *The Tyranny of Numbers* (London: HarperCollins, 2000).

Boyle, Mary. *Schizophrenia: A Scientific Delusion?* (London: Routledge: 2002).

Braun, Bruce and Noel Castree (eds), *Remaking Reality: Nature at the Millennium* (London: Routledge, 1998).

Braver, Lee. *A Thing of This World* (Evanston: Northwestern University Press, 2007).

Brightman, Robert A. *Grateful Prey: Rock Cree Human-Animal Relationships* (Berkeley: University of California Press, 1993).

Brody, Hugh. *The Other Side of Eden: Hunter-Gatherers, Farmers, and the Shaping of the World* (London: Faber, 2001).

Brooke, Roger. *Jung and Phenomenology* (London: Routledge, 1991).

Brown, Norman O. *Life Against Death: The Psychoanalytical Meaning of History* (Middletown, CT: Weslyan University Press, 1959).

Bruner, Jerome S. *Actual Minds, Possible Worlds* (Cambridge, MA: Harvard University Press, 1986).

Buchan, James. *Frozen Desire: An Inquiry into the Meaning of Money* (London: Picador, 1997).

Buchmann, Stephen L. and Gary P. Nabhan. *The Forgotten Pollinators* (Washington DC: Island Press, 1998).

Buck-Morss, Susan. 'Socioeconomic bias in Piaget's theory and its implications for cross-cultural studies'. *Human Development* 18 (1975), 35–49.

Budiansky, Stephen. *Nature's Keepers: The New Science of Nature Management* (New York: Free Press, 1995).

Buell, Frederick. *From Apocalypse to Way of Life* (New York: Routledge, 2003).

Burman, Erica. *Deconstructing Developmental Psychology* (London: Routledge, 2007).

Butler, Judith. *Gender Trouble* (New York: Routledge, 1990).

Butler, Samuel. *Luck, or Cunning, as the Main Means of Organic Modification?* (Whitefish, MT: Kessinger Publishing, 2004 (1886)).

Callicott, J. Baird. 'The wilderness idea revisited: The sustainable development alternative'. *The Environmental Professional* 13 (1991), 235–47.

Carpenter, Edmund. *Oh, What a Blow That Phantom Gave Me!* (St Albans: Paladin, 1973).

Carrette, Jeremy. *Foucault and Religion* (London: Routledge, 1999).

Casey, Edward. *Getting Back Into Place: Toward a Renewed Understanding of the Place-World* (Bloomington: Indiana University Press, 1993).

Cathcart, Thomas and Daniel Klein. *Plato and a Platypus Walk into a Bar: Understanding Philosophy through Jokes* (New York: Abrams Image, 2007).

Cheney, Jim. 'Truth, knowledge, and the wild world'. *Ethics and the Environment* 10(2), Autumn 2005, 101–35.

Childs, Craig. *The Secret Knowledge of Water* (Boston: Little, Brown, and Co., 2000).

Chiu, Lian-Hwang. 'A cross-cultural comparison of cognitive styles in Chinese and American children'. *International Journal of Psychology* 7 (1972), 235–42.

Chomsky, Noam. *Keeping the Rabble in Line: Interviews with David Barsamian.* (Monroe, ME: Common Courage Press, 1994).

Clarke, Desmond M. *Descartes: A Biography* (Cambridge: Cambridge University Press, 2006).

Cohen, Stanley. *States of Denial* (Cambridge: Polity Press, 2001).

Coleman, Les. 'The age of inexpertise'. *Quadrant* 42(5), May 1998, 63–7.

Collier, Andrew. *Critical Realism: An Introduction to Roy Bhaskar's Philosophy* (London: Verso, 1994).

Connerton, Paul. *How Societies Remember* (Cambridge: Cambridge University Press, 1989).

Connerton, Paul. *How Modernity Forgets* (Cambridge: Cambridge University Press, 2009).

Crary, Jonathan. *Suspensions of Perception: Attention, Spectacle, and Modern Culture* (Cambridge: MIT Press, 2000).

Crist, Eileen. *Images of Animals: Anthropomorphism and the Animal Mind* (Philadelphia: Temple University Press, 1999).

Crist, Eileen. 'Can an insect speak? The case of the honeybee dance language'. *Social Studies of Science* 34(1), 2004, 7–43.

Crist, Eileen. 'Beyond the climate crisis: A critique of climate change discourse'. *Telos* 141, Winter 2007, 29–55.

Cronon, William (ed.), *Uncommon Ground: Rethinking the Human Place in Nature* (New York: Norton, 1996).

Crossley, Michele L. *Rethinking Health Psychology* (Buckingham: Open University Press, 2000).

Cushman, Philip. 'Why the self is empty: Toward a historically situated psychology'. *American Psychologist* 45(5), May 1990, 599–611.

Daston, Lorraine (ed.), *Biographies of Scientific Objects* (Chicago: University of Chicago Press, 2000).

Davison, Aidan. *Technology and the Contested Meanings of Sustainability* (Albany: State University of New York Press, 2001).

Dawkins, Richard. *The Extended Phenotype: The Gene as the Unit of Selection* (Oxford: Oxford University Press, 1982).

Dawkins, Richard. 'Viruses of the mind'. In *A Devil's Chaplain* (London: Phoenix Press, 2003).

Deacon, Terrence. *The Symbolic Species* (London: Penguin, 1997).

Deely, John. *Descartes and Poinsot: The Crossroads of Signs and Ideas* (Scranton: University of Scranton Press, 2008).

DeGrandpre, Richard. *Ritalin Nation* (New York: Norton: 2000).

Den Outen, Bernard and Marcia Moen (eds), *The Presence of Feeling in Thought* (New York: Peter Lang, 1991).

Derrida, Jacques. *Limited Inc.* (Evanston: Northwestern University Press, 1988).

Derrida, Jacques. *Memoirs of the Blind* (Chicago: Chicago University Press, 1993).

Descartes, René. *Meditations on First Philosophy* (Cambridge: Cambridge University Press, 1986 (1641)).

De Shazer, Steve. *Keys to Solution in Brief Therapy* (New York: Norton, 1985).

Desjarlais, Robert. *Body and Emotion: The Aesthetics of Illness and Healing in the Nepal Himalayas* (Philadelphia: University of Pennsylvania Press, 1992).

Desjarlais, Robert, Leon Eisenberg, Byron Good, and Arthur Kleinman. *World Mental Health* (New York: Oxford University Press, 1995).

Devaney, M. J. *Since At Least Plato ... and Other Postmodernist Myths* (Basingstoke: Macmillan, 1997).

Dijksterhuis, Ap and Loran F. Nordgren. 'A Theory of Unconscious Thought'. *Perspectives on Psychological Science* 1(2), June 2006, 95–109.

Donald, Merlin. *The Origins of the Modern Mind: Three Stages in the Evolution of Culture and Cognition* (Cambridge, MA: Harvard University Press, 1991).

Donald, Merlin. *A Mind So Rare* (New York: Norton, 2001).

Drengson, Alan. 'The life and work of Arne Naess: An appreciative overview'. *The Trumpeter* 21(1), 2005, 5–47.

Duncan, David James. 'When compassion becomes dissent'. *Orion Magazine*, Jan/Feb 2003.

Dupuy, Jean-Pierre. *The Mechanisation of the Mind: On the Origins of Cognitive Science* (Princeton: Princeton University Press, 2000).

Dworkin, Ronald W. 'The medicalization of unhappiness'. *The Public Interest*, Summer 2001.

Dworkin, Ronald W. *Artificial Happiness: The Dark Side of the New Happy Class* (New York: Carroll and Graf, 2006).

Eagleton, Terry. *Literary Theory: An Introduction* (London: WileyBlackwell, 1983).

Eaton, David. 'Human and Animal: Thinking and Feeling a Way Toward Liberation'. Unpublished doctoral dissertation, Nottingham Trent University, 2008.

Edwards, Derek. *Discourse and Cognition* (London: Sage, 1997).

Ehrenreich, Barbara. 'Farewell to a fad'. *The Progressive*, March 1999, 17–18.

Eidelson, R. J. and J. I. Eidelson. 'Dangerous ideas: Five beliefs that propel groups toward conflict'. *American Psychologist* 58(3), March 2003, 182–92.

Elkin, A. P. *Aboriginal Men of High Degree* (St Lucia: University of Queensland Press, 1977).

Elster, Jon. *Sour Grapes: Studies in the Subversion of Rationality* (Cambridge: Cambridge University Press, 1983).

Enwegbara, Basil. 'Toxic colonialism'. *The Tech* 121(16), April 6 2001.

Epstein, Seymour. 'Integration of the cognitive and the psychodynamic unconscious'. *American Psychologist* 49(8), August 1994, 709–24.

Evans, Leonard. *Traffic Safety and the Driver* (New York: Van Nostrand, 1991).

Evernden, Neil. *The Natural Alien* (Toronto: University of Toronto Press, 1985).

Ewen, Stuart. *All Consuming Images: The Politics of Style in Contemporary Culture* (New York: Basic Books, 1988).

Fairbairn, W. R. D. *Psychoanalytic Studies of the Personality* (London: Routledge, 1952).

Farrelly, Elizabeth. *Blubberland: The Dangers of Happiness* (Cambridge, MA: MIT Press, 2008).

Favareau, Donald. 'The evolutionary history of biosemiotics'. In Marcello Barbieri (ed.), *Introduction to Biosemiotics: The New Biological Synthesis* (Dordrecht: Springer, 2007).

Feist, Gregory J. *The Psychology of Science and the Origins of the Scientific Mind* (New Haven: Yale University Press, 2006).

Feynman, Richard. 'What is science?'. *The Physics Teacher* 7(6), (1969).

Fichte, Johann. *The Vocation of Man*, trans. William Smith (LaSalle: Open Court, 1955).

Foucault, Michel. *The History of Sexuality. Volume 1: An Introduction*. Trans. Robert Hurley (New York: Vintage Books, 1990).

Foucault, Michel and Richard Sennett. 'Sexuality and solitude'. *London Review of Books* 3(9), 21 May 1981, 3–7.

Fox, Warwick. *Toward a Transpersonal Ecology* (Boston: Shambhala, 1990).

Franklin, Adrian. *Nature and Social Theory* (London: Sage, 2001).

Freud, Sigmund. 'Fragment of an analysis of a case of hysteria'. In James Strachey (ed.) *The Standard Edition of the Complete Psychological Works of Sigmund Freud* (London: Hogarth, 1964), Vol. VII.

Freud, Sigmund. 'Civilisation and Its Discontents'. In James Strachey (ed.), *The Standard Edition of the Complete Psychological Works of Sigmund Freud* (London: Hogarth, 1964), Vol. XXI.

Freund, Peter and George Martin. *The Ecology of the Automobile* (Montréal: Black Rose Books, 1993).

Frumkin, Howard. 'Beyond toxicity: Human health and the natural environment'. *American Journal of Preventive Medicine* 20(3), 2001, 234–40.

Gade, Daniel. *Nature and Culture in the Andes* (Madison: University of Wisconsin Press, 1999).

Galilei, Galileo. *Dialogue Concerning the Two Chief World Systems – Ptolemaic and Copernican*. Trans Stillman Drake (Berkeley: University of California Press, 1953 (1630)).

Gamarra, Javier G. P. and Solé, Ricard V. 'Bifurcations and chaos in ecology: lynx returns revisited'. *Ecology Letters*, 2000, 3, 1–8.

Geary, David C. *The Origin of Mind: Evolution of Brain, Cognition, and General Intelligence* (Washington, DC: American Psychological Association, 2005).

Gedo, John E. *The Artist and the Emotional World* (New York: Columbia University Press, 1996).

Geertz, Clifford. *The Interpretation of Cultures* (New York: Basic Books, 1973).

Geertz, Clifford. *Local Knowledge* (New York: Basic Books, 1985).

Gellner, Ernest. *Plough, Sword, and Book: The Structure of Human History* (London: Collins Harvill, 1988).

Gendlin, Eugene. 'How philosophy cannot appeal to experience'. In David Michael Levin (ed.), *Language Beyond Postmodernism: Saying and Thinking in Gendlin's Philosophy* (Evanston: Northwestern University Press, 1997).

Gergen, Kenneth. *The Saturated Self* (New York: Basic Books, 1992).

Geuter, Ulfried. *The Professionalisation of Psychology in Nazi Germany* (Cambridge: Cambridge University Press, 1992).

Gibbons, Ann. 'Paleoanthropology: Bone sizes trace the decline of man (and woman)'. *Science* 276. no. 5314, 9 May 1997, 896–7.

Gibson, William. *Neuromancer* (London: HarperCollins, 1995).

Giddens, Anthony. *Modernity and Self-Identity: Self and Society in the Late Modern Age* (Cambridge: Polity Press, 1991).

Gilder, George. *The Spirit of Enterprise* (New York: Viking, 1984).

Gillson, Lindsey. 'Testing non-equilibrium theories in savannas: 1400 years of vegetation change in Tsavo National Park, Kenya.' *Ecological Complexity* 1 (2004), 281–98.

Gitlin, Todd. 'The politics of communication and the communication of politics'. In James Curran and Michael Gurevitch (eds), *Mass Media and Society* (London: Hodder Arnold, 1991).

Goodman, Paul. *Growing Up Absurd* (New York: Vintage Books, 1960).

Goux, Jean-Joseph. *Symbolic Economies: After Marx and Freud* (Ithaca: Cornell University Press, 1990).

Gray, John. *Straw Dogs: Thoughts on Humans and Other Animals* (London: Granta, 2002).

Gray, Mitchell. 'Urban surveillance and panopticism: Will we recognise the facial recognition society?' *Surveillance and Society* 1(3), 2003, 314–30.

Gross, David. *Lost Time: On Remembering and Forgetting in Late Modern Culture* (Amherst: University of Massachusetts Press, 2000).

Guntrip, Harry. *Schizoid Phenomena, Object Relations, and the Self* (London: Hogarth, 1968).

Hacking, Ian. *The Social Construction of What?* (Cambridge, MA: Harvard University Press, 1999).

Hall, A. Rupert. *Philosophers at War: The Quarrel Between Newton and Leibniz* (Cambridge: Cambridge University Press, 1980).

Hallowell, A. Irving. *The Ojibwa of Berens River, Manitoba*, Jennifer S. H. Brown (ed.) (Fort Worth: Harcourt Brace Jovanovich, 1992).

Hannigan, John. *Environmental Sociology: A Social Constructionist Perspective* (London: Routledge, 1995).

Hansen, James, Makiko Sato, Pushker Kharecha, Gary Russell, David W. Lea, and Mark Siddall. 'Climate change and trace gases', *Philosophical Transactions of the Royal Society* A 365, 2007, 1925–54.

Haraway, Donna. 'A Cyborg Manifesto: Science, Technology, and Socialist-Feminism in the Late Twentieth Century.' In *Simians, Cyborgs and Women: The Reinvention of Nature* (New York; Routledge, 1991).

Hardin, Garrett. 'The Tragedy of the Commons,' *Science* 162 (1968), 1243–8.

Hare, David. 'Why fabulate?'. *The Guardian*, 2 February 2002.

Hardt, Michael and Antonio Negri. *Empire* (Cambridge, MA: Harvard University Press, 2000).

Hargrove, Eugene. 'Taking environmental ethics seriously'. In Dorinda G. Dallmeyer and Albert F. Ike (eds), *Environmental Ethics and the Global Marketplace* (Athens, GA: University of Georgia Press, 1998).

Havel, Václav. 'The power of the powerless'. In John Keane (ed.), *The Power of the Powerless: Citizens Against the State in Central-Eastern Europe* (London: Hutchinson, 1985).

Hay, Louise. *You Can Heal Your Life* (Carlsbad, CA: Hay House, 2004).

Hay, Peter. *A Companion to Environmental Thought* (Edinburgh: Edinburgh University Press, 2002).

Hayles, Katherine. *Chaos Bound: Orderly Disorder in Contemporary Literature and Science* (Ithaca: Cornell University Press, 1990).

Hayles, Katherine (ed.), *Chaos and Order: Complex Dynamics in Literature and Science* (Chicago: University of Chicago Press, 1991).

Hayles, Katherine. *How We Became Posthuman* (Chicago: University of Chicago Press, 1999).

Healy, David. *The Antidepressant Era* (Cambridge, MA: Harvard University Press, 1997).

Henneberg, Maciej. 'Evolution of the human brain: is bigger better?'. *Clinical and Experimental Pharmacology and Physiology* 25(9), Sept 1998, 745–9.

Hern, Warren M. 'Why are there so many of us? Description and diagnosis of a planetary ecopathological process'. *Population and Environment* 12(1), Fall 1990, 9–39.

Hillman, James. *The Essential James Hillman: A Blue Fire* (Routledge, 1990).

Hillman, James and Michael Ventura. *We've Had 100 Years of Psychotherapy and the World's Getting Worse* (New York: HarperCollins, 1993).

Hobsbawm, Eric. *The Age of Extremes* (New York: Pantheon, 1994).

Hoffmeyer, Jesper. *Signs of Meaning in the Universe*, Trans. Barbara J. Haveland (Bloomington: Indiana University Press, 1993).

Hoffmeyer, Jesper. *Biosemiotics: An Examination into the Life of Signs and the Signs of Life* (Chicago: University of Chicago Press, 2009).

Holling, C. S. 'The resilience of terrestrial ecosystems: local surprise and global change'. In W. C. Clark and R. E. Munn (eds), *Sustainable Development of the Biosphere* (Cambridge: Cambridge University Press, 1986).

Holling, C. S. and Sanderson, Steven. 'Dynamics of (dis)harmony in ecological and social systems.' In Susan Hanna, Carl Folke, and Karl-Göran Mäler (eds), *Rights to Nature: Ecological, Economic, and Political Principles of Institutions for the Environment* (Washington, DC: Island Press, 1996).

Holmwood, Leigh. 'BBC Gaza appeal row: Unions protest'. *The Guardian*, 26 January 2009.

Hornborg, Alf. *The Power of the Machine: Global Inequalities of Power, Technology, and Environment* (Walnut Creek: Altamira Press, 2001).

Hosinski, Thomas E. *Stubborn Fact and Creative Advance: An Introduction to the Metaphysics of Alfred North Whitehead* (Lanham, MD: Rowman and Littlefield, 1993).

House, Richard and Nick Totton (eds), *Implausible Professions: Arguments for Pluralism and Autonomy in Psychotherapy and Counselling* (Ross-on-Wye: PCCS Books, 1997).

Hrdy, Sarah. *Mother Nature: Natural Selection and the Female of the Species* (London: Chatto and Windus, 1999).

Hull, R. Bruce and David P. Robertson. 'The Language of Nature Matters: We Need a More Public Ecology'. In Paul H. Gobster and R. Bruce Hull (eds), *Restoring Nature: Perspectives from the Social Sciences and Humanities* (Washington, DC: Island Press, 2000).

Hume, David. *A Treatise of Human Nature* (London: Penguin Classics, 2004).

Humphrey, Michael. 'Reconciliation and the therapeutic state'. *Journal of Intercultural Studies* 26(3), August 2005, 203–20.

Hutchins, Edwin. *Cognition in the Wild* (Cambridge, MA: MIT Press, 1995).

Ingold, Tim (ed.), *Key Debates in Anthropology* (London: Routledge, 1996).

Ingold, Tim. 'The optimal forager and economic man'. In: Philippe Descola and Gísli Pálsson (eds), *Nature and Society: Anthropological Perspectives* (London: Routledge, 1996).

Ingold, Tim. *The Perception of the Environment* (London: Routledge, 2002).

Jackson, Michael. *Paths Toward a Clearing: Radical Empiricism and Ethnographic Enquiry* (Bloomington: Indiana University Press, 1989).

Jacoby, Russell. *The End of Utopia: Politics and Culture in an Age of Apathy* (New York: Basic Books, 1999).

James, Oliver. 'They muck you up: Developmental psychology as a basis for politics'. *The Psychologist* 16 No. 6, (June 2003).

James, William. *Some Problems in Philosophy: A Beginning of an Introduction to Philosophy* (Lincoln, Nebraska: University of Nebraska Press, 1999 (1911)).

Johannes, R. E. *Words of the Lagoon: Fishing and Marine Lore in the Palau District of Micronesia* (Berkeley: University of California Press, 1992).

Johnson, Mark. 'The emergence of meaning in bodily experience'. In Bernard den Outen and Marcia Moen (eds), *The Presence of Feeling in Thought* (New York: Peter Lang, 1991).

Joy, Bill. 'Why the future doesn't need us'. *Wired*, 8.04, April 2000.

Kahn, Peter H. *The Human Relationship with Nature* (Cambridge, MA: MIT Press, 2002).

Kahn, Peter H. and Friedman, B. 'Environmental views and values of children in an inner-city Black community'. *Child Development* 66 (1995), 1403–17.

Kahn, Peter H. and Stephen Kellert (eds), *Children and Nature: Psychological, Sociocultural, and Evolutionary Investigations* (Cambridge, MA: MIT Press, 2002).

Kalland, Arne. 'Indigenous knowledge: Prospects and limitations'. In Roy Ellen, Peter Parkes, and Alan Bicker (eds), *Indigenous Environmental Knowledge and its Transformations* (London: Routledge, 2000).

Kant, Immanuel. *Critique of Pure Reason* (Trans. N. K. Smith). Second Edition (London: Macmillan, 1933).

Kasser, Tim. *The High Price of Materialism* (Cambridge, MA: MIT Press, 2003).

Kauffman, Stuart A. *The Origins of Order: Self-Organisation in Evolution* (New York: Oxford University Press, 1993).

Kauffman, Stuart A. *At Home in the Universe: The Search for Laws of Self-Organisation and Complexity* (London: Viking, 1995).

Kauffman, Stuart A. *Reinventing the Sacred* (New York: Basic Books, 2008).

Kay, Lily. 'A book of life? How the genome became an information system and DNA a language'. *Perspectives in Biology and Medicine* 41(4), Summer 1998, 504–28.

Kaye, Joel. *Economy and Nature in the Fourteenth Century* (Cambridge: Cambridge University Press, 1998).

Keeney, Bradford. *Aesthetics of Change* (New York: Guilford Press, 1983).

Keep, Ewart and Mayhew, Ken. 'The assessment: Knowledge, skills, and competitiveness'. *The Oxford Review of Economic Policy* 15(1), 1999, 1–15.

Kellert, Stephen R. and Edward O. Wilson (eds), *The Biophilia Hypothesis* (Washington, DC: Shearwater Press, 1993).

Kenrick, Douglas T. 'Contrast effects and judgements of physical attractiveness: When beauty becomes a social problem'. *Journal of Personality and Social Psychology* 38 (1980), 131–40.

Kleinman, Arthur. *Rethinking Psychiatry: From Cultural Category to Personal Experience* (New York: Free Press, 1988).

Kleinman, Arthur. '"Everything that really matters": Social suffering, subjectivity, and the remaking of human experience in a disordering world.' *Harvard Theological Review* 90(3) (1997), 315–35.

Kleinman, Arthur, Veena Das, and Margaret Lock. 'Introduction to "social suffering"'. *Daedalus* 125 (1996).

Kleinman, Arthur, Veena Das, and Margaret Lock (eds), *Social Suffering* (Berkeley: University of California Press, 1997).

Kohlberg, Lawrence. *The Psychology of Moral Development* (New York: Harper and Row, 1984).

Kretch, Shepard III. *The Ecological Indian: Myth and History* (New York: Norton, 1999).

Lacan, Jacques. *The Seminar of Jacques Lacan, Book XX, Encore 1972–1973*. Jacques-Alain Miller (ed.). Translated by Bruce Fink (New York: Norton, 1998).

Laing, Ronald D. *The Divided Self: An Existential Study in Sanity and Madness* (Harmondsworth: Penguin, 1960).

Laing Ronald D. and A. Esterson. *Sanity, Madness, and the Family* (Harmondsworth: Penguin, 1967).

Lakoff, George and Mark Johnson. *Philosophy in the Flesh: The Embodied Mind and Its Challenge to Western Thought* (New York: Basic Books, 1999).

Lane, Robert E. *The Loss of Happiness in Market Democracies* (New Haven: Yale University Press, 2000).

Lasch, Christopher. *The Culture of Narcissism* (New York: Norton, 1979).

Lasch, Christopher. 'A response to my critics'. In 'A symposium: Christopher Lasch and the culture of narcissism'. *Salmagundi*, Fall 1979.

Latour, Bruno and Steve Woolgar. *Laboratory Life: The Social Construction of Scientific Facts* (London: Sage, 1979).

Lave, Jean. *Cognition in Practice* (Cambridge: Cambridge University Press, 1988).

Le Breton, David. 'Dualism and Renaissance: Sources for a modern representation of the body'. *Diogenes* 36 No. 142, (1988), 47–69.

Leder, Drew. *The Absent Body* (Chicago: University of Chicago Press, 1990).

Legler, Gretchen. 'Body politics in American nature writing'. In Richard Kerridge and Neil Sammells, *Writing the Environment: Ecocriticism and Literature* (London: Zed Books, 1998).

Lewicki, Pawel, Thomas Hill, and Maria Czyzewska. 'Nonconscious acquisition of information'. *American Psychologist* 47(6), June 1992, 796–801.

Lichtman, Richard. 'The illusion of the true self'. *Capitalism, Nature, Socialism* 9(3), 1998, 113–31.

Locke, John. *An Essay Concerning Human Understanding*. A. D. Woozley (ed.), (Glasgow: Collins, 1964).

Long, Jonathan, Aregai Tecle, and Benrita Burnette. 'Cultural foundations for ecological restoration on the White Mountain Apache reservation'. *Conservation Ecology* 8(1), 2003, Article 4.

Lovelock, James. *Gaia: A New Look at Life on Earth* (Oxford: Oxford University Press, 1979).

Luke, Tim. 'Green consumerism: Ecology and the ruse of recycling'. In Jane Bennett and William Chaloupka (eds), *In the Nature of Things* (Minneapolis: University of Minnesota Press, 1993).

Luke, Tim. 'Cyborg enchantments: Commodity fetishism and human/machine interactions'. *Strategies: Journal of Theory, Culture, and Politics* 13(1), May 2000, 39–62.

Lyng, Stephen and David D. Franks. *Sociology and the Real World* (Lanham, MD: Rowman and Littlefield, 2002).

Mackay, Nigel. 'Psychotherapy and the idea of meaning'. *Theory and Psychology* 13(3), 2003, 359–86.

Maffi, Luisa (ed.), *On Biocultural Diversity* (Washington, DC: Smithsonian Institution, 2001).

Mann, Charles. 'Three trees: Inescapable blendings of the human and the natural'. *Harvard Design Magazine*, Winter/Spring 2000, 31–5.

Manno, Jack P. *Privileged Goods: Commoditization and its Impact on Environment and Society* (Boca Raton: CRC Press, 2000).

Marais, Eugene. *The Soul of the White Ant* (New York: New York University Press, 2009).

Marcuse, Herbert. *One-Dimensional Man: Studies in the Ideology of Advanced Industrial Society* (London: Routledge and Kegan Paul, 1964).

Marin, Peter. 'The new narcissism'. *Harper's*, October 1975, 45–56.

Martin, Brian. 'The contexts of environmental decision-making'. *Australian Quarterly*, 50(1), April 1978, 105–18.

Martín-Baró, Ignacio. *Writings for a Liberation Psychology*. Ed. Adrianne Aron and Shawn Corne (Cambridge, MA: Harvard University Press, 1996).

Marx, Karl. 'Private property and communism'. In David McLellan (ed.), *Karl Marx: Selected Writings* (Oxford: Oxford University Press, 1977).

Masson, Jeffrey. *The Assault on Truth: Freud's Suppression of the Seduction Theory* (New York: Farrar, Straus, and Giroux, 1984).

Maturana, Humberto and Francisco Varela. *Autopoiesis and Cognition: The Realization of the Living* (Dordrecht: Reidel, 1980).

Mayr, Otto. *Authority, Liberty, and Automatic Machinery in Early Modern Europe* (Baltimore: John Hopkins University Press, 1986).

McGilchrist, Iain. *The Master and His Emissary: The Divided Brain and the Making of the Modern World* (New Haven: Yale University Press, 2009).

McQueen, Humphrey. *The Essence of Capitalism* (London: Profile Books, 2001).

McNickle, D'Arcy. *The Surrounded* (Albuquerque: University of New Mexico Press, 1978 (1936)).

Mead, George H. *Mind, Self, and Society from the Standpoint of a Social Behaviorist* (Chicago: University of Chicago Press, 1962 (1934)).

Medin, Douglas L. and Scott Atran (eds), *Folkbiology* (Cambridge, MA: MIT Press, 1999).

Merchant, Carolyn. *The Death of Nature: Women, Ecology, and the Scientific Revolution* (New York: HarperCollins, 1980).

Messer, Ellen and Michael Lambek (eds), *Ecology and the Sacred: Engaging the Anthropology of Roy A. Rappaport* (Ann Arbor: University of Michigan Press, 2001).

Michael, Mike. *Reconnecting Culture, Technology, and Nature: From Society to Heterogeneity* (London: Routledge, 2000).

Midgley, Mary. *Science as Salvation: A Modern Myth and its Meaning* (London: Routledge, 1992).

Midgley, Mary. *Science and Poetry* (London: Routledge, 2001).

Miles, Irene. 'Psychological benefits of volunteering for restoration projects'. *Ecological Restoration* 18(4), Winter 2000, 218–27.

Miller, David and Greg Philo, 'Silencing dissent in academia: The commercialisation of science'. *The Psychologist* 15(5), May 2002, 244–6.

Miller, Mark C. 'Free the Media'. *The Nation*, 3 June 1996.

Miller, Stanley L. 'A production of amino acids under possible primitive earth conditions'. *Science* 117, May 15 1953, 528–9.

Mingers, John. *Self-Producing Systems: Implications and Applications of Autopoiesis* (New York: Springer, 1994).

Minsky, Marvin. 'Virtual molecular reality'. In Markus Krummenacker and James Lewis (eds), *Prospects in Nanotechnology* (New York: Wiley, 1995).

Mirowski, Philip. *More Heat Than Light: Economics as Social Physics, Physics as Nature's Economics* (Cambridge: Cambridge University Press, 1989).

Mitchell, Stephen A. *Relational Concepts in Psychoanalysis* (Cambridge: Harvard University Press, 1988).

Mithen, Steven. *The Prehistory of the Mind* (London: Thames and Hudson, 1996).

Monbiot, George. *Captive State: The Corporate Takeover of Britain* (London: Macmillan, 2000).

Monbiot, George. 'Environmental feedback: A reply to Clive Hamilton'. *New Left Review* 45, May/June 2007.

Monbiot, George. 'Leave it in the ground'. *The Guardian*, 11 December 2007.

Morrison, Toni. 'The Site of Memory'. In William Zinsser (ed.), *Inventing the Truth: The Art and Craft of Memoir* (New York: Houghton Mifflin, 1987).

Munro, Alice. *Lives of Girls and Women* (London: Women's Press, 1971).

Munro, Donald J. *The Concept of Man in Early China* (Stanford: Stanford University Press, 1969).

Murdoch, John. 'The analytic character of late medieval learning: Natural philosophy without nature'. In Lawrence D. Roberts (ed.), *Approaches to Nature in the Middle Ages* (Binghampton, NY: Center for Medieval and Early Renaissance Studies, 1982).

Myers, Gene. *Children and Animals* (Boulder: Westview Press, 1998).

Myers, Norman and Julian Simon (eds), *Scarcity or Abundance? A Debate on the Environment* (New York: Norton, 1994).

Nabhan, Gary P. *The Desert Smells Like Rain* (Tucson: University of Arizona Press, 1982).

Nabhan, Gary P. 'Cultural parallax in viewing North American habitats'. In Michael E. Soulé and Gary Lease (eds), *Reinventing Nature? Responses to Postmodern Deconstruction* (Washington, DC: Island Press, 1995).

Nabhan, Gary P. *Cultures of Habitat: On Nature, Culture, and Story* (Washington, DC: Counterpoint, 1997).

Nabhan, Gary P. 'Cultural perceptions of ecological interactions'. In Luisa Maffi (ed.), *On Biocultural Diversity* (Washington, DC: Smithsonian Institution, 2001).

Nadeau, Robert L. *The Wealth of Nature: How Mainstream Economics Has Failed the Environment* (New York: Columbia University Press, 2003).

Nadeau, Robert L. *The Environmental Endgame* (New Brunswick: Rutgers University Press, 2006).

Naess, Arne. *Ecology, Community, and Lifestyle: Outline of an Ecosophy*, trans David Rothenberg (Cambridge: Cambridge University Press, 1989).

Nettle, Daniel and Suzanne Romaine. *Vanishing Voices: The Extinction of the World's Languages* (Oxford: Oxford University Press, 2000).

Nora, Pierre. 'Between memory and history: Les lieux de memoire'. *Representations* 26 (Spring 1989).

Odling-Smee, F. John, Kevin N. Laland, and Marcus W. Feldman. *Niche Construction: The Neglected Process in Evolution* (Princeton: Princeton University Press, 2003).

O'Flaherty, Coleman A. (ed.), *Highways: The Location, Construction, and Maintenance of Road Pavements*, 4th edition (Oxford: Butterworth Heinemann, 2002).

O'Hanlon, Bill. 'Not strategic, not systemic: Still clueless after all these years'. In Steffanie O'Hanlon and Bob Bertolino (eds), *Evolving Possibilities: Selected Papers of Bill O'Hanlon* (Philadelphia: Brunner/Mazel, 1999).

Öhman, Arne and Susan Mineka, 'The malicious serpent: Snakes as a prototypical stimulus for an evolved module of fear'. *Current Directions in Psychological Science* 12(1), February 2003, 5–9.

Olson, Richard. *Scottish Philosophy and British Physics, 1750–1880: A Study in the Foundations of the Victorian Scientific Style* (Princeton: Princeton University Press, 1975).

Orwell, George. *A Collection of Essays* (New York: Doubleday Anchor Books, 1954).

Oyama, Susan. *The Ontogeny of Information*. Second, revised edition (Durham, NC: Duke University Press, 2001).

Oyama, Susan. 'Speaking of Nature'. In Yrjö Haila and Chuck Dyke (eds), *How Nature Speaks: The Dynamics of the Human Ecological Condition* (Durham, NC: Duke University Press, 2006).

Padel, Felix. *The Sacrifice of Human Being: British Rule and the Konds of Orissa* (Delhi: Oxford University Press, 1995).

Pallasmaa, Juhani. *The Thinking Hand: Existential and Embodied Wisdom in Architecture* (Chichester: Wiley, 2009).

Pauly, Philip J. Review of Anne Harrington, *Medicine, Mind, and the Double Brain*. *Science* 239, 22 January 1988, 422.

Payne, Jonathan. 'All things to all people: Changing perceptions of "skill" among Britain's policy makers since the 1950s and their implications'. (University of Warwick: SKOPE research paper No. 1, August 1999).

Peterson, Anna. 'Toward a materialist environmental ethic'. *Environmental Ethics* 28(4), Winter 2006, 375–93.

Phillips, John. *Contested Knowledge: A Guide to Critical Theory* (London: Zed Books, 2000).

Philo, Greg and David Miller (eds), *Market Killing: What the Free Market Does and What Social Scientists Can Do About It* (Harlow: Pearson, 2001).

Piaget, Jean. *The Psychology of Intelligence* (London: Routledge and Kegan Paul, 1971).

Pimm, Stuart. *The Balance of Nature?: Ecological Issues in the Conservation of Species and Communities* (Chicago: University of Chicago Press, 1991).

Pinch, Trevor J. and Harry M. Collins. 'Private science and public knowledge: The committee for the scientific investigation of the paranormal and its use of the literature'. *Social Studies of Science* 14 (1984), 521–46.

Plato, *Phaedo*, trans. G. M. A. Grube (Cambridge, MA: Hackett Publishing, 1977).

Plumwood, Val. *Feminism and the Mastery of Nature* (London: Routledge, 1993).

Polanyi, Karl. *The Great Transformation* (Boston: Beacon Press, 1944).

Pomeroy, Anne. *Marx and Whitehead: Process, Dialectics, and the Critique of Capitalism* (Albany: State University of New York Press, 2004).

Postman, Neil. *Amusing Ourselves to Death: Public Discourse in the Age of Show Business* (London: Methuen, 1985).

Potter, Jonathan. *Representing Reality* (London: Sage, 1996).

Powell, Jason. *Jacques Derrida: A Biography* (London: Continuum, 2006).

Prigogine, Ilya and Isabelle Stengers. *Order out of Chaos: Man's New Dialogue with Nature* (London: Fontana, 1985).

Pullman, Bernard. *The Atom in the History of Human Thought* (New York: Oxford University Press, 1998).

Raine, Kathleen. *Farewell Happy Fields* (London: Hamish Hamilton, 1973).

Rappaport, Roy A. *Ritual and Religion in the Making of Humanity* (Cambridge: Cambridge University Press, 1999).

Reiss, Timothy J. *The Discourse of Modernism* (Ithaca: Cornell University Press, 1982).

Renfrew, Colin. 'Mind and matter: Cognitive archaeology and external symbolic storage.' In Colin Renfrew and Chris Scarre (eds), *Cognition and Material Culture* (Cambridge: McDonald Institute, 1998).

Richards, Robert J. *The Romantic Conception of Life: Science and Philosophy in the Age of Goethe* (Chicago: University of Chicago Press, 2002).

Ridington, Robin. *Trail to Heaven: Knowledge and Narrative in a Northern Native Community* (Vancouver: Douglas and McIntyre, 1988).

Rival, Laura. 'The growth of family trees: Understanding Huaorani perceptions of the forest'. *Man*, 28(4) (1993), 635–52.

Robins, Kevin. 'Cyberspace and the world we live in'. In Mike Featherstone and Roger Burrows (eds), *Cyberspace/Cyberbodies/Cyberpunk: Cultures of Technological Embodiment* (London: Sage, 1996).

Rodman, John. 'The liberation of nature?' *Inquiry* 20 (1977), 83–145.

Rogers, Carl R. *Client Centered Therapy: Its Current Practice, Implications, and Theory* (London: Constable, 1984).

Rogers, Raymond A. *Solving History: The Challenge of Environmental Activism* (Montreal: Black Rose Books, 1998).

Rogers, Raymond A. *Nature and the Crisis of Modernity* (Montreal: Black Rose Books, 1995).

Rolston, Holmes III. *Environmental Ethics* (Philadelphia: Temple University Press, 1988).

Rolston, Holmes III. *Conserving Natural Value* (New York: Columbia University Press, 1994).

Romanyshyn, Robert D. *Psychological Life: From Science to Metaphor* (Austin: University of Texas, 1982).

Romanyshyn, Robert D. *Technology as Symptom and Dream* (London: Routledge, 1989).

Rosch, Eleanor. 'Universals and cultural specifics in human categorization'. In Richard Brislin, Stephen Bochner, and Walter Lonner (eds), *Cross-Cultural Perspectives on Learning* (New York: Sage, 1975).

Rose, Jacqueline. *The Last Resistance* (London: Verso, 2007).

Ross, Lee. 'The intuitive psychologist and his shortcomings: Distortions in the attribution process.' In L. Berkowitz (ed.), *Advances in Experimental Social Psychology*, Vol. 10 (New York: Academic Press, 1977), 173–220.

Rozin, Paul and James Kalat. 'Specific hungers and poison avoidance as adaptive specializations of learning.' *Psychological Review* 78 (1972), 459–86.

Russell, Keith C. 'What is wilderness therapy?' *Journal of Experiential Education*, 24(2), 2001, 70–9.

Ryan, Robert L. 'A people-centered approach to designing and managing restoration projects: Insights from understanding attachment to urban natural areas'. In Paul H. Gobster and R. Bruce Hull (eds), *Restoring Nature: Perspectives from the Social Sciences and Humanities* (Washington, DC: Island Press, 2000.

Sahlins, Marshall. *Islands of History* (Chicago: University of Chicago Press, 1985).

Sampson, Edward E. 'Cognitive psychology as ideology'. *American Psychologist* 36(7), 1981, 730–43.

Sass, Louis A. *Madness and Modernism: Insanity in the Light of Modern Art, Literature, and Thought* (Cambridge, MA: Harvard University Press, 1992).

Schaffer, William M. 'Stretching and folding in lynx fur returns: Evidence for a strange attractor in nature?' *American Naturalist* 124 No. 6, (1984), 798–820.

Schaffer, William M. and Kot, M. 'Do strange attractors govern ecological systems?' *BioScience* 35 No. 6, (1985), 342–50.

Scheibe, Karl. 'Mirrors, masks, lies, and secrets'. In Joseph de Rivera and Theodore R. Sarbin (eds), *Believed-In Imaginings: The Narrative Construction of Reality* (Washington, DC: American Psychological Association, 1998).

Scheper-Hughes, Nancy and Margaret Lock. 'The mindful body: A prolegomenon to future work in medical anthropology'. *Medical Anthropology Quarterly*, New Series, 1(1), March 1987, 6–41.

Scheper-Hughes, Nancy and Sargent, Carolyn (eds), *Small Wars: The Cultural Politics of Childhood* (Berkeley: University of California Press, 1998).

Schumaker, John F. *The Age of Insanity: Modernity and Mental Health* (Westport, CT: Praeger, 2001).

Scott, David and Fern K. Willits. 'Environmental attitudes and behavior'. *Environment and Behavior* 26(2), 1994, 239–60.

Searle, John. *The Rediscovery of the Mind* (Cambridge, MA: MIT Press, 1992).

Searles, Harold F. *The Nonhuman Environment in Normal Development and Schizophrenia* (New York: International Universities Press, 1960).

Seed, Patricia. *Ceremonies of Possession in Europe's Conquest of the New World 1492–1640* (Cambridge: Cambridge University Press, 1995).

Shapiro, Robert. *Origins: A Skeptic's Guide to Life on Earth* (London: Penguin, 1986).

Shiva, Vandana. *Biopiracy: The Plunder of Nature and Knowledge* (Boston: South End Press, 1997).

Shiva, Vandana. *Monocultures of the Mind: Biodiversity, Biotechnology, and Scientific Agriculture* (London: Zed Books, 1998).

Shweder, Richard. ' Menstrual pollution, soul loss, and the comparative study of emotions'. In Arthur Kleinman and Byron Good (eds), *Culture and Depression: Studies in the Anthropology and Cross-Cultural Psychiatry of Affect and Disorder* (Berkeley: University of California Press, 1985).

Silverstein, Richard. 'Land of the free?'. *The Guardian*, 10 October, 2007.

Simon, Herbert. *Models of My Life* (San Francisco: Basic Books, 1991).

Simon, Julian. *The Ultimate Resource* (Oxford: Martin Robertson, 1981).

Simon, Julian. *Good Mood: The New Psychology of Overcoming Depression*. (La Salle, IL: Open Court, 1993).

Simon, Julian. *The Ultimate Resource 2* (Princeton: Princeton University Press, 1996).

Sluzki, Carlos E. and Donald C. Ransom (eds), *Double Bind: The Foundation of the Communicational Approach to the Family* (New York: Grune and Stratton, 1976).

Smail, David. *Taking Care: An Alternative to Therapy* (London: Dent, 1987).

Smail, David. *The Nature of Unhappiness* (London: Robinson, 2001).

Smail, David. *Power, Interest, and Psychology* (Ross-on-Wye: PCCS Books, 2005).

Smith, Adam. *An Inquiry into the Nature and Causes of the Wealth of Nations*. R. H. Campbell, A. S. Skinner, and W. B. Todd (eds), (Indianapolis: Liberty Press, 1982 (1776)).

Smith, Donald B. *From the Land of Shadows: The Making of Grey Owl* (Seattle: University of Washington Press, 1990).

Smith, Neil. *Uneven Development: Nature, Capital, and the Production of Space*. Third Edition (Athens, GA: University of Georgia Press, 2008).

Snow, John. *Snow on Cholera* (New York: Commonwealth Fund, 1936).

Sokal, Alan and Jean Bricmont, *Intellectual Impostures* (London: Profile Books, 1998).

Sommer, Robert. 'Trees and human identity'. In Susan Clayton and Susan Opotow (eds), *Identity and the Natural Environment: The Psychological Significance of Nature* (Cambridge: MIT Press, 2003).

320   *Bibliography*

Soon, Chun Siong, Marcel Brass, Hans-Jochen Heinze and John-Dylan Haynes. 'Unconscious determinants of free decisions in the human brain'. *Nature Neuroscience* 11, 13 April 2008, 543–5.

Soper, Kate. 'Nature and culture: The mythic register'. In Paul Sheehan (ed.), *Becoming Human: New Perspectives on the Human Condition* (Westport: Praeger, 2003).

Soto, Ana M. and Carlos Sonnenschein. 'Emergentism by default: A view from the bench'. *Synthese* 151, (2006), 361–76.

Spence, Donald. *Narrative Truth and Historical Truth* (New York: Norton, 1982).

Spirn, Anne W. *The Language of Landscape* (New Haven: Yale University Press, 1998).

Spirn, Anne W. 'Restoring Mill Creek: Landscape literacy, environmental justice, and city planning and design.' *Landscape Research* 30(5), July 2005, 359–77.

Steffens, Henry J. *James Prescott Joule and the Concept of Energy* (Folkestone: Dawson, 1979).

Steinberg, Michael. *The Fiction of a Thinkable World: Body, Meaning, and the Culture of Capitalism* (New York: Monthly Review Press, 2005).

Stern, Daniel. *The Interpersonal World of the Infant* (London: Karnac Books, 2000).

Stewart, Ben. 'Drax trial held in a climate of injustice'. *The Guardian*, 3 July 2009.

Stoller, Paul. *Sensuous Scholarship* (Philadelphia: University of Pennsylvania Press, 1997).

Storr, Anthony. *The Dynamics of Creation* (London: Secker and Warburg, 1972).

Strang, Veronica. *Uncommon Ground: Cultural Landscapes and Environmental Values* (Oxford, Berg, 1997).

Sulloway, Frank. *Freud: Biologist of the Mind* (London: Burnett Books, 1979).

Taussig, Michael. *The Devil and Commodity Fetishism in South America* (Chapel Hill: University of North Carolina Press, 1980).

Taylor, Mark C. *The Picture in Question: Mark Tansey and the Ends of Representation* (Chicago: University of Chicago Press, 1999).

Taylor, Shelley and Jonathan D. Brown. 'Illusion and well-being: A social psychological perspective on mental health'. *Psychological Bulletin* 103(2), March 1988, 193–210.

Taylor, Shelley and Jonathan D. Brown. 'Positive illusions and well being revisited: Separating fact from fiction'. *Psychological Bulletin* 116(1), July 1994, 21–7.

Temperton, Vicky M. and Richard J. Hobbs. 'The search for ecological assembly rules and its relevance to restoration ecology'. In Vicky M. Temperton, Richard J. Hobbs, Tim Nuttle, and Stefan Halle (eds), *Assembly Rules and Restoration Ecology* (Washington, DC: Island Press, 2004).

Terdiman, Richard. *Present Past: Modernity and the Memory Crisis* (Ithaca: Cornell University Press, 1993).

Thaler, Richard H. and Cass R. Sunstein. *Nudge: Improving Decisions About Health, Wealth, and Happiness* (New York: Penguin, 2009).

Thomas, Julian. 'Some problems with the notion of external storage, and the case of neolithic material culture in Britain'. In Colin Renfrew and Chris Scarre (eds), *Cognition and Material Culture* (Cambridge: Cambridge University Press, 1998).

Thoreau, Henry David. *Walden* (Ware, Hertsfordshire: Wordsworth Editions, 1995).

Todorov, Tzvetan. *The Conquest of America: The Question of the Other* (New York: Harper and Row, 1984).

Tomasello, Michael. *The Cultural Origins of Human Cognition* (Cambridge, MA: Harvard University Press, 2001).

Tonnies, Ferdinand. *Community and Society* (Mineola, NY: Dover Publications, 2003 (1887)).

Tooby, John, and Cosmides, Leda. 'The psychological foundations of culture'. In Jerome H. Barkow, Leda Cosmides, and John Tooby (eds), *The Adapted Mind: Evolutionary Psychology and the Generation of Culture* (New York: Oxford University Press, 1992).

Totton, Nick. 'The baby and the bathwater: "Professionalisation" in psychotherapy and counselling'. *British Journal of Guidance and Counselling*, 27(3), August 1999, 313–24.

Tuan, Yi-Fu. 'Discrepancies between environmental attitudes and behaviour: Examples from Europe and China'. *Canadian Geographer* 12 (1968), 176–91.

Tuck-Po, Lye. *Changing Pathways: Forest Degradation and the Batek of Pahang, Malaysia* (Lanham: Lexington Books, 2004).

Turner, Frederick. *Beyond Geography: The Western Spirit Against the Wilderness* (New Brunswick: Rutgers University Press, 1983).

Turner, Jack. *The Abstract Wild* (Tucson: University of Arizona Press, 1996).

Tversky, Amos and Kahneman, Daniel. 'Extensional versus intuitive reasoning: The conjunction fallacy in probability judgement.' *Psychological Review* 90 (1983), 293–315.

Twenge, Jean. *Generation Me* (New York: Free Press, 2006).

Ulanowicz, Robert. *Ecology, the Ascendant Perspective* (New York: Columbia University Press, 1997).

Ulanowicz, Robert. *A Third Window: Natural Life Beyond Newton and Darwin* (West Conshohocken, PA: Templeton Foundation Press, 2009).

Vallentin, Antonina. *Einstein: A Biography* (London: Weidenfeld and Nicholsen, 1954).

Van der Ploeg, Jan D. 'Potatoes and knowledge'. In Mark Hobart (ed.), *An Anthropological Critique of Development: The Growth of Ignorance* (London: Routledge, 1993).

Vayda, Andrew P. *Methods and Explanations in the Study of Human Actions and Their Environmental Effects* (Jakarta, Indonesia: Center for International Forestry Research and World Wide Fund for Nature, 1996).

Vera, Franz. *Grazing Ecology and Forest History* (Wallingford: CABI Publishing, 2000).

Vera, Franz. 'The Shifting Baseline Syndrome in restoration ecology'. In Marcus Hall (ed.), *Restoration and History: The Search for a Usable Environmental Past* (London: Routledge, 2010).

Vogel, Steven. 'Environmental philosophy after the end of nature'. *Environmental Ethics* 24(1), Spring 2002, 23–39.

Vogel, Steven. 'The nature of artifacts'. *Environmental Ethics* 25(2), Summer 2003, 149–68.

Walker, Susan, J. Bastow Wilson, John B. Steel, G. L. Rapson, Benjamin Smith, Warren McG. King, and Yvette H. Cottam, "Properties of ecotones: Evidence from five ecotones objectively determined from a coastal vegetation gradient". *Journal of Vegetation Science* 14(4), August 2003, 579–90.

Watzlawick, Paul, Janet B. Bavelas, and Don Jackson, *Pragmatics of Human Communication* (New York: Norton, 1967).

Webley, Paul, Carole B. Burgoyne, Stephen E. G. Lea, and Brian M. Young. *The Economic Psychology of Everyday Life* (Hove: Psychology Press, 2001).

Wetherick, Norman. 'Against cognitive psychology'. *The Psychologist* 16(1), January 2003, 22–3.

Whitehead, Alfred N. *Modes of Thought* (London: Macmillan, 1938).

Whitmire, Tim. 'Blue Ridge Parkway pegs value of vistas'. *Los Angeles Times*, 14 September 2003.

Wilkinson, Richard and Kate Pickett. *The Spirit Level* (London: Penguin, 2010).

Williams, Bernard. *Truth and Truthfulness* (Princeton: Princeton University Press, 2002).

Williamson, Robert G. 'The Arctic habitat and the integrated self'. In Michael Aleksiuk and Thomas Nelson (eds), *Landscapes of the Heart* (Edmonton: NeWest Press, 2002).

Winnicott, Donald W. *The Maturational Process and the Facilitating Environment* (London: Hogarth Press, 1965).

Winnicott, Donald W. *Playing and Reality* (London: Tavistock, 1971).

Worster, Donald. *Nature's Economy: The Roots of Ecology* (San Francisco: Sierra Club Books, 1977).

Worster, Donald. 'The ecology of order and chaos'. *Environmental History Review* 14(2), 1990, 2–17.

Wynn, Thomas. *The Evolution of Spatial Competence* (Urbana: University of Illinois Press, 1989).

Index